D1212670

TIME, THE PHYSICAL MAGNITUDE

BOSTON STUDIES IN THE PHILOSOPHY OF SCIENCE

EDITED BY ROBERT S. COHEN AND MARX W. WARTOFSKY

VOLUME 99

OLIVIER COSTA DE BEAUREGARD

Institut Henri Poincaré, Paris

TIME, THE PHYSICAL MAGNITUDE

D. REIDEL PUBLISHING COMPANY

A MEMBER OF THE KLUWER  ACADEMIC PUBLISHERS GROUP

DORDRECHT / BOSTON / LANCASTER / TOKYO

Library of Congress Cataloging-in-Publication Data

Costa de Beauregard, O. (Olivier)
Time, the physical magnitude / Olivier Costa de Beauregard.
p. cm. — (Boston studies in the philosophy of science ; v. 99)
Bibliography: p.
Includes indexes.
ISBN 90-277-2444-X : $79.00 (U.S.)
1. Space and time. 2. Quantum theory. 3. Physics—Philosophy. 4. Irreversible processes. I. Title. II. Series.
Q174.B67 vol. 99
[QC173.59.S65]
001′.01 s—dc 19
[530.1′.1]

Published by D. Reidel Publishing Company,
P.O. Box 17, 3300 AA Dordrecht, Holland

Sold and distributed in the U.S.A. and Canada
by Kluwer Academic Publishers,
101 Philip Drive, Norwell, MA 02061, U.S.A.

In all other countries, sold and distributed
by Kluwer Academic Publishers Group,
P.O. Box 322, 3300 AH Dordrecht, Holland

Printed in The Netherlands

To

Louis de Broglie,
who introduced me to
theoretical physics

TABLE OF CONTENTS

PART 3 LAWLIKE TIME SYMMETRY AND FACTLIKE IRREVERSIBILITY

EDITORIAL PREFACE

In an age characterized by impersonality and a fear of individuality this book is indeed unusual. It is personal, individualistic and idiosyncratic — a record of the scientific adventure of a single mind. Most scientific writing today is so depersonalized that it is impossible to recognize the man behind the work, even when one knows him. Costa de Beauregard's scientific career has focused on three domains — special relativity, statistics and irreversibility, and quantum mechanics. In *Time, the Physical Magnitude* he has provided a personal *vade mecum* to those problems, concepts, and ideas with which he has been so long preoccupied.

Some years ago we were struck by a simple and profound observation of Mendel Sachs, the gist of which follows. Relativity is based on very simple ideas but, because it requires highly complicated mathematics, people find it difficult. Quantum mechanics, on the other hand, derives from very complicated principles but, since its mathematics is straightforward, people feel they understand it. In some ways they are like the *bourgeois gentilhomme* of Molière in that they speak quantum mechanics without knowing what it is. Costa de Beauregard recognizes the complexity of quantum mechanics. A great virtue of the book is that he does not hide or shy away from the complexity. He exposes it fully while presenting his ideas in a non-dogmatic way.

That is perhaps one of the finest features of this book. So often scientists claim impartiality and disinterestedness, while actually arguing in the most intolerant and parochial way. Costa de Beauregard's treatment of the Einstein—Podolsky—Rosen paradox is eminently fair to both Einstein and Bohr as well as their innumerable spiritual descendants. He even has the courage to consider the paranormal. So his book is a scientific adventure. It is also a scientific autobiography which shows how an imaginative and honest mind grapples with the philosophical concerns of time and physics.

* * *

Costa de Beauregard wrote on these themes before, notably in his two monographs, *Le second principe de la science du temps* and *La notion de temps — equivalence avec l'espace.* He meditates, muses, wanders down interesting byways that seem far from his main path and then turn out to be urgent for understanding as well as exploring. His papers invariably stimulate his admiring (but often irritated) readers for they are the results of an original and extraordinarily intelligent philosopher-physicist, one who thinks long and deeply before setting his thoughts to paper. We recall, among his many contributions which will interest readers of the present book, the following:

'Two Lectures on the Direction of Time', in *Hans Reichenbach: Logical Empiricist*, ed. W. Salmon (Dordrecht, 1977), pp. 341—366.

'Discussion on Temporal Asymmetry in Thermodynamics and Cosmology', reported by Costa de Beauregard, *Proc.* (*Cardiff*) *Intern. Conf. Thermodynamics*, ed. P. T. Landsberg (London, 1970).

'Irréversibilité quantique, phénomène macroscopique', *Louis de Broglie, Physicien et Penseur*, ed. A. George (Paris, 1952), pp. 401—412.

'*CPT* Invariance and Interpretation of Quantum Mechanics', *Foundations of Physics* **10** (1980), 513—530.

'Lorentz and *CPT* Invariances and the Einstein—Podolsky—Rosen Correlations', *Phys. Rev. Letters* **50** (1983), 867—869.

RONALD NEWBURGH
ROBERT S. COHEN

PREFACE

Although this book contains many historical references, it is not a book in the history of science; and although it contains quite a few equations, it is not a book in theoretical physics. Historical references and equations are brought forward as aids, the book being one in the philosophy of science.

Time is a concept needing a multidisciplinary approach, as members of the 'International Society for the Study of Time', to which I belong, know well. And of course physics is a full participant at the conference table where time is discussed.

The main aspects discussed here in *Time, the Physical Magnitude* are:

1. Time as an entity measurable with reference to space via motion, motion being formalized in 'universal' equations: initially, those of the Galileo—Newtonian dynamics; but now, those of the 1905 special relativity theory.

At the October 1983 Session of the General Conference for Weights and Measures it was decided that the velocity of light *in vacuo, c*, is an absolute constant by definition, and that the standard of length is thus no longer a primary standard, but an 'alter ego', so to speak, of the standard of time. Thus the physical 'equivalence' between space and time, which is the essence of the relativity theory, is legalized.

2. Irreversibility, or dissymmetry between past and future, is another fundamental aspect of physical time, appearing in the theories of wave propagation, thermodynamics, the probability calculus, statistical mechanics, and information theory. All these aspects of irreversibility are interconnected, being manifestations of one single fundamental universal irreversibility. This irreversibility is 'factlike, not lawlike', meaning that it resides not in the equations but in the solutions selected as significant. Therefore intrinsic reversibility of physical time needs an in-depth discussion, the origins of which are found in Laplace's 1774 thoughts on the probability calculus, in Loschmidt's 1876 well-known reversibility argument, and in the cybernetical discovery of a mutual convertibility of 'negentropy' and 'information'.

So 'arrowed (directional) causality', where 'cause precedes effect', turns out to be 'factlike, not lawlike', statistical in its nature, and nothing else than one more aspect of physical irreversibility.

3. Relativistic quantum mechanics tightly binds these two main aspects of time. It is a 'wavelike probability calculus', displaying paradoxical interference and beating of 'probability amplitudes'. It is endowed with an extended relativistic invariance where time reversal is allowed. Thus '*CPT* invariance', as it is called (*C*, particle—antiparticle exchange; *P*, space reversal; *T*, time reversal) supplants Loschmidt's *T* reversal. *CPT* invariance, together with Born's wavelike probability calculus, are, according to me, the two pillars of a true understanding of the paradoxical phenomenon known as the 1935 'Einstein—Podolsky—Rosen (EPR) correlations', which display direct, long-range, and arrowless spacetime connections. Here again causality shows up as arrowless at the microlevel.

4. Although the general relativity and relativistic cosmology theories are outside my professional field of inquiry, I have found it necessary to include a brief discussion of each of them, as they have direct bearing upon the characterization of physical time.

And so this book, which finally grew out of my lifelong interest as a theoretical physicist in the aspects and the nature of physical time, is intended to nurture the meditations of all who are interested in Time: physicists certainly, who very often are led to reflect on time as do philosophers of science; philosophers of science mainly interested in physics; scholars of other disciplines needing (or wishing) to see what physicists have to say on time; and finally, of course, interested laymen who are amateurs in such matters.

As a final touch of controversy, I might add that the existence of lawlike time reversal, especially in the form of the negentropy—information reversibility, together with time extendedness in relativity theory and arrowless causality in the EPR correlations, have convinced me that there might very well be some truth in 'the claims of the paranormal'.

So, as I have mentioned, I see the heart of my book in the epistemological interpretation of relativistic quantum mechanics. The reader is advised that he will not be led there along a straight, or even a conveniently arranged, path. Climbing Mount Everest is no affair of a travel agency. Indeed, seeing its pyramidal top in the distance — or not seeing it if there are clouds — from various places in the plains helps inspire one to undertake the long, arduous journey to the very side of the peak, and the final assault on it. Colonel Everest's geodesic surveys had to precede the toils of the climbers of the Chomo Lungma. Somewhat similarly, in this book *time, the physical magnitude* will be presented from a number of not unconnected vantage points. And the reader is advised that there may be some really arduous climbing and some delicate holds on the way up.

Practical advice for use of the book is as follows:

Chapters are numbered in the form $p \cdot q$, sections in the form $p \cdot q \cdot r$, and equations in the form $(p \cdot q \cdot r)$; reference to an equation inside the same section is often abbreviated in the form (r), and reference to a section inside the same chapter in the form q.

Notes, numbered in the form ^{n}, are found at the end of Part 5. A numbered note may be referred to more than once inside the same chapter.

Bibliographical details, referred to in the text by author(s) and date(s) of publication, are placed in alphabetical order at the end of the volume.

Of course I am aware that quite a few books both on time in general and on physical time do exist, of which I know some, but not all. By fear of doing an injustice, I choose to mention none of them here. Time is sufficiently important and intricate a question that one need not 'apologize for one more book on time', and I hope that this one will arouse the interest of the reader.

ACKNOWLEDGEMENTS

Thanks are due to Professor Robert Cohen for having invited me to write a book on Time in the prestigious Boston Series; to Mrs Kuipers, of the Acquisition Department of the Reidel Company, for the very helpful and pleasant correspondence we have exchanged; to two anonymous referees for their helpful comments; to my friend Ronald Newburgh who, as it has turned out, prefers straightening out bad American English to producing good English from the French; and (last but not least) to the typist, my wife Nicole, who has, for three years, been a true Penelope!

Dr Giacomo, Director of the BIPM, has allowed me to make a long quotation from the Document containing the new 1983 definition of the Metre; Prof. Alain Aspect has granted me permission to reproduce Figures illustrating his well-known published EPR work, as also has Dean Robert Jahn for his PK work at Princeton; I express warm thanks to all three of them. Also, the late, and much lamented, Banesh Hoffmann has allowed me to quote him from a book of which he owned the copyright.

The following Publishing Companies are thanked for written permission for quotations, all of which are duly credited:

McGraw Hill Book Company
Physical Society of Japan
Plenum Publishing Company
University of California Press
John Wiley and Sons
American Physical Society

PART 1

GENERALITIES

CHAPTER 1.1

INTRODUCTORY REMARKS

1.1.1. MODELISM OR FORMALISM?

Mathematical physics unites two contrasting natures. Therefore, in its progress, it oscillates between two opposite ideals: modelistic theories resembling engineers' toys and formalistic theories resembling mathematicians' games. It proceeds somewhat like a tightrope dancer, guarding himself on his left and on his right.

As a well-known example of this we have the thermodynamical and statistical group of theories.

Phenomenological thermodynamics, as conceived by Carnot, Clausius, Kelvin and Planck, was a highly efficient, abstract and austere system of principles. In its later days, in the hands of such workers as Duhem and Ostwald, it developed an imperialistic tendency, pretending under the name of 'energetics' to rule over physics, rejecting as fantasies all concrete pictures, even the very concepts of atoms or molecules.

Atomism, however, since the days of Democritus and Lucretius, had been proceding at its slow pace. The idea that 'heat' may really be the 'internal energy' of 'atomic agitation' crept up again and again, in the thinking of Daniel Bernouilli, Davy, Rumford, Lavoisier, Laplace and others. A century ago Clausius, Maxwell and Boltzmann constructed their statistical theory of gases, conceived as a tremendous three-dimensional billiards game — the very example of a modelistic theory. It was an extremely successful one, quantitatively explaining quite a few phenomena, such as the value of the heat capacity of gases, or their viscosity, which were completely beyond the range of phenomenological thermodynamics. Was this a final triumph of modelism over formalism? It was no more than a peripateia.

Gibbs, entering the scene, proposed a powerful and abstract scheme of statistical mechanics, no less austere than energetics had been. Such it has remained to this day, where Jaynes has reduced it to the bare skeleton of an axiomatization inside information theory. In this he goes back to the very spirit of the early theory of probabilities.

What Jaynes has done, by discarding a lot of inessential discourse and extricating the mathematical essentials — unveiling the statue, so to speak — is reminiscent of what Einstein did with the relativity theory, as it existed just before him in the minds of Lorentz and of Poincaré. Lorentz and Poincaré were busy with a modelistic theory of relativity implying the 'ether' concept. But the 'ether' was merely a hindrance which Einstein discarded, thus producing the true relativity theory — an overwhelming triumph of formalism over modelism.

Today, in the field of quantum mechanics, some physicists are more or less trying to repeat the story of statistical mechanics: 'explain' its abstract, probabilistic and very operational rules, by things occurring at a 'subquantal' level. Of this nothing will be said in the present book. The complexity of the tentative 'hidden variables theories' is disheartening to me, too reminiscent of the intricacies of the 'luminiferous ether' theories. Therefore we shall essentially stick to the formalism as it exists, striving to produce a coherent interpretation of it in such arduous question as the Einstein—Podolsky—Rosen correlations. In this I am definitely not denying that there may well be a 'subquantal level'; but I deem it unlikely that it should obey rules resembling those of the old, pre-Bornian, probability calculus.

The implication of what has been said is that, in theoretical physics, the switches from modelism to formalism are far more important, and have much more far-reaching consequences, than those from formalism to modelism.

1.1.2. PARADOX AND PARADIGM

For the word 'paradox' most dictionaries give, as the fundamental definition, something akin to 'a surprising, but perhaps true statement'. It is the etymological meaning. For example, Copernicus's heliocentrism was such a 'paradox' in its days.

When the pickaxe of scientific labor hits a hard paradoxical fact of this sort, which can be neither uprooted nor bypassed, a new appropriate 'paradigm' (in Kuhn's (1970) wording) must be invented to surmount it.

First a 'mathematical recipe' must be found, expressing the surprising facts. Thus Copernicus, replacing the Earth as a reference frame by axes centered at the Sun and piercing the celestial vault in fixed directions, marvellously simplified celestial kinematics; just as Lorentz and

Poincaré, defining a new group for exchanging spatio-temporal reference frames, resolved a whole complex of riddles in electromagnetic kinematics.

But a 'scientific revolution' is achieved only when a conceptual discourse, closely fitting the mathematical formalism, is produced. This is what Einstein did in 1905 and Minkowski in 1908 for the relativity theory. By reducing the conceptual wording to its essentials, and tailoring it strictly to the mathematical body, the designer of the new paradigm dresses the explorer in a suit appropriate for further progression.

This is a triumph of formalism over modelism, the essence of a true 'scientific revolution'. Let it be recalled that, before Kuhn, Duhem (1954, pp. 121—131 and 144—164) had very clearly outlined this 'paradox and paradigm' [1] game.

Most often the new paradigm, consisting jointly of the mathematical recipe and of the description fitting it, itself looks 'paradoxical' for some time — and sometimes even a long time. Thus it is a 'surprising but successful explanation'.

Usually also it is felt by many that it is 'no explanation at all', but merely a sort of mathematical pun. This has been said of both Newton's action-at-a-distance theory and of Einstein's relativistic kinematics. As pointed out by Mermin (1981) this is presently said of the quantal rendering of the Einstein—Podolsky—Rosen correlations.[2] Such objectors suffer from a longing for modelism.

As thus outlined, as if from the top of an 'ivory tower', a 'scientific revolution' along the Duhemian and Kuhnian scheme may not look very dramatic. The truth is, however, that it usually is full of 'sound and fury'. For the discoverer it is a sort of storm on Sinai, where the new truth flashes. And part of the 'intelligentsia' finds that the Moses coming down with the tables of the new law is worrisome and disturbing. Both the Galileo affair and the discussions on the relativity theory just after World War I, are examples of such a peripateia.

1.1.3. UTILITY OF DIMENSIONAL ANALYSIS. UNIVERSAL CONSTANTS

The constants inside a law in mathematical physics are said to be 'universal' if they do not depend on the matter considered. For example, the fundamental Galileo—Newton formula $\mathbf{F} = m\mathbf{a}$ is 'universal' in this sense; the multiplicative constant implicitly contained in it is defined as unity, with zero dimension, by a combined choice of the standards of force, mass and acceleration. This formula holds for any 'point particle'.

If two physical magnitudes, differing in their mode of perception, show up homogeneously in a universal formula, they are said to be 'mutually equivalent' and can then, by a joint redefinition of the standards, be expressed in the same unit. Such magnitudes, notwithstanding their *a priori* dissimilarity, are thus of the same nature. That this is true of $\mathbf{F}$ and $m\mathbf{a}$ is shown by the phenomenon of 'inertial forces'.

The discovery of a new universal constant (or the discovery that a magnitude already known is a universal constant) is a corollary to the invention of a new synthetic theory, uniting classes of facts previously distinct. Thus Planck's constant h 'universally' relates a frequency to an energy, or a 4-frequency to a momentum energy — this being the heart of de Broglie's wave mechanics. And thus the velocity of light *in vacuo* is Einstein's 'equivalence coefficient' between length and time (and also, by its square, Einstein's equivalence coefficient between mass and energy).

Should we say, for example, that the self- and mutual inductance coefficients of the electrostatics and the electrokinetics of the vacuum, the common dimension of which is a length, are lengths? Certainly we can, no more and no less than we do with the coordinates of a barycenter: *they* also depend only on the geometry of the system.

Here is another classical example, going crescendo. Boyle and Mariotte find the isothermal compression law of gases, Charles and Gay-Lussac the universal dilatation law of perfect gases, all this leading, via the redifinition of the zero of temperatures, to the universal law $pv = RT$, and then, via Avogrado's and Ampère's molecular hypothesis, to $pv = NkT$. Maxwell's and Boltzmann's kinetic theory interprets k as an equivalence coefficient between a temperature and a kinetic energy;[3] and Gibbs's formalism of statistical mechanics later leads to the interpretation of $k \ln 2$ as an equivalence coefficient between an information and a negentropy.

One of the most important universal constants is Newton's gravitational constant G. In 1898 Planck noticed (before his discovery of the quantum theory, and before Einstein's discovery of the relativity theory) that the three constants G, c, and h (the latter deduced from Wien's α and β coefficients) are dimensionally independent (in the M,

L, T system of units), and thus yield natural standards of mass, length and time. Let us look at the matter more closely, because it is most interesting.

The Planck mass M is related to G, c and h (or to $\hbar \equiv h/2\pi$) via the formula

$$GM^2 = c\hbar$$

which is quite similar to the formulas

$$e^2 = (1/137)c\hbar$$

and

$$g^2 = 13.4\, c\hbar$$

respectively holding in quantum electrodynamics and in nuclear physics. It is thus extremely natural (Costa de Beauregard, 1961) to interpret M as a basic gravitational charge ($c\hbar/G$ being its square). In terms of elementary particle physics this mass standard is 'tremendous', having the value 2.16×10^{-5} grams! Thus the preceding formulas show that, as expressed in terms of the mass quantum M, gravitation is a strong, not a weak, field![4]

The Planck length L is related to the Planck mass M via an equation that has a form that is familiar in general relativity, namely

$$GM = c^2 L.$$

It is exceedingly small, 1.6×10^{-33} cm. Finally, of course, Planck's time is

$$T = L/c,$$

which equals 5.3×10^{-43} s.

1.1.4. 'VERY LARGE' AND 'VERY SMALL' UNIVERSAL CONSTANTS

Strictly speaking, to say that a 'universal constant' is very large or very small makes no sense of course. What makes sense is to notice that it is *expressed* as very large or very small in units we find 'convenient'.

'Convenience' of the units chosen reflects in some way our existential situation-in-the-world. For example, the velocity of light in the vacuum, c, is very large when expressed in terms of meters and seconds. We find it convenient to choose the meter and the second as associated length and time units because they are adapted to the human size, and because the velocity of our nervous impulse is not a very large multiple of the meter per second.

So, Einstein's c is 'very large'; Planck's h (or $\hbar = h/2\pi$) and Boltzmann's k (or $k' = k \ln 2$) are 'very small'. This very clearly says that the phenomena of relativity, of quantum mechanics, and those pertaining to the information interpretation of entropy, are largely outside the domain of our everyday experience. Modern physics, so to speak, like today's astronautics — or like the 16th-century seafarers — is heading outside the realm of near familiarity into the deep surrounding unknown.

When a universal constant is extremely large or small, it is a time honored exercise to let it go formally to infinity or to zero, and examine what happens: by letting $c \to \infty$, one loses Einstein and recovers Newton; by letting $h \to 0$, one loses Einstein's photon or de Broglie's matter wave, thus recovering Maxwell's electromagnetism and classical mechanics, respectively; and by letting $k \to 0$, one loses cybernetics and falls back on the theory of 'epiphenomenal consciousness'.[5] All this will be examined in due time.

1.1.5. TODAY'S SCIENTIFIC HUMANISM

Physics, in this century, has renewed its themes with such vigor and organizing power that it may well be that *now* its great classical age begins. Humanism, as it seems, has not yet fully perceived how the world view has been radically transformed. No fundamental problem in physics (of course), or biology, or psychology, or even philosophy, can be posed today with the same perspective as it was not very long ago.

When Magellan's companions reappeared from the east having left sailing west, everybody at home had to experience for himself the intellectual adventure of those thoughtful Greek observers who had convinced themselves that the Earth is spherical.

Today, by using cathode-ray tubes, electronic microscopes, photocells, and whatnot, thousands of technicians the world over are coping with problems embodying the quantum and relativistic theories, which they must have at their finger tips. But one has just to open a journal in theoretical physics to see articles written (and, of course, thought out) from beginning to end in terms of four-dimensional geometry and of Born's wavelike probability calculus.

Inevitably the whole world of new ideas, two examples of which have just been given, will penetrate the cultural environment. New educational techniques will have to render them familiar to young people. Each of us has learnt, on his mother's knees, that our Earth is a sphere rotating on itself and circulating around the Sun, and has then easily believed it. Very soon school children will be accustomed to reason in terms of four-dimensional diagrams. And no thinker, whatever his discipline, will be allowed to ignore such fundamental breakthroughs.

1.1.6. EPISTEMOLOGY AS UNDERSTOOD IN THIS BOOK

Epistemology, as understood in this book, consists essentially of a reflection and a comment upon the operational scientific theories. Thus it is exactly the opposite of a systematic enumeration of *a priori* categories. Where the scientific formalism penetrates, the epistemological discourse must penetrate also, whatever the difficulties along the way. Epistemology, as understood here, is a sort of Socratic dialogue the physicist has with himself, aiming at resolving the paradoxes encountered along the route — which of course are not *essential* paradoxes if Nature lives with them, and if a mathematical recipe controls them.

PART 2

LAWLIKE ‘EQUIVALENCE’ BETWEEN TIME AND SPACE

CHAPTER 2.1

THE MORE THAN TWO MILLENNIA OF EUCLIDEAN GEOMETRY

2.1.1. 'EUCLIDEAN THEORY OF SPACE'

Euclid's 'Elements' of geometry are regarded as a masterpiece of scientific writing. This lucid systematic exposition of the discoveries of the Greek geometers can be taken as the first example of a physical theory, the 'Euclidean theory of space', as Emile Picard (1905, p. 10) put it.

Three-dimensional Euclidean geometry remained unchallenged until the middle of the 19th century, when Lobachewski and Bolyai published their independent discovery of a non-Euclidean geometry. Before them Gauss, the discoverer of the intrinsic theory of curved surfaces, had of course hit upon the idea, but had not published his results. It is said, however, that he undertook a geodesic experiment, testing whether or not the Euclidean sum rule of the angles of a large triangle were violated. Nothing of this sort was found, but the implication clearly was that Gauss considered Euclidean geometry as a 'falsifiable' physical theory. Today we know that the phenomenon Gauss was interested in does exist, but on a much larger scale; that is, the cosmological scale. The spacetime of Einstein's general relativity theory (1915) is no longer Euclidean and neither are its three-dimensional spacelike sections. Indeed, the Gaussian-like phenomenon does show up in the bending of light rays passing near a massive body such as the Sun.

Anyway, during the more than two millennia of its reign, Euclidean geometry has not only been a wonderful playground providing numerous ingenious and elegant properties of figures, but also the technical background with which the discoverers and users of classical kinematics, dynamics, electromagnetics, crystallography, and whatnot, formulated their discoveries. Even the advent of the special relativity theory, around 1905, did not put an end to Euclid's ruling: three-dimensional space, in the special relativity theory, is no longer 'absolute', in the sense that once the inertial frame is changed, it no longer maps into itself, but it does remain Euclidean. Also, the

Poincaré—Minkowski spacetime of special relativity is 'flat' or 'pseudo-Euclidean'. It is, in fact, the true sucessor of Euclid's geometry as the theater of physical phenomena.

2.1.2. 'IS IT FALSE THAT OVERNIGHT EVERYTHING HAS DOUBLED IN SIZE?'

This is the way Schlesinger (1964) puts a question addressed to us by Clifford, Riemann and Poincaré. Nothing guarantees, they speculated, that the length of a solid rule does not depend upon its place in space, its orientation, or the epoch — provided, of course, that all relevant 'physical perturbations' depending on constraints, temperature, and whatnot, have been eliminated or taken care of. The list of possible relevant perturbations has never ceased to increase, of course, which is a clear indication that an appropriate universal postulate must underlie this whole problem.

Riemannian arbitrariness of the space metric (inside spaces admitting such a metric) implies that the dependence we are speaking of be entirely geometric in its nature. This is perfectly all right, but there remains the experimental fact that so-called solid bodies do exist, and that — provided again that all relevant physical precautions have been taken care of — two solid rules found congruent to each other when once placed side by side, are again found congruent when such a comparison is repeated at another place, another time, and in a different orientation. This is why the two qualifications of 'solid' and 'indeformable' are largely synonymous with each other.

And so, the expression of the space metric which is based on the assumption — that is, on the definition — that solid bodies are indeformable (provided, of course, etc., . . .) is a privileged one. It is all the more privileged since experimentation does show that the metric thus defined is, to a very high degree of precision, Euclidean.

More exactly, it is the combined experimental phenomenology of solid bodies and of light rays that turns out to be, to a very high degree of approximation, Euclidean. Light rays are found to be rectilinear when the paradigm of solid bodies is adhered to, and, conversely, a consistent use of light rays allows us to set up a whole 'solid' background. A consistent use of telescopes and goniometers, the pieces of which are 'solid' and the theory of which is formalized inside Euclidean geometry, has consistently validated the paradigm of

Euclidean geometry inside the classical context — including, as previously reported, the experiment by Gauss.

2.1.3. ABSOLUTE TIME AND CLASSICAL KINEMATICS

Classical kinematics was born from the union of Euclidean geometry with the concept of an 'absolute time' t, later clarified by Newton. It consists largely of the consideration of relative motions of solid bodies or figures, sliding one over the other in the two-dimensional case. Thus have emerged, in the Euclidean plane, the concept of the 'instantaneous center of rotation', the motion being, at each instant t, 'tangent' to a rotation; and, in the three-dimensional case, the concept of the 'instantaneously tangent helical motion', as formalized by the 'wrench' concept.

All these clever and elegant concepts have been rejected by the special theory of relativity where, strictly speaking, motions of such sort are impossible and where the very concept of a solid or indeformable body is not accepted. Nevertheless, the concepts and theories of classical kinematics remain useful as a sufficient, and very clarifying, approximation in the mechanical technology of, say, planetary gears and things of that sort. So, although the classical kinematics has been outlawed when the new, relativistic, kinematics has become the ruler, nevertheless quite a faithful clientèle has followed her in her exile, where the old game continues to be played and is still found to be pleasant.

2.1.4. THE CLASSICAL 'PRINCIPLE OF RELATIVE MOTION'

Belonging essentially to classical kinematics, this principle states that any motion can be legitimately referred to the 'absolute time' t on the one hand, and to any of the 'solid reference frames', as accepted in this kinematics, on the other hand. For example, according to classical kinematics, it is equivalent to use as a 'solid reference frame' either the village square, or the merry-go-round turning on the village square; either the village square or the merry-go-round could be thought of ideally as being at rest. Let us remember that this argument had been opposed to Galileo by his judges, the rotating Earth being the merry-go-round, and the center of the Sun, together with the configuration of the distant stars, being the 'village square'. Also, a rifle bullet in helical motion is considered as a valid 'solid reference frame' in classical kinematics.

The next chapter will relate how, among other things, the dynamics of Galileo and Newton did not accept the kinematical 'principle of relative motion'. It conferred an 'absolute' meaning, supported by observable facts, to accelerations and rotations of 'solid reference frames'. Therefore it defined a 'restricted principle of relativity' where only uniform relative velocities of solid reference frames are allowed, the privileged family thus defined being termed the family of 'Galilean reference frames'.

So there existed a discrepancy inside the realm — or, should we rather say, the confederation — of classical kinematics and dynamics, as each of them had its own relativity principle.

This state of affairs persisted until the rise of the relativity theory. Then the dynamical, or restricted, relativity principle, appropriately modified, became the universal relativity principle. An entirely new kinematics, based on the restricted relativity principle, became the ruler over the entire realm of physics.

CHAPTER 2.2

THE THREE CENTURIES OF NEWTONIAN MECHANICS: UNIVERSAL TIME AND ABSOLUTE SPACE

2.2.1. REMARKABLE APHORISMS BY ARISTOTLE

"Time numbers motion in reference to before and after" is an often quoted sentence of Aristotle. For him motion meant what we call change; to him, our motion was 'local motion'. So Aristotle's aphorism referred to both the reversible aspect of time as manifested in spatial motion and as applied in the technologies of chronometry, *and* to the irreversible aspects of time which Aristotle discussed under the wording of 'generation' and 'corruption'. The physicist is concerned with both aspects, as already mentioned in the Introduction — and as will be more evident in later chapters.

There is, however, another Aristotelian aphorism, not unrelated to the former, very relevant to the physicist, and indeed very far reaching, notwithstanding its apparent circularity: "we measure time by means of motion and motion by means of time". Let us take 'motion' as spatial motion. What we have here is the very definition of chronometry, either celestial or terrestrial, as rendered operational by Kepler, Huygens, Harrison and their followers.

So do sleep in peace, old Aristotle, till the days of the Renaissance. Many cycles of years and days will number Time, many generations and corruptions will occur in not quite repetitive succession; also, many craftsmen's devices such as sandglasses, clepsydras and rudimentary clockworks will appear before the days of Galileo and Newton, the days of the scientific and technological breakthrough of *measurable time*. Of course, both Ptolemy's and Kepler's laws are mathematical recipes for celestial kinematics, and both the medieval and Huygens's devices are clockworks; but, in between, a radical change in spirit has occurred. There are examples akin to this in biological evolution, when an entirely new functional organ is derived from a pre-existing one that had a quite different use.

So it seems as if the old Aristotelian dream of measurable time, dormant for many centuries, suddenly germinated and sprouted as if triggered by a sudden inspiration.

2.2.2. KEPLER (1571—1630) AND GALILEO (1564—1642): CELESTIAL AND TERRESTRIAL MECHANICS

Copernicus (1473—1543), basing his reasoning upon his own astronomical observations, resurrected Aristarchus's beautiful hypothesis, and placed the Sun at the center of our planetary system; he also referred the angular motions of the planets not to the Earth but to the distant array of the so-called fixed stars — an ominous background the gravity of which, according to Mach, may well rule the local laws of our dynamics.

So, having King Sun rather than Mother Earth as the ruler, the pieces of the puzzle immediately fitted much better. Ptolemy's epicycles became useless, and, as a zeroth-order approximation, all planets, the Earth included, move along concentric circles with uniform velocities.

Then Kepler stepped in, and, using Tycho Brahé's measurements, produced his three famous laws. The second of these states that the areas swept by the radius-vector joining any planet to the Sun are proportional to time, and the third one that the squared revolution durations are proportional to the third power of the major axes.

Let us stop at this solemn event in the history of science. If there were only one planet, there would be no 'third law'; and, then, the 'second law' would not really be a law, but a simple decree stating how we should 'number time'. But, as there is more than one planet, Kepler's second law expresses a mutual coherence of motions such that the planetary system is indeed a big clockwork, 'numbering time' consistently. Then the 'third law' gains its full status. Truly, this is a brand new celestial kinematics! As numbered by this planetary clock, Time will soon be termed 'universal time' by Newton.

Now, as we have said, the 'motions' of this clock are referred to a certain spatial frame, having its center at the center of the Sun and axes piercing the celestial vault in fixed directions. Is this an 'absolute spatial reference frame'? This is a big question, needing serious thought.

So let us get down to Earth with Galileo. Not that Galileo was uninterested in celestial 'motions' ('local' or otherwise): satellites of Jupiter, sunspots, and what not. But his outstanding discoveries were in the fundamentals of dynamics, and he arrived at them by terrestrial experimentation — in fact, mainly via cleverly devised thought experiments, the intellectual tool used again and again by creative theoretical physicists.

Galileo's most brilliant discovery is that of the *principle of inertia*, which is intimately tied with the *dynamical principle of relativity of motion* and is a first approximation to the *universal principle of special relativity*, as it came to be named later. There had been significant medieval hints in that direction, but Galileo was the first to grasp the idea plainly. His principle states that a particle endowed with mass, and free from all external influence, will be either at rest or in uniform motion — these two statements being essentially equivalent.[1]

So let us consider first two different spatio-temporal frames of reference, in the sense of the classical kinematics discussed in the previous chapter. Using of course the same time scale in both, we see that the condition for invariance of the principle of inertia under change of the reference frame is that both frames are in uniform motion with respect to each other. This is a very drastic restriction to the *kinematical principle of relative motion* discussed in Chapter 3, according to which any two solid reference frames were considered equivalent for the description of motions! On the other hand, the new Galilean principle goes far beyond the more or less intuitive idea that there exists a preferred, or 'absolute', spatial reference frame — as implied, for example, in Aristotle's idea that all moving bodies tend spontaneously to come to rest. So the *principle of inertia*, and the closely connected *dynamical principle of relativity*, are something significantly new in physics — something truly needing a 'lynx-eyed' discoverer, for no convincing direct experimental evidence, neither celestial nor terrestrial, was at hand. Kepler's seemingly eternal motions were elliptical, and terrestrial rolling balls do decelerate. So Galileo's principle could be vindicated only by its numerous consequences, most of which were to be unravelled later.

Now, what of the time scale? Clearly, once the spatial question is settled, another condition is required for invariance of the inertial law when changing the reference frame: that the two time scales keep in step; that is, are linearly related to each other.

So, the *dynamical principle of relativity* states first that there exists *a preferred set of equivalent space reference frames, all in uniform motion with respect to each other* — the family of 'Galilean frames', as they came to be named — and *a preferred time scale*, which it is only natural to call Galilean also. In these spatio-temporal frames, and these alone, the principle of inertia holds. If it is then asked which are these space and time reference frames such that the principle of inertia holds, the answer is: these are the Galilean space-and-time frames. We here find an example of the circularity, or 'autofoundation', of all fundamental physical principles, as discussed by Gonseth (1964).

What, then, of the Copernicus and Kepler astronomical reference frame? It turned out later, through Newton's work, that this frame is almost exactly a Galilean frame, because the center of mass of the Solar System is extremely near the center of the Sun. The rotating Earth, however, is not at all a Galilean frame, as demonstrated later by the Foucault pendulum experiment. However, in the days of Galileo's trial all this was far from clear. When Galileo's judges objected that the Earth is just as good a reference frame as the Sun and the stars, they were obviously adhering to the old kinematical relativity principle — all the more so as the Earth is a very good place at which to sit and discuss the matter. On the other hand, when grumbling (as it is said) "e pur si muove", poor Galileo, no less certainly, was adhering to his new dynamical relativity principle, but of course unable to explain this convincingly. It is very sad indeed that this discussion was not an academic one, and that the unjust weight of the argument of authority on one side and the too understandable impatience of the discoverer on the other side (why did he not simply admit that 'everything goes on as if'?) have produced a judiciary error extremely harmful to both.

As for the matter of chronometry, to which we are coming, it also turned out, through Newton's work, that Kepler's celestial clockwork and Galileo's uniformly moving free-point particle do indeed measure one and the same Time.

2.2.3. THE UNIVERSAL GALILEO–NEWTONIAN LAW $\mathbf{F} = m\ddot{\mathbf{r}}$

Newton (1642–1727) was born the very year Galileo died. Co-discoverer, with Leibniz, of the differential and integral calculus, he was able to codify mathematically Galileo's discoveries, to which he added his own concept and formula of universal gravitation.

So Newton's 'second law' relates the acceleration $\ddot{\mathbf{r}}$ of a point particle of mass m acted upon by a force $\mathbf{F}$ according to the universal formula

$$\mathbf{F} = m\ddot{\mathbf{r}}, \qquad (2.2.1)$$

which is obviously 'Galileo-invariant'; that is, unchanged under a change of inertial frame (mass, time scale, and force being unchanged).

What then occurs for those changes of 'solid reference frames' allowed by the kinematical principle of relative motion? An easy calculation yields the expression $-m\ddot{\mathbf{r}}$ of the ordinary inertial force produced by a linear acceleration of the frame, and a more sophisticated one that of the 'Coriolis force' $2\boldsymbol{\omega} \times \mathbf{r}$ produced by a uniform angular rotation of the frame. The Earth is such a rotating body, where the Coriolis force shapes, for instance, the tradewinds and the whirlwinds; it is the basis of the gyroscopic compass, and has been conspicuously demonstrated by the famous Foucault pendulum experiment (1851). All this is 'deduced rationally' by combining Euclidean geometry, the concept of 'universal time', the classical description of the motions of rigid bodies, and the fundamental dynamical formula (1).

By itself this formula — the cornerstone of the whole construct of classical mechanics — is a *universal chronometer* defining, for the first time in history, *time as a measurable magnitude*. This is because, in this universal formula — the explicit expression of which is

(2.2.2) $\mathbf{F} = m\,\mathrm{d}^2\mathbf{r}/\mathrm{d}t^2$,

— the vectors $\mathbf{r}$ in Euclidean space, the forces $\mathbf{F}$, and the masses m, are all measurable magnitudes, their respective additions being validly defined via appropriate procedures. Therefore Time (that is, in this frame of thought, Newton's, and everybody's, universal time) is defined, for the first time in history, as *an* (*indirectly*) *measurable magnitude*. This, indeed, is a great step towards the realization of the old Aristotelian dream: for one-third of the parameters (measurements of $\mathbf{r}$) time is 'spatialized', while, for two-thirds (measurements of $\mathbf{F}$ and of m) it is 'dynamicized'. Of course, the Keplerian elliptical motions, and the (ideal) Galilean inertial uniform motions, directly yield spatialized measurements of Time; but these are *not universal* measurements, because the concept of a universal velocity standard is lacking in classical mechanics.

As for practical clocks the choice is very large in principle, be they astronomical[2] (the circulating or the rotating Earth, for instance) or terrestrial (Galileo's pendulum, as applied by Huygens, for example). Each and all of them are, so to speak, various incarnations of the one and same universal principle expressed in formulas (1) or (2). And if it happens (it always happens) that any one of these material clocks is found to be faulty, it is immediately put aside and replaced by a better one; that is, by a more faithful representation of the ideal clock, as expressed fundamentally in formula (1) or (2). *During the three centuries of the reign of Galilean–Newtonian mechanics,* $\mathbf{F} = m\ddot{\mathbf{r}}$ *has been the universal clock defining time as a measurable magnitude.*

Most historical textbooks quote gunpowder and the compass as technical revolutions having affected the course of history, and this of course is very true (alas for gunpowder). But these textbooks underrate the revolution of exact timing, which has not only drastically changed our whole way of life, but also, by allowing us to compare 'astronomical' and 'terrestrial' time, has rendered possible the measurement of longitudes at sea. Compass, sextant and chronometer have been, till the eve of this 20th century, the triad of fundamental aids to navigation.

Now, what of those imperfections which lead to discarding, one after the other, astronomical or terrestrial clocks? All can be traced back to accidental irregularities and to wear; that is, to statistical irreversibility — the second aspect of time to be discussed later. For example, the rotating Earth, besides being irregular in its motion, is continuously slowing down. So when Galileo, pointing his telescope at the Sun, discovered sunspots he showed the incorruptibility the Greeks had attributed to celestial bodies to be false. And when Newton — by likening the falling of an apple to the attraction to Earth of a Moon that was nevertheless kept from falling by Huygens's centrifugal force — thus opened up the way to the skies of Galileo's mechanics, he also opened upwards the route of both Law and Perturbation of the Law; that is, to the 'Second Law'. The skies themselves are wearing down, but this is another story to be told later.

Concluding this section, we state that *before Galileo and Newton, time was not a measurable magnitude. The very possibility of defining time as a measurable magnitude rests on the principle of inertia; that is, on the dynamical relativity principle.* No 'scientific revolution' has been greater than this one.

2.2.4. 'GREATNESS AND SERVITUDE' OF CLASSICAL MECHANICS

All through the 17th, 18th and 19th centuries a constellation of highly gifted mathematicians — Huygens, Leibniz, the Bernouillis, Euler, Laplace, Lagrange, Hamilton, and quite a few others — systematically derived consequences of the Galilean–Newtonian principles, producing the dynamics of systems of point particles, of solids, of elastic bodies, of fluids, and what not, together with many celestial or terrestrial applications, and also very powerful abstract formalisms termed 'analytical mechanics'. The latter are still under scrutiny today, following the works of Poincaré and Cartan on invariant integrals, and expressed in terms of 'symplectic geometry'. Even the advent of the quantum and the relativity theories has not put an end to technological applications of classical mechanics, which remains in many cases a perfectly valid and useful approximation.

On the whole, 'rational mechanics' (as classical mechanics came to be named) grew in the prestigious and regular style of a Versailles Palace. An overall glance at the foundations of the building is in order now.

Universal time, as codified by Newton, was the main one. Let us recall Newton's own very solemn words (1687, p. 10): "Absolute, true, and mathematical time, of itself, and of its own nature, flows equably without relation to anything external, and by an other name is called duration." So Newton, as it seems, likens the 'flow of time' to that of a continuously running river, carrying from source to sea events witnessed in turn by anyone standing on a bridge. This looks much like some prefiguration of the relativistic spacetime picture, where, just the opposite, the water of the river is considered at rest, but the onlookers — you and I — are thought of as swimming upwards, towards the future.

This Newtonian concept of flow and duration differs somewhat from the man-in-the-street's view, according to which 'only the present exists', or 'is real', 'the past existing no more' and 'the future not yet'. It most certainly differs also from how time was generally understood by the practitioners of classical mechanics. As Bergson (1907, Ch. 1) put it:

> The present state of a system is defined by equations where enter [time derivatives], that is, essentially, actual velocities and actual accelerations. Therefore only the present is at stake, a present, however, taken together with its tendency. So, in fact, the systems upon which science operates are inside an instantaneous present renewing itself ceaselessly, and never inside the real, concrete, duration where the past adheres to the present. When the [classical] mathematician computes the future state of a system after some time t, nothing prevents him from assuming that, in between, the material universe has vanished, only to suddenly reappear later. . . . What flows in between, that is, real time, does not count, and cannot enter the computation. . . . Even when dividing the time

interval in infinitesimal parts by use of the differential dt, he merely implies that he will consider present accelerations and velocities . . . ; always it is a given instant, that is, a 'stopped' one, that is at stake, and not the time which flows. On the whole, *the world upon which the* [classical] *mathematician* [physicist] *operates is a world which dies and springs up at every time instant — the very same one to which Descartes was referring when speaking of continued creation.*

This seems to me a faithful description of what most classical mathematicians and physicists had in mind while at work. That it is a conception metaphysically different from Newton's seems clear. However, a little thinking tends to vindicate Newton's intuition and should correct the erroneous view so aptly described by Bergson. Given at some time t the complete set of positions and velocities of classically interacting point particles we find by integration the complete evolution of the system, from past to future, and thus *the system does have mathematical 'duration'*; so, the 'present instant' t of Newton's — and in fact everybody's — concept of 'universal time' is truly like a bridge crossing the flow.

It should be remembered also that, inside the very formalism of classical mechanics, an elegant and useful algorithm exists, with implications very different from those criticized by Bergson: the *extremum action principle* of Euler, Maupertuis and Hamilton, using *an integral extended over time* from an initial to a final instant of evolution. Moreover, this algorithm is *essentially symmetric in past and future.* Each of these points gave rise to some dreaming, quickly followed by sedative comments. Nevertheless, as we shall see later, there *is* indeed good reason for dreaming of this. . . .

It should be remembered also that Lagrange's programme of analytic mechanics included a treatment of time similar to that of the space coordinates — but this could not go very far, in the absence of a spacetime metric. So, in fact, it is quite true that, as Bergson (1907, Ch. 4) put it "*modern* [that is, classical] *science is mainly characterized by its tendency to use time as an independent variable*". The advent of the relativity theory was to change this, by *not* using time as an independent variable, and placing the four 'coordinates' x, y, z, ct on an equal footing.

As for *space* — Euclidean space — it was termed *absolute*, being changed into itself by both the (old, larger) kinematical and the (new, more restricted) dynamical relativity principles. The transformation formulas for the Galilean group

(2.2.3) $t' = t \qquad x' = x - ut \qquad y' = y - vt \qquad z' = z - wt,$

where u, v, w denote the Cartesian components of the velocity of the primed frame with respect to the unprimed one, leave unchanged time and the Euclidean metric of space, but do produce a time dependent expression of the parametrization of space. Of course, there is also the possibility of producing a more intricate parametrization by making use of the kinematical principle of relativity, but then, dynamically speaking, the inertial forces enter the picture.

Newton's (1687, p. 10) characterization of 'absolute space' is appropriate at this point: "Absolute space, in its own nature, without relation to anything external, remains similar and immovable. Relative space is some movable dimension or measure of absolute space."

Thus it seems that Newton, as in fact everybody else, had in mind something more than what his mathematics was saying: that there should exist a preferred spatial reference frame, with respect to which any motion could be termed 'absolute'; that is, different from rest. Did this stem perhaps from some longing for 'terra firma', a concept shaken by the Copernicus and Galileo earthquake? It may be.

Anyway, as it was quite certain that dynamics by itself could definitely not produce anything of this sort, the 19th-century classical physicists undertook an interminable "Hunting of the Snark" named, in that case, 'luminiferous aether', for they believed that light waves had to be propagated in a 'medium'. Such a 'medium' would, of course, make a quite decent 'absolute spatial frame'. This is a story to be told in the next chapter.

2.2.5. GRAVITATION

This book is concerned only marginally with gravitation. It is impossible, however, not to refer briefly to Newton's great discovery of *universal gravitation*,[3] which he expressed fundamentally as the mutual attractive force between two massive point particles separated by a distance **r**:

(2.2.4) $\mathbf{F} = \pm\, G m_1 m_2 r^{-3} \mathbf{r};$

G denotes the universal gravity constant, the value of which is presently given as

$$G = 6.664 \times 10^{-8}\,\mathrm{cm}^3\,\mathrm{g}^{-1}\,\mathrm{s}^{-2}.$$

What is typical of gravitation is that the masses appearing in formula (4) are the same as those appearing in the inertial formula (1) or (2), so that m cancels out when such formulas are combined. Therefore the gravity field is a field of accelerations. This was clearly understood by Galileo in his thought experiment in which, say, n bricks are allowed to fall freely, either stuck together by cement, or not. Galileo argued that, of course, the n bricks separated will fall with identical motions, which will be also the motion of the n bricks stuck together. This reasoning illustrates the 'equivalence' of inertial and gravitational mass — the starting point of Einstein's thinking about gravitation. Also, this illustrates the additivity of masses, given the additivity of 'weights' via the use of scales.

2.2.6. SYMPLECTIC MANIFOLDS AND ANALYTICAL MECHANICS

Quoting J. M. Souriau (1982, pp. 352—354):

> Symplectic geometry resembles much, at the start, Euclidean geometry ... but the symmetry condition $g_{kj} = g_{jk}$ [on the metric tensor] is replaced by the antisymmetry condition $\sigma_{kj} = -\sigma_{jk}$. ... This is possible only if the manifold is of even parity. The symplectic structure was discovered in 1811 by Joseph-Louis de Lagrange; the covariant and contravariant components of the σ tensor are Lagrange's 'brackets' and 'parentheses'. Their discovery is due to an in-depth investigation of the structure of the equations of dynamics. The manifold to which Lagrange confers a symplectic structure is the set of solutions of the 'equations of motion' of a dynamical system — the space of motions, shall we say. This theory, developed in [his] 'Analytical Mechanics' ... was not fully understood by his contemporaries and followers; Poisson and Hamilton, for example, used it in a restricted form, [thus] occulting together the global and the relativistic properties of dynamics. A century later, extending Henri Poincaré's work on 'integral invariants', Elie Cartan re-invented the symplectic form, ... thus allowing the true dimensions of Lagrange's work to reappear. ... The symplectic structure also shows up in all spaces the points of which are solutions of a variational calculus. ... But the variational formulation of dynamics is a regression from its symplectic formulation.

So speaks a mathematician who is very conversant with theoretical physics. Quoting now Abraham and Marsden (1978, p. 402):

> Some believe that the Lagrangian submanifold approach will give deeper insight into

quantum theories than does the Poisson algebra approach. In any case, it gives deeper insight into classical mechanics and classical field theories.

Other aspects of analytical mechanics are, of course, the variational Hamilton–Jacobi scheme, well suited for displaying the connection between classical mechanics and wave mechanics; and the well-known Hamilton formalism favoring time and energy, thus largely blocking the way towards relativistic dynamics.

CHAPTER 2.3

THREE CENTURIES OF KINEMATICAL OPTICS

2.3.1. FERMAT (1601–1665) AND HUYGENS (1629–1695)

It is interesting that, at its very dawn, kinematical optics started from the idea which, at its zenith, it proclaims with a fanfare: that *time is actually extended*!

For deducing Snell's and Descartes's laws of reflection and refraction Fermat proposed an axiom quite typical of the ideas mathematicians do bring into physics: that light selects the path which makes its transit time as short as possible. As shown later by Hamilton, this transit time may also be the longest possible. Therefore, the general statement of Fermat's principle is that the transit time of light is an 'extremum', or 'stationary'. The calculation, of a sort that came to be termed 'variational calculus', uses *an integral extended over time*, which moreover, is *essentially symmetric with respect to past and future* — a property to be discussed all through this work. Of course, Fermat did not explicitly use the integration algorithm; but, since the days of the Greeks, much differential, integral and variational calculus has been performed without use of explicit algorithms!

As was better understood later, Fermat's principle introduces naturally bunches of light rays, curvilinear, or broken, depending on how the refractive index varies, in space, whether continuously or discontinuously. These curves are orthogonal to families of curved surfaces, the so-called 'wave fronts'. The underlying geometrical theory is that of 'congruences of normal trajectories'. The classical discipline of geometrical optics which, even today, remains useful as an approximation to physical optics, essentially rests on it.

The wave-front concept was introduced by Huygens, who also deduced the Snell—Descartes laws from his scheme.

Both the Fermat and the Huygens concepts — which fit together quite smoothly — predict that the velocity of light is smaller inside a refracting medium than in the vacuum — just contrary to the prediction of Newton's corpuscular theory. Direct experiments by Foucault in 1850 later confirmed the Fermat—Huygens prediction. However, and quite unexpectedly, light corpuscles were resurrected by Einstein, in a manner shown by Louis de Broglie to be consistent with the wave theory.

Huygens, very naturally, believed that his light waves needed a medium to support propagation, as do sound waves. This all pervasive medium (the existence of which, for other reasons, Descartes had also assumed) was named, after the Greeks, the 'ether'.

The 'luminiferous aether', as it came to be named, was just what was needed to carry the hopes of those believing, as did Newton, in the existence of a preferred spatial reference frame, with respect to which motions should be termed 'absolute', and rest, 'absolute rest'. Whence the feeling grew that appropriate experiments in kinematical optics might well uncover effects of the 'ether wind' felt when rushing through the ether.

This can be understood quite easily. Suppose that a point source S, at 'absolute rest', emits light continually in all directions at the 'absolute velocity' c. A detector moving uniformly in a straight line through S, with an 'absolute velocity' v, ($v < c$), will first encounter wave trains of velocity $c + v$, and later wave trains of velocity $c - v$. If not passing through S, this detector will receive wave trains the velocity of which varies continuously from $c + v$ to $c - v$, depending on the direction from which they are received. All this follows from the classical assumption of (vectorial) addition of the velocity of light waves with that of a material body.

What had not been remarked explicitly was that, by conferring on the hypothetical 'ether' the position of an ultimate arbiter of 'absolute' motion or rest, one was *ipso facto* conferring on light propagation a unique and supreme status in the universal problem of measuring lengths, time intervals and motions. That light does indeed have this status, unique among all other physical phenomena, turned out to be true, and it remains to be shown why and how this is so. It was in terms of modelism that the early followers of Fermat and of Huygens were thinking, when they unwittingly tied together the fates of kinematics and of optics, not suspecting that the answer to their riddles was to be couched finally in terms of mathematical formalism. So, a dramatic stage effect lay ahead in the form of *complete identification of fundamental kinematics with the optics of the vacuum.*

Narrating the aborted tracking of the 'ether wind' makes a fantastic detective story: how the 'wanted unknown', named 'ether', each time escaped the prepared snare, so that in the end neither ether nor absolute rest or motion was found. Instead, radically new aspects of an 'absolute' emerged: the invariant velocity c of light in vacuum — a 'universal constant', a 'conversion coefficient' between space and time; and a four-dimensional pseudo-Euclidean spacetime, replacing Euclid's three-dimensional space.

However, while usually detective stories maintain the suspense, so that, in the end, the Great Detective, in a press conference, unravels the whole riddle, I choose here a quite different policy. I shall proceed in the manner of the Greek drama in which the actors, tormented for making their decision, rush into action and play their game, while the chorus, knowing too well what the gods see, and where it all leads, sings, in an atemporal mode, a song that the actors hear not.

So when, in 1676, Roemer performs the very first valid experiment in kinematical optics, the rising curtain displays an early dawn, where two young shoots do not draw the attention they should: Fermat's extremum principle, and Huygens's mutually orthogonal rays and wave fronts.

2.3.2. ROEMER (1676) AND BRADLEY (1728): THE TWO FIRST MEASUREMENTS OF THE VELOCITY OF LIGHT

Ole Roemer, at the Paris Observatory, observed in the year 1676 that the orbital period of Jupiter's satellites, as measured on Earth, is variable: it increases during the six months while the Earth goes away from, and decreases during the six months while it goes towards, Jupiter (as Jupiter's motion is very slow, in one year it has hardly moved). This is easily understood if the light emitted by the Jupiter system and received on Earth has a finite velocity, c. Then the computation of c is easy if the dimensions of the Earth's ecliptic are known.

As for these, it is a matter of geodesy and trigonometry. The passages of Venus in front of the solar disc offer at this end excellent opportunities. There were such passages in 1631 and in 1639, and geodetic measurements of terrestrial meridian arcs were performed by Norwood in 1636 and by Picard in 1665. Using these, plus his own observations, Roemer gave for c a value which, expressed in today's units, would be around 227 000 km/s — to be compared with today's accepted value 299 792.5 ± 0.5 km/s. This was very fine work indeed, and a historical breakthrough.

The formula of the Roemer effect is exactly the same as that of the Doppler effect, discovered in 1842, with the period of a Jupiter moon in place of the oscillation period of the light source. Today astronomers have reversed the procedure: they use the value of c, as obtained in high precision laboratory measurements, and measurements of the Doppler shift of light emitted by stars, for highly accurate measurements of the dimensions of the ecliptic.

Around 1728 Bradley, at Oxford, observed that the apparent positions of stars seen in a direction orthogonal to the ecliptic plane all describe equal ellipses replicating the shape of the ecliptic. This is easily understood by combining the almost orthogonal velocities of the Earth along the ecliptic and of the light coming from the stars. From his observation Bradley derived the value 320 130 km/s for c.

Roemer's and Bradley's methods are very similar to each other.[1] They use a one-way light transit, but require that the receptor — the Earth — be used in at least two different states of motion. In fact, it is used in an infinity of such states, which is a good thing for precision, and allows a closure of the measurement.

What of the ether wind in all this? It simply disappears, at first order in $\beta = v/c$, from the overall formulas, which only display the relative velocities of source and receiver. A happy coincidence, it was then thought, but a first hint towards a universal relativity principle, as we think today. Anyway, as the Earth's velocity is very small compared to that of light (30 km/s against 300 000 km/s) there is no hope that such methods can directly reveal deviations from the classical additive law of velocities. But there is another aspect to the question.

2.3.3. COULD BRADLEY'S DISCOVERY ALLOW A FORMULATION OF THE RELATIVITY THEORY?

This question is put by Yilmaz (1972) whose argument is thus: taking seriously the Fermat—Huygens concepts would imply, as it seems, the assumption of their invariance under a change of the inertial reference frame. So orthogonality of light rays and wave fronts should be preserved in such changes. Then, by a direct calculation, Yilmaz derives the Lorentz transformation.

It can be remarked that orthogonality of light rays and wave fronts implies (via the geometrical theory of envelopes) that the velocity c is the same in all directions. In other words, Yilmaz's argument amounts to the requirement of invariance of the isotropy of c — the very argument used by Einstein in his derivation of the relativity theory.

This looks all very convincing today. But in1728? The very idea of a non-additive law for composing velocities would certainly have been rejected as absurd. Also, the idea that the velocity of light has a privileged position compared with all other velocities was far from clear in those days: why light rather than falling rain, for example, which also suffers 'aberration' when viewed from a running carriage? True, the 'ether' had been likened to an absolute frame, so light, after all, had implicitly been ranked as supreme. But the time had not yet come when physicists had made up their minds to go from modelism to formalism. . . .

2.3.4. A COROLLARY TO BRADLEY'S ABERRATION: PHOTOGRAPHY OF A FASTLY MOVING OBJECT

Penrose's (1959) and Terrell's (1959) publications of a relativistic theory of photography of rapidly moving objects produced quite a stir, and prompted the appearance of confirmatory papers. Here I present a simple classical derivation of the formula in the approximation where the velocity of the object is small compared with that of light, showing the close relation between the Penrose—Terrell effect and Bradley's aberration.

Consider first two stars exactly aligned with the Earth. Via Bradley's effect they follow each other along precisely the same little ellipse, the nearer star being the leader. Therefore, by carefully observing the aberration of a star, it is in principle possible to fix its distance from Earth inside a light-year interval — but alas to no avail, because the integral part of the distance, as expressed in light years, is not known.

Consider now the ideal case of two such stars, aligned exactly on the perpendicular to the ecliptic plane passing through the Sun, at a small distance l from each other. Setting $\beta \equiv v/c$ and $\alpha = \tanh^{-1} \beta$, the distance Δl between the images of the two stars on Bradley's ellipse is $\Delta l = l \sin \alpha$.

Now we come to Terrell's photography of moving objects. Figure 1 shows, when viewed from above, a straight highway with a parallelipipedic boxcar running on it at velocity $v \ll c$; four trees have been added to make up a landscape. A photographer, stationed at O, takes a snapshot

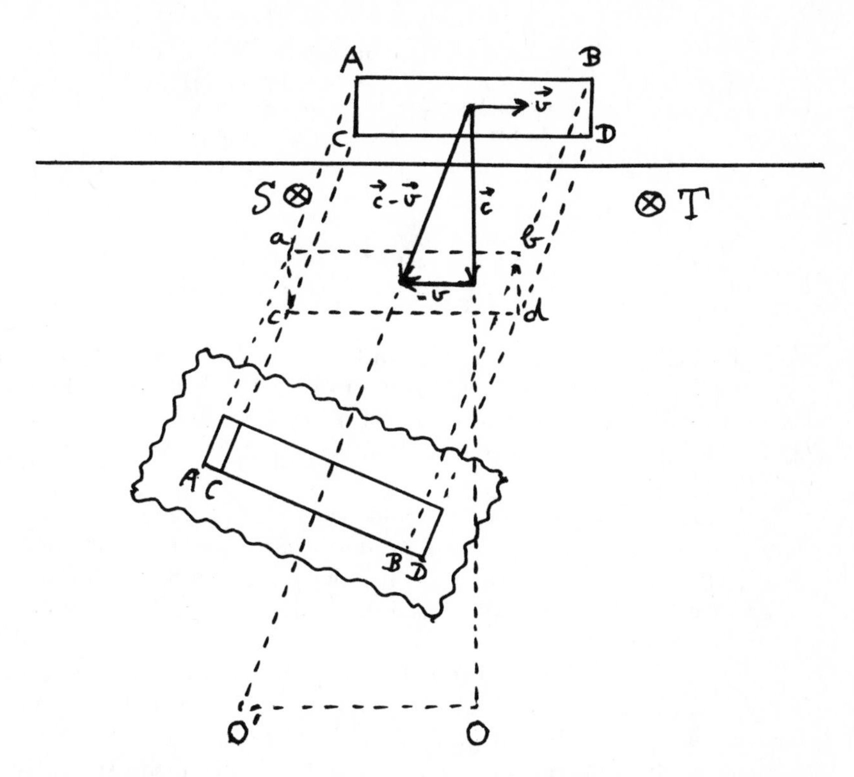

Fig. 1. 'Aberration' in photography of a rapidly moving object (as witnessed from above): viewed from O, the boxcar, running at velocity $v \equiv c\beta$, is seen as rotated by the angle $\tanh^{-1} \beta$. A snapshot of the boxcar is insered: its length is reduced in the 'FitzGerald—Lorentz ratio' $(1 - \beta^2)^{-1/2}$, and its rear side is visible.

of the whole scenery just when he sees the boxcar passing at the foot P of the perpendicular from O to the highway — a point which, in terms of Newton's absolute time, it has left.

The question is: how are the light rays reaching the photographer at O directed? As for the landscape 'there is no problem': the light rays emanating from objects near P, say the trees Q, R, S, T, are parallel to PO. However, the rays emanating from points $ABCD$ of the boxcar must be aimed, according to Bradley's aberration, at an angle $\alpha = \tanh^{-1} \beta$ as pictured (that is, opposite to Bradley's angle proper).

Therefore, as photographed at P, the boxcar is viewed as rotated by the angle α (Figure 1) so that its rear is visible, and projected with a length $\Delta l = l \sin \alpha$, if l denotes its width. Also, its length L is projected as $\Delta L = L \cos \alpha \simeq L(1 - \beta^2)^{1/2}$; that is, reduced by what came to be termed later the Fitzgerald—Lorentz contraction.

This is Terrell's formula for small objects photographed in transverse motion, obtained here in the limiting case where the velocity v is small.

2.3.5. ARAGO'S 1818 EXPERIMENT AND FRESNEL'S VERY FAR-REACHING 'ETHER DRAG' FORMULA

In 1818 Arago measured the refraction by a prism of light emitted by stars near the ecliptic plane; he found, to his surprise, that the deviation is the same as if the source were at rest in the laboratory. To explain this Fresnel promptly created his hypothesis and his formula of an 'ether drag coefficient'.

His formula, from which the 'absolute velocity' of the refracting body drops out, turned out to be a *universal non-additive law* for combining the velocity of light with that of a slowly moving material body, allowing a straightforward derivation of Einstein's relativity theory. I shall come back to this in Chapter 2.5.

It must be admitted that something significant had been overlooked by both Arago and Fresnel: the Doppler effect, discovered later, in 1842. So, in fact, as Mascart pointed out in 1893, Fresnel's reasoning implicitly assumed that the light source is on Earth, like the prism, both rushing together through the ether.

Potier (1874) noticed that this formula is a universal non-additive one for combining the velocities of light and of matter. M. von Laue (1907b) remarked that it is the differential of the exact relativistic formula. Hadamard (1930), using the theory of Lie groups integrated Fresnel's formula, and obtained the relativistic law. Finally, Ramakrishnan (1973) directly produced the mapping, in the sense of Lie groups, of the straight line $-\infty < A < +\infty$ into the finite segment $-c < v < +c$, thus deriving the relativistic formula for combining velocities. All this is explained in Chapter 2.5.

The question is: Could such views have been adduced as early as

1818, considering, among other things, that Lie's group theory was not published before 1887? The answer is *yes*, as will be made clear in Chapter 2.5. The algorithm involved is a simple one, belonging to hyperbolic trigonometry. And the general concept of Lie groups could easily have been bypassed by working the problem directly.

So, the conclusion is the same as the one stated *à propos* Bradley's aberration: the time had not yet ripened for a move from modelism to formalism.

2.3.6. 'NORMAL SCIENCE' IN OPTICS THROUGHOUT THE 19TH CENTURY

Excellent work, both experimental and theoretical, was performed in optics during the 19th century, firmly establishing the correctness of the wave theory of light, and producing the explanations of a host of new phenomena. We are not concerned here with the whole matter (notwithstanding its great interest), but only with a few points directly relevant to our subject.

Of course, the two most prominent theories are those of Fresnel (1788—1827) and of Maxwell (1831—1879).

Fresnel understood light as a periodic transverse oscillation; Fermat's stationary time principle became in his hands a principle of stationary phase — a far-reaching principle in many respects, including the relation between the older 'geometrical' and newer 'physical' optics, the questions of relativistic covariance, and the connection between wave theory and mechanics. Hamilton (1805—1865) produced the fully fledged expression of the Fermat principle, displaying the similarity between it and his own extremum action principle. Finally Maxwell's work, coming after quite a few 'elastic type' approaches, produced the appropriate system of partial derivative equations showing light to be an electromagnetic radiation.

In 1842 Doppler, by observing double stars (there are such coincidences!) discovered the 'Doppler effect', the exact theory of which was given by Fizeau in 1849.

The first measurements of the velocity of light in air were performed by Fizeau, using the toothed wheel method, and then by Fizeau and by Foucault separately, using the rotating mirror method (1849—1850). Velocity measurements by Foucault verified that the velocity of light in a medium is inversely proportional to the refraction index. And in 1851, using an interferometric method, and running water as a refracting medium, Fizeau verified directly Fresnel's ether drag formula.

This being said we come back to the velocity composition problem.

Between 1818 and 1873 a variety of experiments, aimed at detecting the 'ether wind', were performed by Babinet, Airy, Ångström, Hoek, Fizeau, Mascart — all with a null result. Among these I mention those by Ångström and by Mascart using a source, a prism or a grating, and a receiver, all fixed on a solid body of arbitrary orientation at rest in the laboratory. The Doppler effect is thus eliminated, and an ether wind effect, if it existed, would be isolated. This method amounts to comparing the wavelength of a propagating periodic wave to the length of a solid rule — which is predicted and measured as independent of the ether wind to first order in the small quantity $\beta = v/c$.

So, two important ideas progressively emerged: a technical one and a philosophical one.

The technical idea consisted in the remark tht all these experiments were sensitive only to the first order in β; already Fresnel had mentioned that his formula would not give first-order wind effects and, in 1848, Stokes made a similar remark. Finally Veltmann in 1873, and Potier in 1874, using Fermat's principle, demonstrated that suppression of ether-wind effects to the first order in β is a general consequence of Fresnel's formula. Therefore it was only natural to think of experiments designed for detecting second-order ether-wind effects — the task Michelson set for himself.

The philosophical idea was that *the 'principle of relativity' should be a universal law of nature*, so that no physical experiment whatsoever could detect an 'ether wind'. Mascart, the experimentalist who had performed so many first-order ether-wind experiments, put it this way (Mascart, 1874, p. 420):

> . . . The translatory motion of the Earth has no appreciable effect at all on the optical phenomena produced with a terrestrial source or with solar light. These phenomena are incapable of demonstrating the absolute motion of a body. Relative motions are the only ones we can make evident.

And so it happened that, in 1887, at Cleveland, Michelson and Morley performed their famous second-order experiment — again with a null result. Essentially, this experiment used a differential setting of the type of experiments aimed at counting the number of standing light waves carried by a solid rule, to which we shall come back in the next chapter. The two 'solid rules' of the Michelson experiment were orthogonal to each other, and, thus, differently affected by an ether wind.

Very much as in the early days of Arago and of Fresnel, the theorist's answer was an *ad hoc* formula suppressing the ether wind, but now to the second order in β: the FitzGerald (1893)–Lorentz (1895) universal 'contraction law of solid bodies under an ether wind', proportional to $(1 - \beta^2)^{1/2}$. Again, this was a very far-reaching recipe, leading, via group theory, to the famous Lorentz–Poincaré and Einstein transformation formulas. These, incidentally, were already known almost exactly to the crystallographer Voigt in 1887, and had been considered also by Larmor in 1900.

What was philosophically wrong with the otherwise very clever formulas of Fresnel and of FitzGerald–Lorentz was that they assumed the existence of an ether wind only to eliminate it. On the other hand, a group theoretical treatment of either the Fresnel or the FitzGerald–Lorentz formula does dissolve the ether concept into nothingness. Such is the effect of going from modelism to formalism. But the time was not yet ripe.

The sort of thinking going on between 1887 and 1905 is, of course, best represented by the laborious building of the Lorentz and later the Lorentz–Poincaré relativity theory, so concocted as to suppress all observable ether-wind effects, yet nevertheless assuming the existence of an ether! This is all the more paradoxical (in a derogatory sense) in that Lorentz as it seems, and most certainly Poincaré, had become (after Mascart) convinced that the relativity principle is a universal physical principle, valid for all phenomena, regardless of increasing powers of the (usually small) quantity $\beta = v/c$.

In Poincaré's three famous philosophical books *Science and Hypothesis* (1906a), *The Value of Science* (1905), *Science and Method* (1908), so many quotations could support this point that drastic choices are needed. I pick the two following quotations, translated directly from the French.

In *Science and Method*, Part 3, Chapter 2, entitled 'Mechanics and Optics', one reads at the end:

Anyway, it is not possible to avoid the impression that the principle of relativity is a general law of Nature, and that it will never be possible, whatever means are imagined, to detect anything but relative velocities. . . . So many experiments of different kinds have given concordant results that one cannot but be tempted to endow this relativity principle with a value comparable to that of the [thermodynamic] equivalence principle, for example. It is appropriate, nevertheless, to examine which consequences this way of looking at things would lead us to, and then to submit them to an experimental control.

In *The Value of Science*, Chapter 8, subtitle 'The Principle of Relativity', one reads:

. . . All experiments aiming at measuring the speed of the Earth with respect to the ether have yielded negative results . . . . The means employed have been varied, and finally Michelson has pushed precision to its [presently] ultimate limits; all to no avail. In order to explain such an [experimental] obstinacy the mathematicians are today obliged to display all their ingenuity. Their task is not an easy one, and, if Lorentz has succeeded, it is only by piling up hypotheses. [The] most ingenious one is that of local time. Consider two observers who wish to time their watches by [exchanging] optical signals . . . ; as they know that light is not transmitted instantaneously, they use crossed signals. . . . Thus timed, each watch will not display the real time. This does not matter, as we have no means to [find out what is the real time]. . . . As required by the principle of relativity, an observer has no means of knowing if he is at rest or in absolute motion. Unfortunately, even this is not sufficient, and complementary hypotheses are needed; it must be assumed that moving bodies undergo a uniform contraction along their direction of motion. . . .

When reading these, or other analogous quotations, one cannot but wonder how it happened that Poincaré was not a codiscoverer of the relativity theory! But also one gets some hints as to what could have been the psychological blocks inside the mind of so eminent a mathematical physicist and philosopher of science. This is a problem to which Holton (1973, pp. 185—195) and Miller (1981) address themselves.

If one adds that it is Poincaré (1906b) who discovered, in 1905, the geometrical interpretation of the Lorentz transformation, as a 'rotation' of Cartesian axes inside a (pseudo) Euclidean four-dimensional spacetime, also displaying the corresponding geometrical invariants — that is, the very concepts of the 1908 Minkowski paradigm — one is struck again by a feeling of awe and fatality.

There is another author who must be quoted at this point, although his conclusions were not as far reaching as those of Mascart and of Poincaré — and an author who, according to Holton (1973), may well have given a decisive direction to Einstein's thinking: August Föppl, whose pedagogical books were new and widely read in the days when Einstein was a student.

Föppl, in his Introduction to Maxwell's *Electricity Theory*, published in 1894, writes: "We cannot judge *a priori* unimportant the fact that, for example, a magnet moves near an electric circuit at rest, or that the latter is moving while the other is at rest." He continues by explaining that, if both pieces of apparatus are at relative rest, no effect of their 'absolute motion' shows up. Then he extends his reasoning by showing that, in the two cases considered first, only relative motion is significant, the 'induction formula' being the same in both.

This, of course, is the very example put forward by Einstein in his 1905 presentation of the 'theory of relativity'. Holton feels sure, for this and other reasons, that Einstein, consciously or not, borrowed this argument from Föppl.

The formal solution — that is, *the solution* — to the 'no-ether-wind' problem, that is, the *Theory of Special Relativity*, will be discussed in Chapter 2.5.

The Book of Daniel relates how a composite colossus — not a very convincing work of art — was suddenly smashed to pieces by a stone — *Ein Stein* — that had come full speed from elsewhere, and then grew as a huge mountain covering the whole of Earth. This is exactly what happened to the patchwork of Lorentz and Poincaré in electromagnetic theory. To continue with biblical quotations, what Einstein did was to 'uncover the Sense of the Scriptures' — scriptures that Lorentz and Poincaré had written down completely!

It is significant that in his epoch-making article of 1905 Einstein does not mention the second-order Michelson experiment, but refers only to some of the earlier first-order experiments. In other words, he transmutes into mathematics — into already known mathematics — a conviction that had been Mascart's and Poincaré's: that the relativity principle is a universal one. What, then, was the missing link in the previous arguments?

Since Arago's 1818 'null result' and Fresnel's 'ether drift formula,' Fizeau's 1852 experiment, Stokes's 1848 and Potier's 1874 remarks, Michelson's 1887 'null experiment', and FitzGerald's and Lorentz's contraction hypothesis, all through the 19th century kinematical optics had been wandering, meeting from time to time an oasis in the desert, but not suspecting that, far underneath, there was a *universal* water layer. Near the end of the journey, Lorentz and Poincaré, piling up mathematical reasoning, got hold of the Tables, with the Law correctly written down — however, not trusting the Word completely. So they could not enter the New Land, nor taste the wine from the grapes over there. Again, from cup to lip, what was the slip?

It was Einstein who entered the Promised Land, after felling a high wall by the following trumpet blast:

> We will raise . . . the principle of relativity to the status of a postulate, and . . . introduce another postulate . . . only apparently irreconcilable with the former, namely, that light is . . . propagated in empty space with a definite velocity *c*. . . .

That statement, implying non-additivity of the velocity composition law, opened the way to the New Land, by exorcising the ether phantasm in nothingness.

2.3.7. IN ELECTROMAGNETISM ALSO THERE WAS A DORMANT RELATIVITY PROBLEM

Electrostatics and magnetostatics were founded in parallel in the years 1785—1789 by Coulomb. Electric discharges of capacitors, and the generation and study of continuous electric currents were then investigated by various famous people. In 1820, Oersted discovered a coupling between electricity and magnetism, thus triggering Ampère's experimenting and thinking. Then Faraday discovered the induction phenomenon; and then, independently, Maxwell (1864) and L. Lorenz (1867) discovered the general laws of electromagnetism — which are Lorenz invariant; that is, *are essentially relativistic*.

What I want to emphasize is this: electrostatics and magnetostatics each implies its own universal constant, as present in the expression of the force between charges. As is well known, these lead respectively to the so-called electrostatic and electromagnetic systems of units.

As is also well known, the ratio of these two constants — the electric and magnetic 'permeabilities of the vacuum' — has the dimension of a squared velocity, which velocity is then by necessity a universal constant; that is, an absolute conversion coefficient between lengths and times. There was matter for conjecture — even if no hint was given as to how this 'equivalence between space and time' should show up.

The 'absolute electromagnetic measurements' by Gauss (1833) and Neumann (1845) led Kirchhoff to remark, in 1848, that this (formal) velocity equals that of light in the vacuum. Of course, this is the keystone in Maxwell's equivalence of light with an electromagnetic radiation.

Let it be said that joint static measurements of the electric and the magnetic constants were used in 1906 as a precision measurement of the velocity of light, by Rosa and Dorsey.

And finally there is, of course, Einstein's remark pertaining to the *one* formula of Faraday's induction phenomenon: relativistic invariance of Maxwell's equations was his answer to the riddle.

2.3.8. UNEXPECTED END OF THE HUNTING OF THE SNARK

No 'ether wind', no 'medium at absolute rest' were found, but a quite unexpected new 'absolute' was discovered: c, the velocity of light *in vacuo*, isotropic in all inertial frames, the same in all inertial frames if evaluated in its own natural units, the wavelength L and the period T.

Truly, the sinusoïdal waves of coherent light, pure as they are in their definition, regular as they are in their repetition, are the ideal high precision standards for gauging space and numbering time, for duplicating space unto time and time unto space.

Measuring space, measuring time, measuring the velocity of light, are not independent of each other. By fixing, say, the value of c by definition, measuring space and measuring time are rendered essentially identical. Indeed, this is today the very state of the art in metrology and in chronometry, where law has to conform to fact.

'Spatialization of Time', in Bergson's (1907) wording, is the one ultimate conceivable end of the old Aristotelian dream. Time is then measured by the *motion of light*, and referred directly and absolutely to space via the equivalence formula $L = cT$.

So, in the end, the snark was 'killed' — in the sense that it was found never to have been alive.

CHAPTER 2.4

TODAYS NEC PLUS ULTRA OF METROLOGY AND CHRONOMETRY: 'EQUIVALENCE' OF SPACE AND TIME

2.4.1. FUNDAMENTAL SIGNIFICANCE OF THE MICHELSON—MORLEY TYPE OF EXPERIMENT

As previously said, Michelson and Morley's 1887 experiment is a differential arrangement of the type of setup used, since 1893, for comparing a solid length standard and an optical standing wavelength standard, to be discussed in the next section.

Had this experiment detected an 'ether-wind' effect — that is, a direction dependence of the ratio of the two length standards — then, in principle, every comparison between these should have been associated with a measurement of the direction and magnitude of the 'ether wind'. Conversely, the demonstration that no 'ether wind' appears is tantamount to establishing a *de jure* unconditional equivalence between a solid and an optical standing wave standard, the choice between them being only a matter of attainable precision and of reproducibility, as discussed in the next section.

The most precise repetition of the Michelson—Morley experiment *stricto sensu* was performed by Joos in Jena in 1930. It could have detected an 'ether wind' of 1.5 km/s, to be compared with the sensitivity of non-differential metrological measurements in those days: some 50 km/s, corresponding to a relative precision in wavelengths of 10^{-8}.

Good as it was, Michelson's and Morley's result was received at the time as 'paradoxical' by most physicists — but not all, however, as we have seen in Chapter 2.3. Today it is just the contrary: according to de Broglie's wave mechanics, a solid body is nothing more than a (very complex) standing matter wave, and it would be quite embarrassing if an optical and a material standing wave did not display the same kinematical behavior.

Let me mention that in 1932 Kennedy and Thorndike performed a Michelson—like experiment with very unequal arm lengths, also finding a 'null effect'.

In 1964 Townes and coworkers renewed this approach by using the very coherent light produced by lasers and a time rather than a length metrology, not improving much, however, on Joos's result. As the technology thus initiated differs radically from the Michelson one *stricto sensu*, the whole matter is reconsidered in Section 6.

So let it be concluded here that experiments of the Michelson sort have paved the way to the present-day optical metrology.

2.4.2. OPTICAL METROLOGY

As early as 1828, soon after Young and Fresnel had established the wave theory of light, Babinet suggested that a light wavelength be used as a length standard.

In 1893 Michelson and Benoit performed the first direct comparison between the solid platinum length standard kept in Sèvres and the red cadmium wavelength. During the years 1905—1906 Benoit, Pérot and Fabry, using improved methods and apparatus, augmented the precision of the previous determination without modifying its magnitude.

The clever device known as the Pérot—Fabry interferometer was conceived for this work, and has had numerous applications ever since. It consists of a system of two parallel plane mirrors which are lightly silvered so that some light escapes after each reflection. The multi-component emerging wave displays a set of very narrow directional fringes. It is thus an excellent tool for seeking an ether-wind effect without using a spatial differential arrangement — and it came to be used precisely for that, as will be explained in Section 6.

Ever since that time work has been pursued along such lines, and, in 1967, the 11th International Conference of Weights and Measurements decided that the standard of length be defined in terms of one selected narrow line of the atom of krypton 86.

2.4.3. MICROWAVE CHRONOMETRY

Going from the traditional time standards — the rotating or circulating Earth — to the electromagnetic time standard as defined by the period of some radiation of molecular or atomic origin did not require (at first sight) the solution of a preliminary problem, as did the similar problem in metrology. Here also, however, wave mechanics has something significant to say.

Let it be recalled that, in principle, the typical time standard in

Galilean–Newtonian mechanics was the freely moving point particle, the motion of which is rectilinear and uniform by the very virtue of the two tightly interconnected inertia and restricted relativity principles. Similarly, in the wave theory of light, the typical time standard is in principle the period of a plane monochromatic wave propagating in free space. Now, as explained by Einstein's 1905 and 1912 photon theory, these two standards are essentially interconnected via the quantal equivalence formulas $W = h\nu$ and $\mathbf{p} = h\boldsymbol{\kappa}$, relating the photon's energy W and momentum $\mathbf{p}$ and the wave's temporal ν and spatial $\boldsymbol{\kappa}$ frequencies. In 1925 Louis de Broglie extended these formulas to matter waves, thus making a *universal bond between the mechanical and the wave time standards.*

So, in dealing with chronometry, going from the mechanical to the wave time standard implies no radical change, the preference again being merely a question of precision and reproducibility of the standard.

Analogous remarks hold of course in the case of more sophisticated mechanical or electromagnetic clocks, like those traditional astronomical or man-made chronometers initiated by Kepler and by Huygens, or, today, those molecular or atomic mini-clocks used for monitoring the Hertzian chronometers.

The first steps in microwave chronometry were taken in 1946 by Pound and Rebka and in 1947 by Smith and coworkers, using as a frequency standard the inversion line of the ammonia molecule. In 1951 Townes developed his 'maser', a microwave oscillator also stabilized by the ammonia transition. In 1956 Rabi based a frequency standard on an atomic transition of caesium atoms moving in a beam. Work continued along these lines and, in 1955, Essen and Parry in Great Britain and Zacharias in the U.S.A. produced operational time standards of the Rabi sort.

Finally, in 1967, the 13th International Conference on Weights and Measures decided that the time standard be an appropriately chosen line of the caesium atom.

2.4.4. MEASUREMENTS OF THE VELOCITY OF LIGHT

A two way light beam on Earth was used in Fizeau's 1848 measurements using the toothed-wheel method (over a 8.6 km baseline) and in

Fizeau's and Foucault's independent 1850 measurements using Arago's rotating-mirror concept (over a 20 m baseline). The latest determinations of *c* based on this sort of approach have been Karolus's and Mittelstaedt's in 1928, Anderson's in 1941, and Bergstrand's in 1951, using a light beam chopped by a Kerr cell; and Aslakson's in 1949, using a radar.

But, of course, the Michelson null effect suggests a much more radical approach, that of *comparing the wavelength and the period of an optical or a microwave radiation.* The difficulty, however, is that metrology is easier in the optical realm and chronometry in the microwave realm, so that some means for bridging the gap is needed. In fact, since, say, 1950, the history of the measurements of *c* has been that of improving the bridging between the optical measurements of length and the microwave measurements of time.

In 1950 Hansen and Bol used a resonant cavity, the size of which was measured optically and the period generated directly. Essen's 1950 method is very similar. Froome in 1952—1954 and Florman in 1954 used waveguides. Rank in 1954—1955, and Plyler in 1954, measured infrared wavelengths and microwave periods of a molecular spectrum. Bay, Luther and White, at the NBS in Washington, used a helium—neon laser and an evacuated Pérot—Fabry cavity, together with an ingenious bridging method (Bay *et al.*, 1972).

All this led to the present-day 'heterodyning' methods, as initiated in 1972—1974 by Evanson and coworkers (1972) at the NBS in Boulder, and by Blaney and coworkers (1974) at the NPL in Teddington. The former give *c* as 299 792 456.2 ± 1.1 and the latter as 299 792 459 ± 0.8 m/s.

This is extremely impressive, and of course raises the question of fixing the value of *c* by definition, and, then, either the length or the time standard (but not both). But this is a very intricate matter, not independent of improved choices of fundamental wavelength or period standards; that is, choices of lines finer than those previously used, and also decisions concerning the 'secondary standards' to be used in various ranges of measurements.

Thus, at the 1973 October meeting of the CIPM (Comité International des Poids et Mesures), the 'recommended' value of *c* was 299 792 458 m/s, together with recommended values of various wavelengths of the methane molecule and of the iodine atom.

2.4.5. IMMINENT FULFILMENT OF THE OLD ARISTOTELIAN DREAM

"Measuring motion by means of time and time by means of motion", that was the dream. And now *light* fulfills the dream.

Four tightly interrelated types of experiments — so tightly interrelated indeed that their respective techniques and interpretations largely overlap — are at stake: Michelson's null-effect experiments, optical metrology, microwave chronometry, and measurements of c.

This overall phenomenology 'allows and suggests', in Duhem's (1954) wording, that one standard be used for both length and time measurements, with a defined value of c. So, the relativistic assumption that c is a universal constant is equivalent, in Poincaré's (1906a, *passim*) wording, to a 'disguised definition' — which is 'allowed and suggested' by experimental facts. This is typically 'autofoundation' in Gonseth's (1964) wording — the typical circularity inherent in all basic physics — or, should we rather say, the very 'bootstrap nature' of physics.

And so, as in the older days of the Galileo—Newtonian chronometry, we see that the problem of measuring time is intimately bound to the inertia-and-relativity principle — but, of course, to the new version of it, where time and space are directly related via c.

So, while there seemed at first sight to be no radical revolution when considering separately the transition from mechanical to microwave metrology and chronometry there is one indeed when considering both together: *a conceptual revolution underlying a metrological revolution.*

The conversion from the Galilean to the Einsteinian concept of the relativity principle not only implies a very deep plunge towards fundamentals, but also brings back pearls from the abyss: a wonderfully increased precision in basic measurements.

2.4.6. WONDERS OF LASER PHYSICS: THE 1978 BRILLET AND HALL 'REPETITION' OF THE MICHELSON EXPERIMENT

In this beautiful experiment (Brillet and Hall, 1979), yielding an upper limit of some 40 cm/s to the 'ether wind', the solid length etalon was a Pérot—Fabry interferometer, but the measurement itself consisted of a beat frequency measurement. The Pérot-Fabry interferometer was phase locked to a helium—neon laser, both continuously rotating around a vertical axis on a granite slab. The beam was reflected upwards, along the rotation axis, and made to interfere with a reference methane laser fixed in the laboratory. The beat frequency was measured ('optical heterodyning'), and so a time-differential technique instead of Michelson's original space-differential technique was used.

Plotting their results during a period of 238 days, Brillet and Hall assigned an upper limit to the 'ether wind' 4000 times smaller than that of Joos and Townes. This corresponds to a relative frequency shift smaller than 2.5×10^{-15}!

I guess a long time will elapse before some similar experiment is undertaken!

REMARK. In their paper Brillet and Hall refer to a study by Robertson (1949), concluding, from a combined consideration of the Michelson, the Kennedy—Thorndike and the Ives—Stilwell experiments, that, between two inertial frames moving along x, the metric transforms as

$$ds^2 = dt^2 - c^{-2}(dx^2 + dy^2 + dz^2)$$

$$ds'^2 = (g_0\, dt')^2 - c^{-2}[(g_1\, dx')^2 + g_2^2(dy'^2 + dz'^2)].$$

According to this, Joos has demonstrated that $g_2/g_1 - 1 = (0 \pm 3) \times 10^{-11}$, and Brillet and Hall that $g_2/g_1 - 1 = (3 \pm 5) \times 10^{-15}$. This speculation obviously refers to a possible non-congruence of a solid state etalon and the standing wave optical etalon — an extremely unlikely hypothesis, in my opinion.[1]

2.4.7. WONDERS OF LASER PHYSICS: METROLOGY VIA DOPPLER-FREE SPECTROSCOPY

So many clever ideas and devices have been adduced to the field of laser spectroscopy that only a very brief outline can be given here.

Since the invention of the laser itself (Schawlow and Townes in 1958 and Maiman in 1960) the major advance has been that of tunable lasers, as obtained by quite a few different techniques. These are the optical analogues of microwave oscillators, and are available as strong, narrow-band sources in the visible and in the infrared portions of the spectrum.

Another major scientific advance has been the elimination of

Doppler broadening, either in absorption or in emission spectra using two antiparallel laser beams interacting with a perpendicular atomic or molecular beam.

In 1976 the Hänsch and the Bordé groups have independently inaugurated the 'saturated absorption' technique by using a beam chopper. In a later improvement chopping was replaced by adjustable polarizations of each beam.

In 1970 Vasilenko and coworkers proposed a new method for observing two photon transitions free from Doppler broadening. The two photons, of unequal frequencies, raise an atom or molecule from a lower to a higher energy level through an intermediate virtual level. If they are carried on antiparallel laser beams, the overall energy and momentum absorbed does not depend on the motion of the atom or molecule. Nearly simultaneously, in 1974, Biraben—Cagnac—Grynberg, Levenson—Bloembergen and Hänsch—Harvey—Meisel—Schawlow succeeded in producing this remarkable phenomenon.

These are but samples of a sheaf of ingenious ideas that are presently put to work (Baird, 1983; Schawlow, 1982). In the realm of metrology, this means that optical or infrared, atomic or molecular, radiations are available with extremely high definitions as length standards — definitions in fact superior to those attainable with solid length standards, such as Pérot—Fabry resonators. That optical standing plane waves can thus be excited with their nodes positioned at very much higher precision than atomic dimensions is a cause of some wonder, even when perfectly justified via the equations. So, the traditional sort of solid length standard can no longer compete with the immaterial optical wave standard. Matter has yielded to light — Jacob to the angel.

Therefore, as soon as the techniques of optical heterodyning have developed to the point where a safe connection can be established between the 'metrological' and the 'chronometric' frequency ranges, the time standard will also become the length standard, via a chosen value of *c*.

2.4.8. OCTOBER 1983: THE SPEED OF LIGHT AS SUPREME 'MOTION REFEREE', AND THE NEW IMMATERIAL LENGTH STANDARD

In its 1983 October session, the 17th General Conference for Weights and Measures issued a Resolution entitled 'Definition of the Metre' which, with written permission of Dr Giacomo, Director of the BIPM, I quote from its English translation (Documents . . . , 1984; Giacomo, 1983). The reader will appreciate how this official, administrative, document also is a masterpiece in the philosophy of science.

DEFINITION OF THE METRE:

Resolution 1

The Seventeenth Conférence Générale des Poids et Mesures

considering:

— that the present definition does not allow a sufficiently precise realization of the metre for all requirements;

— that progress made in the stabilization of lasers allows radiations to be obtained that are more reproducible and easier to use than the standard radiation emitted by a krypton-86 lamp;

— that progress made in the measurement of the frequency and wavelength of these radiations has resulted in concordant determinations of the speed of light whose accuracy is limited principally by the realization of the present definition of the metre;

— that wavelengths determined from frequency measurements and a given value for the speed of light have a reproducibility superior to that which can be obtained by comparison with the wavelength of the standard radiation of krypton-86;

— that there is an advantage, notably for astronomy and geodesy, in maintaining unchanged the value of the speed of light recommended in 1975 . . . (c = 299 792 458 m/s);

— that a new definition of the metre has been envisaged in various forms all . . . giving the speed of light an exact value, equal to the recommended value, and that this introduces no appreciable discontinuity into the unit of length, taking into account the relative uncertainty of $\pm 4 \times 10^{-9}$ of the best realizations of the present definition of the metre;

— that these various forms, making reference either to the path travelled by light in a specified time interval or to the wavelength of a radiation of measured or specified frequency . . . have been recognized as . . . equivalent and that a consensus has emerged in favour of the first . . . ;

— that the Comité Consultatif pour la Définition du Mètre is now in a position to give instructions for the practical realization of such a definition, instructions which could include the use of the orange radiation of krypton-86 used as standard up to now, and which may in due course be extended;

decides

1. The metre is the length of the path travelled by light in vacuum during a time interval of 1/299 792 458 of a second;
2. The definition of the metre in force since 1960 . . . is abrogated . . . .

Sic transit gloria: "The king is dead. Long live the king." 'Considering' the care with which the metrologists are exercising their art, it is gratifying to the philosopher of science that they have ended up with the corpus of definitions fitting the very nature of Time, Space and Motion.

2.4.9. WONDERS OF LASER SPECTROSCOPY: CHRONOMETRY VIA OPTICAL HETERODYNING

Since 1967 the official definition of the time standard is as follows: "The second is the duration of 9 192 631 770 periods of the radiation corresponding to the transition between the two hyperfine levels of the ground state of the caesium-133 atom." This frequency is in the microwave range. Various caesium primary standards are operating, and used for producing 'atomic time scales' in such laboratories as the U.S. National Bureau of Standards (NBS), the Canadian National Research Council (NRC), the German 'Physikalisch-Technische Bundesanstalt (PTB) and the Washington Naval Observatory (USNO). All are using resonant cavities tuned to a caesium atomic beam. Their accuracy capability is estimated as better than 10^{-12}.

As previously said, the important issue in this field is to connect directly, by optical heterodyning, the caesium frequency to the optical frequencies used in metrology — for example, to the rich arrays of very narrow spectral lines displayed, in the infrared, by the carbon dioxide or the methane molecules, or, in the visible, by the iodine atom.

Systems having a non-linear response may be used to generate harmonics of a sinusoïdal signal, or a beating between two superposed sinusoïdal signals. Multiplication, division, addition, subtraction of frequencies is thus possible — a procedure named 'heterodyning' in radiofrequency technology, but novel in the optical range.

So, by using lasers phase locked to very narrow atomic or molecular lines and, for example, the non-linear device consisting of a point contact metal—oxide—metal diode, quite a few dedicated laboratories in the world have been able to bridge the tremendous distance between the microwave frequency of the caesium time standard and some very narrow lines in the iodine atom or the methane molecule spectra. This amounts to 'an exact count of cyclical events that occur at a rate of over 500 000 000 in a microsecond' (Baird, 1983). The Institute for Semiconductor Physics in Novosibirsk, for example, has announced the first 'optical clock' in which microwaves were locked to the 3.39 micron absorption line in methane.

So much work is going on in this fascinating field of endeavor that it is not possible to quote everything of significance. As striking examples of what has been achieved, I select the recent direct frequency measurements of the iodine 473 and 520 THz (10^{12} Hz) lines, as performed by Evanson and coworkers at the NBS in Boulder.

It is estimated that the stability barrier of the caesium time (or frequency) standard is of the order of 10^{-15}, and its accuracy (that is, the faithfulness of its representation of the atomic line) is just a little less.

So, the universal Shiva Dance of molecules, atoms, and even nuclei, emitting and absorbing electromagnetic radiation from the microwave to the gamma ranges, is ticking time — and time keeps in rhythm with it. Visiting the laboratories where this basic research is going on, such as Bordé's laboratory in Villetaneuse, and seeing the huge series of highly resolved lines of the iodine or the carbon dioxide spectra, is truly fascinating.

2.4.10. MOSSBAUER EFFECT (HEIDELBERG, 1957)

Another frequency standard accurate to one part in 10^{15} has been known since 1957 as the 'Mössbauer effect'. It uses gamma rays emitted or absorbed in nuclear transitions, the nucleus being tightly bound inside a crystal, so that the whole crystalline body (and not only the transiting nucleus) recoils. For example iron-57 thus yields a frequency standard accurate to some 10^{-18} (one part in a million trillion).

No method is presently available for connecting such extremely high frequencies to practical clocks. Therefore, Mössbauer-style measurements can only be 'relative' ones, testing some perturbation affecting either the emitter or the receiver of the radiation. A convenient way of exploring the frequency range is via the Doppler shift, with velocities as low as a few centimeters per second.

A spectacular fundamental Mössbauer experiment has been a laboratory measurement of Einstein's gravitational frequency shift as caused by the Earth's field, with an altitude difference of 22.5 m (Pound and Rebka (1959), Harvard).

2.4.11. APPLIED METROLOGY, TACHYMETRY AND CHRONOMETRY

Having the standard is one thing; using it is another matter.

As for length measurements, the scheme of performing a time measurement and using an accepted value for c is common practice with the radar, and Bergstrand's 'geodimeter' — a useful newcomer in geodetical surveys.

Tachymetry, that is, velocity measurements via Doppler shifts, is common practice in astronomy, and in terrestrial technology.

The measurement of the Earth's velocity via Doppler shifts is the method in use now for measuring the dimensions of the ecliptic, and, thus for evaluating the ratio of the 'parsec' to the meter.

Also, it is by its very definition that the other astronomical length unit, the 'light year', conforms to this scheme.

CHAPTER 2.5

ENTERING THE FOUR-DIMENSIONAL SPACETIME PARADIGM

2.5.1. WALKING THROUGH THE ENTRANCE GATE

The two previous chapters have explained how a long 'theoretic–experimental reasoning', pursued all through the 19th century, finally led to the conclusion that the velocity of light *in vacuo*, c, is measured as isotropic in all inertial frames. This fact 'allows' and 'suggests' that the length and the time standards be chosen as the wavelength and the period of some optical or microwave radiation. Thus, by a joint definition of the length and the time standards, c is made an absolute constant, a 'conversion coefficient' between space and time.

The relation between the squared distance $r^2 = x^2 + y^2 + z^2$ travelled by light during the time t, with all four coordinates x, y, z, t belonging to the same spatio-temporal 'inertial reference frame', is such that

(2.5.1) $x^2 + y^2 + z^2 - c^2t^2 = 0.$

This expression retains its form in all inertial frames, c being an absolute constant.

Since, geometrically speaking, this is the null squared distance from the origin of a 'light class' of point-events in a hyperbolic, 'pseudo-Euclidean', geometry, reference frames are thus interpreted, according to Poincaré [1906] and Minkowski [1907, 1908], as orthonormal Cartesian tetrapods in the *four-dimensional spacetime*. A natural question then is: Are all Cartesian spacetime tetrapods, including those related by reversal of part, or all, of the axes, physically valid as 'inertial frames'? That is an important question, to be pondered from now on, to the end of this book.

In any case, (1) is the equation of a 'null cone', or 'light cone' (Figure 2) — the 'reality' of which is of course a peculiarity of hyperbolic geometries. The non-null vectors of spacetime are such that

(2.5.2) $x^2 + y^2 + z^2 - c^2t^2 = s^2;$

if $s^2 > 0$ (s real) they point outside the light cone and are termed

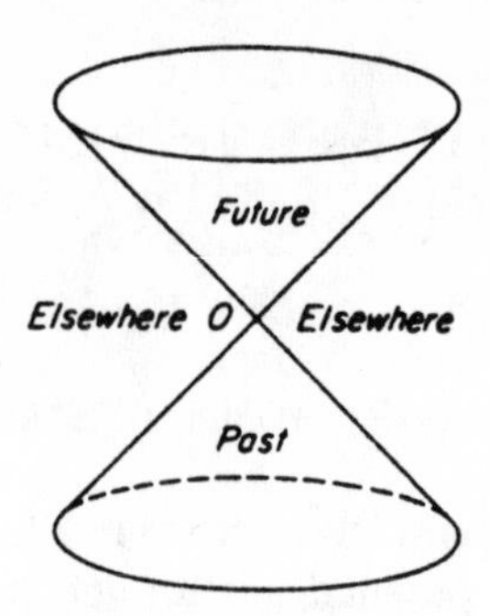

Fig. 2. The 'light cone' of equation $x^2 + y^2 + z^2 - c^2t^2 = 0$ trisecting spacetime in 'past', 'future' and 'elsewhere'.

'spacelike'; if $s^2 < 0$ (s purely imaginary) they point inside the light cone and are termed 'timelike': 'future timelike' if they point 'upwards', 'past timelike' if they point 'downwards' (Figure 3). So (2) is the equation of a hyperboloïd, a one-sheet hyperboloïd if $s^2 > 0$, a two-sheet hyperboloïd if $s^2 < 0$.

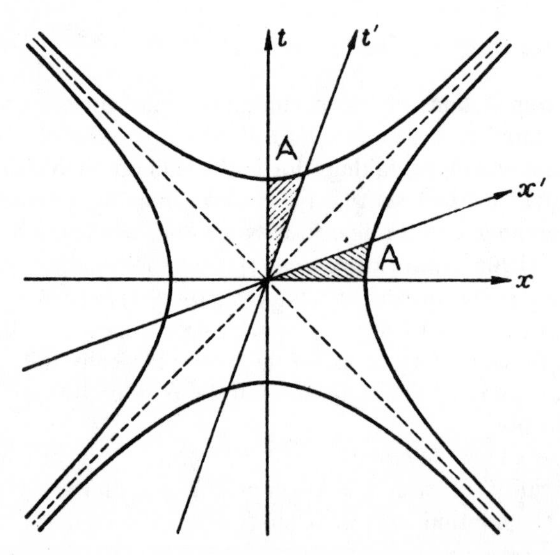

Fig. 3. 'Hyperbolic rotation' in the (x, ct) plane as geometrical rendering of the Lorentz transformation; such a continuous rotation cannot reverse the directions of the space x and the time ct axes.

By rotating a Cartesian tripod inside the ordinary Euclidean space one cannot exchange right and left; so, the complete set of such tripods is obtained via the product of rotations by a space symmetry, termed 'parity reversal', and denoted by P. Somewhat similarly, as visualized in Figure 3, a continuous rotation of the tetrapod can neither reverse the time axis, nor exchange right and left in the x, y, z subspace. So, to exchange 'past' and 'future', the product of a spacetime rotation L_0 by a reversal of the time axis, denoted T, is needed. On the whole, the complete Lorentz—Poincaré group L thus comprises four subgroups: the *continuous rotation group* L_0, which is *orthochronous* and *orthochiral*; and its products by the three *discrete symmetries* P, T and PT.

Does this make sense in physics?

As for the parity, or chirality, reversal P, there are cases where it certainly makes sense; for example, the electrodynamics of current-carrying wires and magnets needs a discussion of parity reversal.

As for the time reversal T, there are of course problems in which it does make sense; for example, the 'microreversibility problem' as discussed by Loschmidt in 1876. In fact the T-reversal question arises in all discussions pertaining to equations describing elementary evolutions.

However, macroscopically speaking, if one requires that the obvious 'irreversibility laws' existing in wave propagation, in thermodynamics, in statistical mechanics and in information theory be relativistically invariant, then the time-reversal operation T must be excluded. This is what Einstein meant when he said that "one cannot telegraph into the past".

Let us conclude this section by stating that *the Lorentz—Poincaré group of macrorelativity is the product of the continuous rotation group* L_0 *by the parity-reversal operation* P. *Time reversal is excluded in macrorelativity.*

2.5.2. PLAYING WITH HYPERBOLIC TRIGONOMETRY

I bring together here known formulas and reasonings usually found in different places. This, I believe, will be found illuminating.

In the present approach, the most natural deduction of the Lorentz—Poincaré—Einstein formulas

$$(2.5.3)\quad x' = \alpha(x + ut), \qquad t' = \alpha(t + ux/c^2), \qquad y' = y, \qquad z' = z;$$
$$\beta \equiv u/c, \qquad \alpha \equiv (1 - \beta^2)^{-1/2};$$

is via hyperbolic trigonometry. The expressions for x' and ct' have been written with a '+' sign in order to make forthcoming formulas more symmetrical, so that $(u, 0, 0)$ denotes the velocity of the unprimed system with respect to the primed one. Obviously, these formulas are those of the continuous, 'orthochiral and orthochronous' subgroup. In the limit $\beta \to 0$, $\alpha \to 1$, they reduce to the Galilean ones (2.2.3).

The crucial novelty in formula 2.5.3 is the contribution of x to t', rendering a Lorentzian inertial frame a reference frame of space *and time*, which are symmetrically (and partially) transformed into each other by relative motion.

Formulas (3) are no more than those of a hyperbolic rotation of the spacetime x and ct axes:

(2.5.4) $$x' = x \cosh A + ct \sinh A, \qquad ct' = x \sinh A + ct \cosh A$$

with

(2.5.5) $$\cosh A = (1 - \beta^2)^{-1/2}, \qquad \sinh A = \beta(1 - \beta^2)^{-1/2}, \\ \tanh A = \beta;$$

as $c\beta = u$ is the relative velocity of the two frames, A has been termed their relative 'rapidity' (Levy-Leblond, 1976).

The relativistic composition law of velocities thus is

(2.5.6) $$\tanh(A + B) = \frac{\tanh A + \tanh B}{1 + \tanh A \cdot \tanh B}$$

or

(2.5.7) $$w = \frac{u + v}{1 + uv/c^2},$$

belonging to a group. By setting $A + B = -C$, we get

(2.5.8) $$\begin{cases} A + B + C = 0 \Rightarrow \\ u + v + w + uvw/c^2 = 0, \end{cases}$$

showing that 'addition of rapidities is composition of velocities'. Setting $C = 0$ implies that $A + B = 0$, thus displaying the existence of the 'neutral element' and of the 'inverse transformation' inside the group.

Let us suppose that, ignoring the whole story of kinematical optics, we happen to know that there exists an upper limit, $\pm c$, to velocities. The natural question then is: Which is the velocity composition law that is isomorphic to the additivity of the reals, $-\infty \leqslant A \leqslant +\infty$? We are looking for a faithful mapping, in the sense of Lie groups, of the real line upon the finite segment $-c \leqslant u \leqslant +c$. An obvious candidate is

$$(2.5.9) \quad u = c \tanh A,$$

the variation of which is displayed in Figure 4. Then we see, via (6), (7) and (8), that this indeed is an answer; and, by the theory of Lie groups, it is *the* answer.[1]

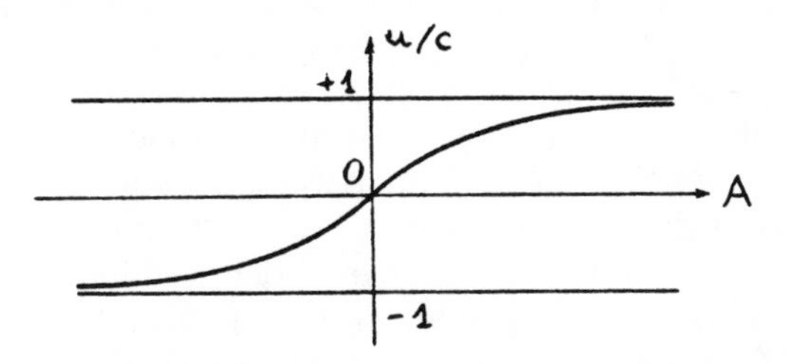

Fig. 4. 'Rapidity' A versus velocity u. Assuming that velocities u along x cannot exceed that of light, $-c \leqslant u \leqslant +c$, the function $A = \tanh^{-1}(u/c)$ is an *a priori* candidate for relating A to u. Indeed, the composition law of the A's is isomorphic to the additive composition law of the reals.

This is an extremely fast and intuitive approach to the relativity theory, the idea of which is found in papers by Hadamard (1923), Abelé (1952), Malvaux (1952) and Ramakrishnan (1973), but not presented by them in so compact a form as here.

Having thus obtained the velocity composition law (7) or (8), we can easily derive the Lorentz formulas (3) or (4). To achieve this we remark that, for timelike vectors, a velocity is the ratio of a length to a time. Therefore setting $x = l \sinh A$, $ct = l \cosh A$, we write down, for timelike vectors, the Lorentz formulas as

$$(2.5.10) \begin{cases} \sinh(A+B) = \sinh A \cosh B + \cosh A \sinh B, \\ \cosh(A+B) = \cosh A \cosh B + \sinh A \sinh B. \end{cases}$$

What then for spacelike vectors? As (6) may be rewritten as

$$(2.5.11) \tanh(A+B) = \frac{\operatorname{cotanh} A + \operatorname{cotanh} B}{1 + \operatorname{cotanh} A \cdot \operatorname{cotanh} B}$$

an analytic continuation is possible where two of the three vectors involved are spacelike, as remarked by Ramakrishnan. For these we write $x = l \cosh A$, $ct = l \sinh A$, so that the two formulas (10) are then exchanged. Since $u = c \operatorname{cotanh} A$ is not a faithful representation of the real axis, the third vector must remain timelike.

There have been quite a few attempts at 'tachyonic' Lorentz transformations aiming at transgressing the latter prohibition, but they have only led to unacceptable consequences (Marchildon *et al.*, 1983).

2.5.3. ON THE GENERAL LORENTZ–POINCARÉ–MINKOWSKI TRANSFORMATION

The general transformation law between Cartesian frames is known as an *orthogonal transformation.*[2] For spacetime its discussion is found in the textbooks, including mine (1966). An interesting remark is that two successive Poincaré rotations, the hyperbolic rotation (x, ct) planes of which do not coïncide, do produce a Lorentz transformation, but one including a rotation of the spatial axes. This phenomenon, known as the 'Thomas precession', has observable consequences.

Let us look at the matter more closely, considering first rotations inside Euclid's 3-space. Denoting u, v, w as a circular permutation of the indexes 1, 2, 3, we find the generators of rotations 'around' the axes of a Cartesian tripod to be

$$\partial \alpha_u = x_v \, \partial_w - x_w \, \partial_v$$

obeying the commutation relation

$$[\partial \alpha_u \, \partial \alpha_v - \partial \alpha_v \, \partial \alpha_u] = \partial \alpha_w .$$

Therefore the result of two such rotations performed in succession depends upon their order; in the infinitesimal case, the 'product' of a u by a v rotation brings out, as a bonus, a w rotation also.

Similar results hold inside the pseudo-Euclidean 3-space y, z, ct, but, then, the (y, ct) and the (z, ct) rotations are hyperbolic Lorentz transformations *stricto sensu*. Two cases are then of interest.

The commutator of the generators of two Lorentz transformations (y, ct) and (z, ct) is the generator of a spatial rotation (y, z): this corresponds to the 'Thomas precession' previously mentioned.

The commutator of the generators of (x, y) and (z, x) spatial rotations is expressible as the generator of either a spatial (y, z) rotation (as previously said) or of an (x, ct) Lorentz transformation. In this sense there is an 'equivalence', or a one-to-one correspondence, between a Lorentz transformation and a spatial rotation; this shows up in the phenomenon known as 'Terrell's photography', discussed in Section 2.5.6.

2.5.4. ON THE GALILEO–NEWTON PARADIGM AS A LIMIT OF THE POINCARÉ–MINKOWSKI ONE

We ask now what sort of relation exists between the new spacetime and the old space-and-time concepts.

As made clear in Chapter 2.3, the relativistic phenomena do appear only when the velocities of material bodies are sufficiently large with respect to c. This amounts to saying that the joint length and time scales effectively usable must be in a ratio not too small with respect to c. Mathematically speaking, of course, the optimum choice is the one equating c to 1 — the one implicit in our formulas and figures, where ct rather then t has been used consistently.

In units that we find 'practical' for existentialistic reasons (say, the meter and the second) c is expressed as extremely large. Therefore, when using them, the light cone is pictured as extremely flat (Figure 5(a)), the spacelike region being squeezed, so to speak, between the 'past' and the 'future' regions. In the limit $c \rightarrow \infty$ the spacelike region is completely squeezed out (Figure 5(b)), so that Minkowski's trisection of spacetime into 'past', 'future' and 'elsewhere' is reduced to Newton's (and everybody else's) dissecting it into a 'past' and a 'future' separated by a 'universal present instant' t.

In other words, the Poincaré–Minkowski spacetime metric is then decomposed into a space metric and a time metric.

It goes without saying that *the very existence of a spacetime metric* (*together with the trisecting light cone*) *basically excludes any world*

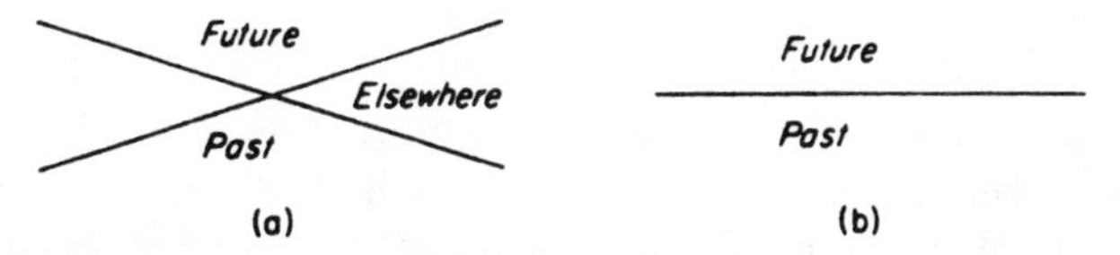

Fig. 5. As expressed in 'practical' space and time units, the velocity of light, c, is very large, and the light cone is then pictured as widely open, with 'elsewhere' squeezed between 'past' and 'future'. Going to the limit $c \rightarrow \infty$ completely expels the 'elsewhere', and the Newtonian dichotomy between 'past' and 'future' is thus recovered.

picture implying a global or universal separation between a 'past' and a 'future'. Therefore being space extended (as emphasized by Descartes), *matter is essentially conceived, in the relativity theory, as also time extended*, both in the past and in the future. So the relativistic matter does indeed possess 'duration'.

This was not necessarily the case for the Newtonian matter (notwithstanding Newton's personal feelings to the contrary).

2.5.5. FRESNEL'S ETHER DRAG LAW AS A VELOCITY COMPOSITION FORMULA

Fresnel's 'ether drag' formula reads

$$(2.5.12)\ c' - c/n = v(1 - 1/n^2)$$

with n denoting the index of a refracting medium moving in the laboratory with the velocity v, and c' the resulting velocity of light. Let it be recalled that this formula was a response to Arago's 1818 experiment aimed at detecting the 'absolute' motion of the Earth. In 1852 Fizeau verified the formula in the laboratory. Putting things together makes clear that (12) is a velocity composition formula, which was Potier's (1874) remark.

Let us verify that Fresnel's formula (12) is nothing more than the differential of the previous formulas (8) and (9). Setting

$$\beta = 1/n, \qquad \beta + \Delta\beta = c'/c,$$

we rewrite (12) as

$$v/c = \Delta\beta/(1 - \beta^2).$$

Then, as v, the velocity of matter, is small with respect to c/n, c' and $c(1 - \beta)$ (all of the order of c) we consider $\Delta\beta$ as a differential $\mathrm{d}\beta$, and v/c also as a differential which we denote $\mathrm{d}A$. Thus we get

$$\mathrm{d}A = \mathrm{d}\beta/(1 - \beta^2), \qquad \text{Q.E.D.}$$

These calculations would have been easy for Fresnel or Potier. That Lie's group theory was not published until 1887 is no objection, because, as we have seen, there is no need of an explicit recourse to the general theory of Lie groups to go through the reasoning.

But, in 1818 and even in 1874, a serious proposal that the law for combining velocities might not be the additive one certainly would have been rejected as 'crazy' — and 'crazy' its consequence as well, that there is an upper bound to velocities.

2.5.6. TERRELL'S RELATIVISTIC PHOTOGRAPHY REVISITED

We have seen in Section 2.3.4 that use of the classical additive law for combining velocities shows that a small object moving transversely at a relative velocity $v = c\beta$ is viewed as rotated by an angle $\alpha = \tanh^{-1} \beta$. In other words, the small angles $i\beta$ in the (x, ct) plane and β in the (y, z) plane are in a one-to-one association.

If, say, the 'moving object' is a parallelepipedic boxcar of width l and length L, its rear side is visible, projected as $l \sin \beta$, and its length is seen as reduced to $L \cos \beta \simeq L(1-\beta^2)^{1/2}$ (Figure 1, p. 33).

Penrose's (1959) and Terrell's (1959) relativistic reasoning show that this result holds exactly, regardless of the value of $\beta < 1$. My book on special relativity (1966, pp. 45—48) presents a very clever geometrical argument due to Becker deriving this conclusion, and displaying the close relationship between Bradley's aberration and the Penrose—Terrell phenomenon.

There is, of course, no reasonable hope of taking a snapshot of an object of recognizable shape moving at a velocity such that the effect would show up, but it is nevertheless interesting to think about the matter.

Going back (see Figure 1, p. 33) to the example of the boxcar running along a speedway, we consider the directions of the light rays reaching the observer stationed at O. The light rays emanating from the environment in the vicinity of the boxcar (say, the four trees pictured) are directed parallel to PO, while those emanating from the boxcar are aimed at the Bradley angle β. Therefore one can mimic the Penrose—Terrell phenomenon by superposing two snapshots: one of the environment taken from O, and one of a boxcar stationed at P taken from O'. Thus the complete picture consists of two incongruent parts, somewhat like those 'double exposures' unwittingly produced by old cameras when one forgot to roll the film.

2.5.7. TIME DILATATION AND THE 'TWINS PARADOX'

According to the formulas recalled in Section 1, a time interval T as measured by some clock moving at uniform velocity u with respect to some inertial reference frame $O_0x_0y_0z_0ct_0$ (say, the period separating two beatings of the clock) is measured in the O_0 system as $T_0 = T(1 - \beta^2)^{-1/2}$. Between the corresponding frequencies the relation $\nu_0 = \nu(1 - \beta^2)^{1/2}$ holds; this is called the 'transverse Doppler shift', and has been measured in 1938 by Ives and Stillwell, using light emitted by an atomic beam.

A large manifestation of this effect is displayed by the so-called π-mesons of the cosmic radiation, spontaneously decaying with a 'proper mean lifetime' of 2.5×10^{-8} s (as measured when they are at rest). Thanks to the relativistic time dilatation, and to their great velocity, they can traverse the whole terrestrial atmosphere, some 10 km thick, after having been generated in its upper layers by the primary cosmic radiation.

These phenomena are 'coordinate effects', somewhat analogous to the 'perspective effects' occurring in Euclidean geometry.

But what if the 'moving clock' C follows a curvilinear path in spacetime, intersecting at two points 1 and 2 the straight path of a clock C_0 following the axis O_0ct_0? *An assumption is needed concerning the behavior of the accelerated clock C.* It is indeed a very natural assumption, quite similar to the one underlying the working of odometers. An odometer following a curved path in ordinary 3-space measures the 'curvilinear abscissa' along the path, as tangent at each point C to the unit t vector of the so-called 'comoving tripod' $Ctnb$, the normal unit vector n of which points along the acceleration (when the velocity is constant in magnitude). Similarly, we assume that *the accelerated clock C measures the curvilinear abscissa along the* (*timelike*) *trajectory, that is its own 'duration'* or 'proper time' — very much as an automobile's odometer measures its 'mileage'. 'Aging' is the common fate of both.

Consider now two brand new identical automobiles leaving, say, Chicago with mileage zero, and meeting in Denver, the one, C_0, having travelled a direct route, and the other one, C, having followed an extremely tortuous route through New Orleans and Dallas. When both automobiles are finally parked side by side in Denver their odometers display very different mileages; say 1000 miles for C_0 and 2000 miles for C.

It is very much the same with the two twins we now consider. These twins — two true twins — have lived side by side, thus remaining practically identical to each other, until some point-instant 1 where one, C_0, decides to stay home, and the other, C, decides to travel in a spaceship undergoing strong accelerations. Let it be remarked that *acceleration* is precisely *curvature* of the timelike worldline. 'Home' is likened to an inertial rest frame and thus, strictly speaking, should not be located on Earth. However, if we imagine that C undertakes a journey very far out into the Galaxy before coming back, the difference between the reference system of C_0 and an inertial frame is negligible.

When both former twins meet again at point-instant 2, it is found, of course, that both have aged, but the 'accelerated twin' much less than the 'sedentary twin'. The calculation is fairly simple: integrating the previous formula we get

$$T = \int (1 - \beta^2)^{1/2} \, \mathrm{d}t_0$$

as the age of the accelerated former twin, the age of the sedentary one being $\int \mathrm{d}t_0 > T$!

Were we not saying that there is a strict analogy between the case of the two cars and that of the two brothers? And now, the conclusion is that the wandering car has aged more than the steadily travelling one, but that the wandering brother has aged less than the steadily travelling one! Well, the analogy is perfectly true; the difference stems entirely from the fact that space is Euclidean while spacetime is pseudo-Euclidean: *for timelike trajectories, the chord is longer than any arc it subtends.*

Can it then be said that the brother C has converted all the vast space he traversed, while in his accelerated cell, into time gained for living? If so, the meaning would be similar for someone put in deep freeze and later recalled to life. Truly, this man would not have lived longer. Simply, he would have been given the opportunity to jump into the future, and to become a contemporary of his grandsons.

For the two cars, of course, it was just the contrary: the car with the straighter route will still be alive when the other one is dead.

Besides the numerous verified consequences of relativistic time dilatation, there has been a very nice experiment quite convincing to the layman.

In 1971 Hafele and Keating (1972) flew around the Earth, from Washington to Washington, by commercial airplanes, one westward and one eastward, together with two high-precision atomic clocks. Everybody knows that a westbound sonic aircraft almost 'stops the sun', while an eastbound one makes days and nights terribly short. In other words, a westbound aircraft C_0, by counterbalancing the rotation of the Earth, follows in spacetime an almost straight path, while an eastbound aircraft C doubles, so to speak, the rotation of the Earth, and thus follows a helix more twisted than, say, the one of Washington.

We of course do not mention the intermediate landings and takeoffs nor gain and loss of speeds and altitudes. We are also omitting something significant: the dependence of the rate of a clock upon the gravity potential of the Earth, as formalized by Einstein's 'general relativity' of 1916.

This being said, and due account being taken of everything, Hafele's and Keating's experiment has quite nicely vindicated the relativistic evaluation of the so-called 'twins paradox'. The difference in aging was calculated and measured as some 300 nanoseconds, while the flight durations were of the order of 45 hours.

If, before leaving Washington, the two pilots had been two true twins equally freshly shaved, and if some device existed that could measure the length of beards to a relative precision of some 10^{-10}, then the 'twins paradox' could be tested that way. This is the same as saying that a wristwatch is anybody's spacetime odograph.

Another kind of clock would measure the fuel consumption of the airplane. If measured to a relative precision of 10^{-10}, this also would display the phenomenon.

But the fuel consumption of an airplane also measures (due account being taken of everything else) the length of the path travelled over lands and seas. Therefore, strictly speaking, the westbound and the eastbound crews do not attribute the same length to the Washington parallel. This is an other typically relativistic effect, and one having to do with the 'paradoxes' of rotating solid bodies.

2.5.8. THE HARRESS (1912) AND SAGNAC (1913) EFFECTS

These effects, in which two light beams go in opposite directions along a closed contour fixed on a rotating disc, are very analogous to the Hafele and Keating effect. They are 'absolute' effects, implying two intersections of two spacetime trajectories. These, however, were 'null' lines in Sagnac's experiment, where the light travelled *in vacuo*. In the Harress experiment the light travelled in a material medium glass. In each case the measurement was interferometric.

It is found (Costa de Beauregard, 1966, p. 35), both theoretically and experimentally, that the time lag has the expression

$$\Delta t = 4c^{-2}\Omega\mathscr{A},$$

Ω denoting the angular velocity of the disc, and $\mathscr{A}$ the area of the circuit.[3]

Therefore, like the Foucault pendulum experiment, the Harress and Sagnac experiments do display the 'absolute' character of rotations, thus confirming the 'restricted relativity principle' against the old, kinematical, 'principle of relative motions'.

2.5.9. THE PROBLEM OF ACCELERATING A SOLID BODY

It is easily seen that the Poincaré—Minkowski four-dimensional geometry does not admit the concept of a 'rigid body' remaining undeformed under acceleration, either linear or angular (except in the one case of Born's (Fokker, 1965, pp. 68—70) hyperbolic motion of a sphere).

Therefore the question of accelerated solid bodies cannot be treated independently from that of a relativistic elasticity theory.

This remark will end our rapid survey of the basic relativistic phenomenology which displays, as we have seen, quite a few curiosities.

2.5.10. KINEMATICS IDENTIFIED WITH VACUUM OPTICS. THE RESTRICTED RELATIVITY PRINCIPLE AS A KINEMATICAL PRINCIPLE

The very way we have entered the spacetime paradigm shows that relativistic kinematics is so closely connected with the optics of the vacuum that these are indeed just the two faces of a single coin. When Huygens and his followers started the whole affair by identifying their 'luminiferous aether' with an 'absolute reference frame', they were far from foreseeing the final resolution of the adventure.

It was not suspected either that the classical kinematical 'principle of relative motion', as discussed in Chapter 2.1, would have to yield to the 'restricted relativity principle' *on the grounds of kinematics proper.*

Of course, *conferring a 'universal' character to the restricted relativity principle amounts to stating that it is a kinematical principle* — because lengths, times and pure numbers are the *universal* language of physics.

The characteristic magnitudes of optics — wavelengths, periods and phases — are precisely lengths, times and pure numbers, so that the marriage of kinematics and optics is a perfectly suitable one.

It remains however to be understood what is so special with the motion of light. A clear answer is provided by Louis de Broglie's wave mechanics, to be discussed at the end of the next chapter.

That the velocity of light *in vacuo* truly is a *conversion coefficient* between space and time measurements is evident in formulas (2) and (3). The analogy of the (quadratic) formula (2) with the (linear) expression of the 'equivalence' law between work and heat

$$\mathrm{d}E = \mathrm{d}W - J\,\mathrm{d}Q$$

is fairly clear.

This analogy may be pursued. Restricting the complete Lorentz–Poincaré group to its orthochronous subgroup is some sort of an irreversibility principle, a 'second law' of the macrorelativity theory. However, as shown by Loschmidt in 1876 and by Zermelo in 1896, reversibility rules in microphysics. So *one should remain alert to the possibility that the full Lorentz–Poincaré group, including time and space reversals, is the one expressing relativistic invariance at the microlevel* — that is, the one characterizing the 'microrelativity theory'.

CHAPTER 2.6

THE MAGIC OF SPACETIME GEOMETRY

2.6.1. INTRODUCTION

Spacetime geometry, as the new playground of physics, stems from Poincaré's seminal remarks (1906b) expanded in Minkowski's famous articles (1907; 1908a, b). Another source is an elegant paper by Hargreaves (1908) where, however, the concept of the spacetime metric is not found.

The first Treatise using this presentation is by M. von Laue (1911) of which Einstein said, 'tongue in cheek', that he could hardly follow it. Other early Treatises of this style are due to Pauli (1958), and to Fokker (1965), both translated later into English. Since then, the four-dimensional style has become more and more dominant; quite a few treatises of that sort exist, among which I select Rindler's 1969 book, and for convenience, my own 1966 book, to which I shall often refer in this chapter.

As interesting dissonant curiosities I first mention the 1966 and 1972 books by Arzelies containing a lot of useful information and very thorough bibliographies; in these, the exploration proceeds in the pre-Minkowskian pedestrian style. I mention also the 1968 book by Shadowitz based on a trick, termed the Loedel or the Brehme diagram, which I leave as a surprise to be discovered by the reader.

The Poincaré—Minkowski formalization of spacetime geometry provides a bird's eye view, directly displaying the essentials of relativistic problems, and thus avoiding the numerous treacherous pitfalls obstructing the pedestrian approaches. There are cases where only the 'manifestly covariant' ways of reasoning and computing have been able to produce the right answer to intricate problems — for example, in quantum electrodynamics.

The four-dimensional presentation of the evolution of a physical process displays it all at once, past, present and future together, *the expressions 'all at once' and 'together' being of course not synonymous with 'at the same time'*. Therefore, the qualification 'relativistic physics' (so easily understood when viewed along the historical development of the subject) suddenly looks utterly inappropriate; *it really is an 'absolute' presentation, unmasking the deceitful 'appearances'*. However, the fold in the discourse draping the mathematics has been so strongly ironed that it seems impossible to flatten it. Thus one must satisfy oneself with the wording 'relativistic physics' — without being thus misled.

Technically speaking, *the four-dimensional view of a physical evolution transposes the three-dimensional static presentation of a permanent regime*, as occurring in many hydrodynamical or electrodynamical problems for example. Quite a few instances of that sort will be presented in this chapter. In other cases the classical analogy that comes to mind is that of 'direct connections at spacetime distance', similar to those showing up between points in the three-dimensional Euclidean space. Fokker's excellent book (1965) is full of examples of this sort. Its underlying philosophy has come out as quite operational in, say, the Wheeler—Feynman electrodynamics, which I present in Section 2.6.14 and in the Feynman formalization of quantum electrodynamics, which I present in Chapter 4.7.2.

A few words are now in order concerning a technical point.

If pictures of spacetime connections are drawn in, say, the (x, ct) plane, all coordinates must be real, so the metric to be used is the real hyperbolic metric of Equation (2.5.2). We have already met quite a few examples of this.

However there is a drawback in this, because the 'covariant' and 'contravariant' coordinates of a vector (or of the generalizations of vectors known as 'tensors') are equal to each other only when the metric is *stricto sensu* Euclidean — with '+' signs before all squares. For elucidation of this point I refer to any Treatise; for example, mine (1966). This is why Poincaré and Minkowski used the trick of a purely imaginary time coordinate, denoting any point-event as $x^i = x_i$ with $x^4 = x_4 = ict$ so that, as $i^2 = -1$, the squared spacetime distance comes out as

$$s^2 \equiv x^i x_i = x_1^2 + x_2^2 + x_3^2 + x_4^2;$$

x^i and x_i denote the 'contravariant' and 'covariant' coordinates of a point-event, and, according to a convenient Einstein convention, automatic summation is implied over repeated indexes (covariant and contravariant ones).

By definition, a 'tensor' $T^{ij\cdots}{}_{kl\ldots}$ transforms as a product of vectors

$A^i B^j \ldots C_k D_l \ldots$. Its 'rank' n, the number of its indexes, is lowered to $n - 2$ by 'contracting' a covariant and a contravariant index; that is, equating them, and summing.

Eventually, under permutation of two indexes, tensors are symmetric ($T^{ij} = T^{ji}$) or skew symmetric ($T^{ij} = -T^{ji}$), which, if necessary, can be denoted as $T^{(ij)}$ or $T^{[ij]}$, respectively.

The Poincaré—Minkowski trick of setting $x^4 = ict$ is used by many authors, including me, but is by no means compulsory. Quite a few authors do not use it — and in fact, for obvious reasons, everybody working in general relativity. Anyway, one way or the other, all authors feel free to illustrate their papers with cavalier sketches which are quite easily understood. This I will do in the following pages.

Up to this point I have refrained from writing down many formulas. In this chapter I must relax this decision, as chosen formulas will be the only means for conveying what is at stake. But, then, I must apologize twice: first to the mathematical expert, who will feel that I am explaining trivialities and, even worse, doing so with insufficient rigor; and also apologize to the reader not very conversant with mathematics, who will think that I should be able to explain the matter in words.

So, well aware that both sides will be dissatisfied, I nevertheless must proceed, adding to the main, classical, parts of the building a few selected examples of matters less widely known, but well illustrating the style and spirit of the relativity theory.

2.6.2. INVARIANT PHASE AND 4-FREQUENCY VECTOR

The familiar expression of the phase φ of a plane monochromatic light wave propagating in the vacuum in the direction $\mathbf{k}/k$ with the space frequency $\mathbf{k}$ and the time frequency $\nu \equiv ck$ displays a 'manifest' spacetime symmetry:

$$(2.6.1) \quad \varphi = 2\pi(\nu t - \mathbf{k} \cdot \mathbf{r}).$$

As it is invariant under 3-space rotations and the parity reversal, one would guess that it is also invariant under 4-space hyperbolic rotations and the time reversal.

This argument can be sharpened. From Yilmaz's (1972) remark (see Section 2.3.3) that invariance of the Fermat principle implies that of the isotropy of c, the Lorentz-invariance of formula (1) is deduced. Also, from the classical presentation of the photography of moving objects as discussed in Sections 2.3.4 and 2.5.6, we learn that Bradley's aberration, expressible as a hyperbolic rotation $i\beta$ in the (x, ct) plane, is compensated by a spatial rotation of angle β in the (x, y) plane, y being the direction along which the light propagates. All this requires that the phase φ be a relativistic scalar, Lorentz, P and T invariant.

Therefore **k** and ν combine to form a spacetime 4-vector k^i or k_i with $i = 1, 2, 3, 4$ and $k^4 = k_4 = i\nu/c$; thus the phase shows up as the four-dimensional scalar product

$$(2.6.2) \quad \varphi = -2\pi(k^i x_i),$$

with the restriction $k^i k_i = 0$ in the present case of light propagation.

The classical expressions $\Delta\nu/\nu = \beta$ of the Doppler effect, and $\Delta\alpha = \beta$ of Bradley's aberration, are then recognized as differential forms of the Lorentz transformation of the 4-vector k^i.

2.6.3. THE 4-VELOCITY CONCEPT

As used in Section 2.5.2, with components $c \tanh A$ and c in the (x, ct) plane, the velocity concept was not coordinate independent — that is, not 'covariant'. But a modified velocity concept, with components $c \sinh A$ and $c \cosh A$, is a 4-vector of length c — and is 'covariant'.

So, using the imaginary time component, we write in general, with $u = 1, 2, 3$, and $\beta = v/c$,

$$(2.6.3) \quad V^u = c\beta^u(1-\beta^2)^{-1/2}, \qquad V^4 = ic(1-\beta^2)^{-1/2}.$$

Up to the factor c, V^i thus is a timelike unitary 4-vector:

$$(2.6.4) \quad V_i V^i = -c^2.$$

2.6.4. INTEGRATION AND DIFFERENTIATION IN SPACETIME

In relativistic physics two sorts of curvilinear integrals do appear: timelike ones and spacelike ones, the latter usually closed:

$$\int_1^2 a \, \mathrm{d}x^i, \qquad \oint a \, \mathrm{d}x^i.$$

In such covariant integrals a must be a tensor (or a scalar) with or without an index contracting with i.

In double integrals, the integration element $[dx^i\, dx^j]$ is a skew symmetric 6-component object, $[dx^i\, dx^j] \equiv -[dx^j\, dx^i]$, with three 'spacelike components' $[dx^u\, dx^v]$ (u, v = 1, 2, 3) generalizing the familiar surface element, and three 'timelike components' $[dx^u\, c\, dt]$. It is often convenient to use the so-called 'dual' 6-component skew-symmetric tensor

$$ic\, ds^{ij} = \pm [dx^k\, dx^l],$$

the sign being '+' if the permutation of indexes (i, j, k, l) is even. Both spacelike and timelike double integrals

$$\iint a\, ds^{ij}$$

do occur.

In threefold integrals, the integration element is a skew-symmetric 4-component object, that is, the dual of a 4-vector

$$ic\, du^i = [dx^j\, dx^k\, dx^l],$$

the fourth component of which generalizes the familiar Euclidean volume element $[dx^1\, dx^2\, dx^3]$, while the three others are fluxlike, with expressions $ic[ds^u\, dt]$. Of course, the three-dimensional 4-vector volume element is orthogonal to the integration domain to which the three vectors in the [ ] are tangent; this is like the orthogonality, in Euclidean space, of the vector 'surface element' **ds** to the surface.

Both sorts of integrals, with du^i timelike and du^i spacelike, do occur.

When du^i is timelike — that is, when the integration domain is spacelike — we denote it as $\mathscr{C}$ and call it a 'cap' when it is associated with the evolution of a fluid of 4-velocity V^i (see Figure 6). When du^i is spacelike, we denote it as $\mathscr{W}$ and call it a 'wall' when associated with the evolution of a fluid. In that case (see Figure 6) we take as the 'wall' the current lines passing through the contour of $\mathscr{C}$, so that, on the wall, $V^i\, du_i \equiv 0$.

Spacelike integrals such as

$$A = \iiint_{\mathscr{C}} a\, du^i$$

(with or without an 'i contracting index' in a) generalize the usual

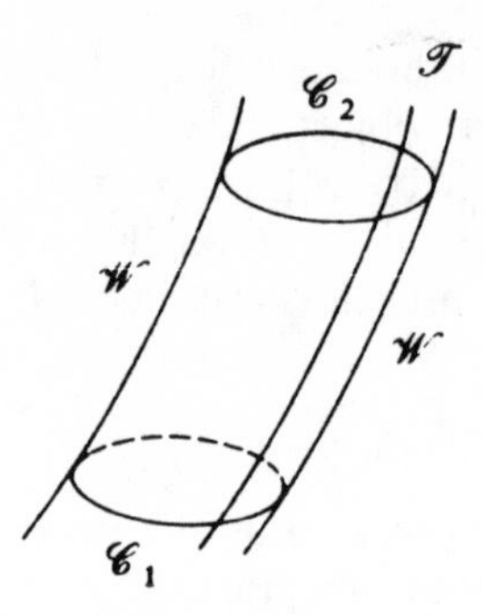

Fig. 6. Relativistic kinematics of a fluid: a bunch of timelike current lines or 'world-tube', its 'wall' $\mathscr{W}$, and two spacelike 'caps' $\mathscr{C}_1$ and $\mathscr{C}_2$ or 'states' of a fluid drop.

relation between an integrated magnitude and its density. Two important remarks are in order at this point.

The first one is that, in spacetime geometry, *an integrated magnitude and the corresponding density 'a' do not have the same tensorial rank*; the difference between the two ranks is ± 1, according as there is not, or is, in a an 'i contracting' index. I find it useful to give in Table I a few important examples of this important property (brackets denote skew symmetry).

TABLE I

	Density	*Integral magnitude*
Electric charge	4-vector j^i	Scalar Q
Force	4-vector f^i	Skew-symmetric tensor $F^{[ij]}$
Torque	Skew-symmetric tensor $\mu^{[ij]}$	Tensor $M^{[ij]k}$
Momentum energy	Tensor T^{ij}	4-vector p^i
Angular momentum	Tensor $\sigma^{[ij]k}$	Skew-symmetric tensor $S^{[ij]}$

The second remark is that, for generality and for covariance, the spacelike curvilinear 'cap' $\mathscr{C}$ is taken arbitrarily, not orthogonal to the time axis of any Poincaré tetrapod (except, sometimes, at a transitory stage, either in research or in pedagogy, for making clear the contact with prerelativistic physics). Therefore, a threefold integral contains in fact four terms, one formally similar to the familiar volume expression, and three being flux like contributions.

Timelike integrals of the form

$$A = \iiint_{\mathscr{W}} a \, \mathrm{d}u^i$$

essentially are of a flux type, the three main components of $\mathrm{d}u^i$ being $ic[\mathrm{d}s^u \, \mathrm{d}t]$. Of course there is also the (usually small) fourth component $[\mathrm{d}x^u \, \mathrm{d}x^v \, \mathrm{d}x^w]$ which, classically, was absent.

Finally we must consider the fourfold sort of integrals

$$A = \iiiint a \, \mathrm{d}\omega$$

with

$$ic \, \mathrm{d}\omega \equiv [\mathrm{d}x^1 \, \mathrm{d}x^2 \, \mathrm{d}x^3 \, \mathrm{d}x^4].$$

These are often taken inside the contour '$\mathscr{C}_2 - \mathscr{C}_1 + \mathscr{W}$' consisting (Figure 6) of two 'caps' $\mathscr{C}_1$ and $\mathscr{C}_2$, representing two 'successive states' of the same portion of fluid, and the 'wall' $\mathscr{W}$, made of fluid lines joining the contours of $\mathscr{C}_1$ and $\mathscr{C}_2$. The reason for the negative sign before $\mathscr{C}_1$ is that one finds it convenient to confer the same 'past timelike orientation' on the vectors $\mathrm{d}u^i$ on both $\mathscr{C}_1$ and $\mathscr{C}_2$. The reason why the orientation of $\mathrm{d}u^i$ on $\mathscr{C}$ is 'past' rather than 'future' timelike is that a factor $i^2 = -1$ comes out in the product of two fourth components of spacetime vectors; of course, that must be corrected for a smooth 'correspondence' between the classical and the relativistic formulas.

Two often useful expressions of the four-dimensional volume element are as follows. One, occurring in the case of an evolving fluid, is

$$\mathrm{d}\omega = \mathrm{d}x^i \, \mathrm{d}u_i.$$

The other one, displaying a 'correspondence' with prerelativistic physics, is

$$\mathrm{d}\omega = \mathrm{d}v \, \mathrm{d}t.$$

2.6.5. INVARIANT OR SCALAR VOLUME ELEMENT CARRIED BY A FLUID

This simple and useful concept has been introduced independently by

McConnell (1929), Synge (1934) and myself (1966, p. 50). The definition is:

$$(2.6.5)\quad u \equiv \iiint_{\mathscr{C}} V^i \, \mathrm{d}u_i, \qquad 0 \equiv \iiint_{\mathscr{W}} V^i \, \mathrm{d}u_i$$

so that, using Green's formula,

$$(2.6.6)\quad u_2 - u_1 = \iiiint \partial_i \, V^i \, \mathrm{d}\omega$$

with, in the metric such that $x^4 = ict$, $\partial_i = \partial^i \equiv \partial/\partial x^i = \partial/\partial x_i$.

Incidentally, the formula

$$(2.6.7)\quad V^i \, \partial_i a \equiv \mathrm{d}a/\mathrm{d}\tau \equiv a',$$

with τ denoting the 'proper time' along a fluid line, will be used again and again in the following pages.

2.6.6. THE GREEN- AND STOKES-LIKE INTEGRATION TRANSFORMATION FORMULAS

The general formula of this sort is derived in the relativity theory treatises, and for it I refer to my book (1966, p. 19). The few specifications of this formula needed in this chapter are simple, and I hope the reader will trust me when I write them down 'impromptu'.

2.6.7. RELATIVISTIC ELECTROMAGNETISM AND ELECTRODYNAMICS

Like classical optics, Maxwell's electromagnetism really was already relativistic. Thus it is that Einstein's answer to Föppl's riddle concerning Faraday's induction formula was merely a display of the relativistic covariance of Maxwell's equations. In his 1908 famous work Minkowski acted as a tailor, not as a 'body builder'.

As optics was the root of the relativity theory, electromagnetism appears as its trunk. In its use of the energy, force, torque and heat concepts, it clearly indicates which tensorial variances should be attributed to these magnitudes and to their densities, thus allowing a straightforward derivation of the relativistic dynamics and relativistic thermodynamics.

All textbooks explain how the electric **E** and magnetic **H** fields partially transform into each other under a change of reference frame, and how both combine into a 6-component skew-symmetric tensor, the electromagnetic field E^{ij} or its 'dual', the 'magneto-electric field' $H^{kl} = iE^{ij}$. The textbooks explain also that H^{ij} is the spacetime curl $\partial^i A^j - \partial^j A^i$ of the '4-vector potential' uniting the classical **A** and V potentials ($A^4 = iV$). If there are electric **P** and magnetic **M** polarizations — that is, presence of an electric **D** = **E** + **P** and a magnetic **B** = **H** + **M** 'induction field' — two spacetime tensors appear: $B^{ij} = (\mathbf{B}, \mathbf{E})$ and $H^{ij} = (\mathbf{H}, \mathbf{D})$. Then

(2.6.8) $B^{ij} = \partial^i A^j - \partial^j A^i.$

The classical electric current **j** and charge q densities combine to produce a '4-current density' j^i ($j^4 = icq$), the conservation of which is expressed as

(2.6.9) $\partial_i j^i = 0$

Then the electric charge Q is a scalar, defined as

(2.6.10) $Q = \iiint_{\mathscr{C}} j^i \, \mathrm{d}u_i \quad \text{or} \quad Q = \iiint_{\mathscr{W}} j^i \, \mathrm{d}u_i;$

the latter expression is a charge flux through a timelike wall.

In a 'purely convective' case the current j^i is timelike and collinear with the 4-velocity V^i of the fluid. Then

(2.6.11) $Q_2 - Q_1 = \iiint_{\mathscr{C}_2 - \mathscr{C}_1 + \mathscr{W}} j^i \, \mathrm{d}u_i = \iiiint \partial_i j^i \, \mathrm{d}\omega = 0;$

the charge of a 'fluid cap' is conserved. If there is conduction, the second integral (10) is not zero, and, according to formula (11), the charge variation between the 'states' $\mathscr{C}_1$ and $\mathscr{C}_2$ is due to the flux through $\mathscr{W}$.

In the convective case the Lorentz force density **f** and the power density **f** · **v** imparted to the fluid combine as

(2.6.12) $f^k = B^{kl} j_l$

with

$$(2.6.13)\, f^4 = ic^{-1}\mathbf{f} \cdot \mathbf{v}$$

requiring

$$(2.6.14)\; V_i f^i = 0,$$

so that f^i is a spacelike 4-vector. If there is conduction, $V_i f^i$ is non-zero, and precisely equals the rate at which Joule heating is produced (my book, 1966, p. 61).

No less important than the scheme of a continuous electric current density is that of a point charge Q. Then the expressions $\mathrm{d}\mathbf{p} = \mathcal{F}\, \mathrm{d}t$ of the momentum, and $\mathcal{F} \cdot \mathbf{v}\, \mathrm{d}t$ of the work, or energy, imparted to the charge combine as

$$(2.6.15)\, \mathrm{d}p^i = QB^{ij}\, \mathrm{d}x_j \equiv F^{ij}\, \mathrm{d}x_j$$

with

$$(2.6.16)\, \mathrm{d}p^4 = ic^{-1}\, \mathcal{F} \cdot \mathbf{v}\, \mathrm{d}t$$

entailing

$$(2.6.17)\; V_i\, \mathrm{d}p^i = 0.$$

There is a close parallelism between the three latter and the three former formulas. A reasonable guess concerning the relation between the second rank force tensor F^{ij} and the corresponding force density f^i is

$$(2.6.18)\, \mathrm{d}F^{ij} = f^i\, \mathrm{d}u^j - f^j\, \mathrm{d}u^i.$$

In fact, multiplying both sides by $\mathrm{d}x_j$, one gets

$$(2.6.19)\, \mathrm{d}x_j\, \mathrm{d}F^{ij} = f^i\, \mathrm{d}x_j\, \mathrm{d}u^j = f^i\, \mathrm{d}\omega, \qquad \text{Q.E.D.}$$

Finally the 'elastic style' Maxwell and Maxwell—Minkowski tensors will be considered.

Maxwell has shown that the Coulomb plus Laplace force densities, later termed the Lorentz force density, can be cast in the form $f^u = \partial_v M^{uv}$, the 'elastic tensor' M^{uv} being symmetric ($M^{uv} = M^{vu}$) or not, according to the absence or presence of electric and/or magnetic polarizations. In the latter case $\mu^{uv} \equiv M^{uv} - M^{vu}$ is none other than the sum of the electric and magnetic torque densities, which is just what the elasticity theory required.

Besides this, a field energy density w showed up, together with an 'energy flux density' $\mathbf{E} \times \mathbf{H}$ and a 'momentum density' $\mathbf{D} \times \mathbf{B}$ termed the 'Poynting vectors'. Poynting and Poincaré have respectively discussed energy and momentum transfer and conservation through the electromagnetic field. Minkowski, in his 1908 paper, showed that all these formulas combine so as to form a spacetime M^{ij} 'elastic style tensor', and demonstrated the formula

$$f^i + \tfrac{1}{4} B^{kl}[\partial^i]H_{kl} = \partial_j M^{ij}$$

where the Lorentz 4-force density f^i is supplemented by a Stern—Gerlach style force density; $[\partial_i] \equiv \underset{\rightarrow}{\partial}_i - \underset{\leftarrow}{\partial}_i$ denotes a partial differential operator operating once towards the right and once towards the left — a mathematical object showing up more than once in theoretical physics.

Thus, Faraday's concept of a field transmitting energy fluxes and momentum densities through the vacuum, and storing an energy density, has found a mathematical expression. For a discussion of other possible elasticlike tensors, and related questions in electrodynamics, I refer to my paper (1975).

A final remark is that *the twin interpretations of the Poynting vector as an energy flux and as a momentum density entail the existence of a mass—energy equivalence in the ratio* c^2.

2.6.8. ENTERING RELATIVISTIC DYNAMICS

All approaches to relativistic dynamics are, one way or another, inductive. Here is, I believe, an easily understandable one.

Inside a classical fluid flowing in permanent regime the divergence of the energy flow, $\boldsymbol{\partial} \cdot (w\mathbf{v})$, where $\mathbf{v}$ denotes the velocity and w the energy density, is related to the density of applied power $\mathbf{f} \cdot \mathbf{v}$ via $\mathbf{f} \cdot \mathbf{v} = -\boldsymbol{\partial} \cdot (w\mathbf{v})$ Arguing that a three-dimensional permanent regime is a faithful analog of a four-dimensional evolution, we transcribe this formula in spacetime language as

$$(2.6.20)\ V_i f^i = -\partial_j (wV^j)$$

with V^i denoting the 4-velocity defined in (3) and (4). The implication is that there is in f^i a contribution $f^i_{\parallel}$, the expression of which is

$$(2.6.21)\ f^i_{\parallel} = c^{-2} V^i\, \partial_j (wV^j).$$

Now, the fundamental dynamical equation of a classical fluid of mass density ρ is, $\mathbf{v}'$ denoting the acceleration, $\mathbf{f} = \rho\mathbf{v}'$. This holds in particular in a permanent regime, and therefore, arguing as before, we transcribe this formula into

$$(2.6.22)\; f^i_\perp = \rho V'^i \equiv \rho V^j\, \partial_j V^i$$

because, according to (3), $V_i V'^i = 0$.

Thus we have found two orthogonal components, $f^i_\parallel$ and $f'^i_\perp$, of the 4-vector

$$(2.6.23)\; f^i = f^i_\parallel + f'^i_\perp,$$

and so, by adding (21) and (22), we must obtain a geometrically covariant equation. This is possible if and only if

$$(2.6.24)\; w = c^2\rho,$$

which displays a *universal law of physical equivalence between mass and energy*, in the 'tremendous' ratio c^2. Then, the fundamental dynamical laws of the relativistic fluid are

$$(2.6.25)\; f^i = \partial_j(\rho V^i V^j) \equiv V^i\, \partial_j(\rho V^j) + \rho V'^i$$

together with (21), (22) and (23).

Formula (24) expresses *the big discovery of Einsteinian dynamics, its very cornerstone, and the one missing in Newtonian dynamics.* It has meaning to the general public, who know it as 'Einstein's formula'.

The universal law formalized as (24) says that all energy has mass, and that all mass implies the existence of an energy — perhaps a 'dormant' one that can be awakened . . . .

Concerning the electromagnetic field discussed in the previous section, the Maxwell–Minkowski 'elastic tensor' thus distributes not only energy but also mass throughout space. Therefore, the so-called 'vacuum' is not so empty after all: if not matter, there is at least radiation in it; and this radiation has mass and momentum.

Of course, the ρ and the w appearing in formulas (20) to (25) are relativistic scalars, or invariants. Those showing up in the scaffolding formulas quoted in the text are not. Finding the relations between the classical ρ and w and the relativistic ones is left to the sagacity of the interested reader.

Just for completion I now write down the following integral formulas, analogous to some encountered previously.

The momentum energy of a fluid cap is

$$(2.6.26)\ p^i = \iiint_{\mathscr{C}} \rho V^i V^j \, \mathrm{d}u_j \equiv \iiint_{\mathscr{C}} \rho V^i \, \mathrm{d}u$$

and, using (25), we get

$$(2.6.27)\ p_2^i - p_1^i = \iiint_{\mathscr{C}_2 - \mathscr{C}_1} \rho V^i V^j \, \mathrm{d}u_j = \iiiint f^i \, \mathrm{d}\omega$$

displaying f^i as a source of momentum energy.

The 'proper energy' of a fluid cap is

$$(2.6.28)\ W = \iiint_{\mathscr{C}} w V^i \, \mathrm{d}u_i.$$

Using (20) we then get

$$(2.6.29)\ W_2 - W_1 = \iiint_{\mathscr{C}_2 - \mathscr{C}_1} w V^i \, \mathrm{d}u_i = \iiiint V_i f^i \, \mathrm{d}\omega$$

displaying $V_i f^i$ as a 'proper energy source'.

2.6.9. FLUID MOVED BY A SCALAR PRESSURE: A QUICK LOOK AT RELATIVISTIC THERMODYNAMICS

Very many works have been devoted to relativistic thermodynamics but none, so far as I know, has inquired as to what should replace, in a basically continuum thermodynamics, the two faithful plough horses harnessed by Clausius: integrability of energy and entropy. In other words, which concept corresponds, in fluid dynamics, to an exact differential? *Exact divergence* is the answer. In a short article (1984b) which I summarize here, I have shown how *exact energy and entropy divergences* can be used as cornerstones for a relativistic thermodynamics. This approach avoids introducing a heat current density or a heat stress—energy tensor, which concepts obviously revive the late 'caloric theory'.[1]

The accepted relativistic dynamical equation of a fluid moved by a scalar pressure ϖ is

$$(2.6.30)\ \partial^i\varpi = \partial_j[(\rho + c^{-2}\varpi)V^iV^j] = c^{-2}\partial_j(\sigma V^iV^j).$$

It is obtained by replacing in (25) f^i by $\partial^i\varpi$, thus generalizing the prerelativistic formula $\mathbf{f} = \boldsymbol{\partial}\varpi$, and adding the contribution $c^{-2}\varpi$ to the mass density ρ, as required by the mass—energy equivalence, as ϖ is known to be an energy density. The quantity

$$(2.6.31)\ \sigma \equiv w + \varpi$$

transposes Gibb's thermodynamical 'enthalpy density'.

Multiplying Equation (30) by V_i and contracting, we get

$$(2.6.32)\ \partial_i(wV^i) = -\varpi\partial_i V^i, \qquad \partial_i(\sigma V^i) = V^i\partial_i\varpi \equiv \varpi',$$

'corresponding' to the thermodynamical differentials of energy and enthalpy

$$\mathrm{d}U = -\varpi\ \mathrm{d}v, \qquad \mathrm{d}F = v\ \mathrm{d}\varpi.$$

This correspondence is expressed in the first four columns of the following table. The four latter columns have been added by analogy, T denoting a typical 'intensive' magnitude, S a typical 'extensive' magnitude and s^i its density

Classical	d	v	$\mathrm{d}v$	$v\mathrm{d}$	T	$\mathrm{d}S$	$S\ \mathrm{d}T$	$\mathrm{d}(TS)$
Relativistic	∂_i	V^i	$\partial_i V^i$	$V^i\ \partial_i$	T	$\partial_i s^i$	$s^i\ \partial_i T$	$\partial_i(Ts^i)$

From the classical thermodynamical differentials of energy and generalized enthalpy,

$$\mathrm{d}U = \sum \mathbf{F}\cdot \mathrm{d}\mathbf{r} - \varpi\ \mathrm{d}v + T\ \mathrm{d}S + \sum \mu\ \mathrm{d}n$$

$$\mathrm{d}\mathscr{F} = \sum \mathbf{F}\cdot \mathrm{d}\mathbf{r} + v\ \mathrm{d}\varpi - S\ \mathrm{d}T - \sum n\ \mathrm{d}\mu$$

(where $\mathbf{F}\cdot\mathrm{d}\mathbf{r}$ denotes an external work, $-\varpi\ \mathrm{d}v$ the pressure work, $T\ \mathrm{d}S$ the absorbed heat, $\mu\ \mathrm{d}n$ a 'Gibbs chemical work') we derive immediately, using the above table, the 'corresponding' relativistic formulas

$$(2.6.33)\ \partial_i(\varpi V^i) = V^i f_i - \varpi \partial_i V^i + T\,\partial_i s^i + \sum \mu\,\partial_i \nu^i,$$

$$(2.6.34)\ \partial_i(\sigma V^i) = V^i f_i + V^i \partial_i \varpi - s^i \partial_i T - \sum \nu^i\,\partial_i \mu.$$

Similarly, we rewrite the 'state equation' of perfect gases,

$$\mathrm{d}(\varpi v - knT) = 0,$$

where k denotes Boltzmann's constant, as

$$(2.6.35)\ \partial_i(\varpi V^i - kT\nu^i) = 0.$$

We notice, in the expression of $\mathrm{d}U$, the presence of 'integrating factors' $1/\varpi$ for the pressure work, $1/T$ for the heat, $1/\mu$ for a chemical work, and the same ones in the relativistic formula (33).

Let us call (30) the external and (32) the internal dynamical equation of a fluid. We see that *there exists a strict isomorphism between the classical and the internal thermodynamics of a fluid.* This amounts to saying that, apart from explicit covariance, nothing physically new emerges.

It is, of course, quite different with the external thermodyamics of a fluid — an intricate matter to which quite a few papers have been devoted. I shall not pursue the matter here.[2] (See Gariel, 1986.)

2.6.10. DYNAMICS OF A POINT PARTICLE

The relativistic expression of the Galileo—Newtonian fundamental equation $\mathbf{F} = \dot{\mathbf{p}}$ has already been obtained as (15), that is

$$(2.6.36)\ \mathrm{d}p^i/\mathrm{d}\tau \equiv p'^i = F^{ij} V_j \equiv F^i$$

where, if F^{ij} is skew symmetric,

$$(2.6.37)\ V_i p'^i = 0.$$

Equation (15) has shown us that $\mathrm{d}p^4 \equiv ic^{-1}\,\mathrm{d}W$ is the energy transferred to the point particle. Now we show that if m denotes the so-called 'Maupertuisian mass', we also have $\mathrm{d}p^4 = ic\,\mathrm{d}m$; this merely results from an extension to the fourth dimension of the expression $\mathbf{F} = (m\mathbf{v})'$ of Newton's 'second law'. Thus we rederive Einstein's mass—

energy equivalence in the form

(2.6.38) $W = c^2 m.$

Covariantly speaking, one rewrites the relation between momentum-energy, velocity and mass as

(2.6.39) $p^i = m_0 V^i$

with $m_0 = \text{const}$ by virtue of (37).

According to formulas (3) we see that $m = m_0(1 - \beta^2)^{-1/2}$, displaying the 'velocity dependence of the Maupertuisian mass'; it is such that $m \to \infty$ if $\beta \to 1$. The 'rest mass' m_0 is of course a scalar.

The connection between this presentation and the one expressed by formulas (20) to (25) is straightforward: it goes through (18), (19) and (27), where the world tube is squeezed in the form of a filament.

A generalization of these formulas where the force F^{ij} is no longer skew symmetric, and, thus, the rest mass no longer constant, is possible.

2.6.11. ISOMORPHISM BETWEEN THE CLASSICAL STATICS OF FILAMENTS AND THE RELATIVISTIC DYNAMICS OF SPINNING-POINT PARTICLES

Let us recall that the first equation of the statics of a filament of curvilinear abscissa l subjected to a linear force density $\mathbf{f}$, and thus undergoing a tension $\mathbf{T}$, is

(2.6.40) $\mathrm{d}\mathbf{T} = \mathbf{f} \cdot \mathrm{d}\mathbf{l}.$

If the filament is perfectly flexible $\mathbf{T}$ is tangent to it, and that is the end of the story. In a still more specific case $T \equiv |\mathbf{T}|$ is constant; then $\mathrm{d}\mathbf{T}$ is directed along the 'principal normal unit vector' $\mathbf{n} = \mathrm{d}\mathbf{t}/\mathrm{d}l$ of the 'comoving unit tripod' $\mathbf{t}, \mathbf{n}, \mathbf{b}$.

If the filament is a stiff one, the preceding formula is supplemented by

(2.6.41) $\mathrm{d}\mathbf{S} = \mathbf{g} \cdot \mathrm{d}\mathbf{l} + \mathbf{t} \times \mathbf{T}$

where $\mathbf{g}$ denotes a 'linear torque density' and $\mathbf{S}$ 'an angular tension'.

It so happens that the four-dimensional equations of the dynamics of spinning-point particles, as proposed by Weyssenhof (1947), Weyssenhof and Raabe (1947), and other authors (Corben, 1968; Costa de Beauregard, 1966, p. 87; Papapetrou, 1961; Sciama, 1968) are isomorphic to these equations. They read

$$dp^i = F^{[ij]}\,dx_j$$

$$ds^{ij} = M^{[ij]k}\,dx_k + V^i p^j - V^j p^i,$$

the first one being (36) above. The second one expresses the variation of the internal angular momentum, or 'spin' of the particle, in responding to the applied torque M^{ijk}. Setting

$$F^i \equiv F^{ij} V_j, \qquad M^{ij} \equiv M^{ijk} V_k,$$

and, τ denoting the proper time, we rewrite them as

$$(2.6.42)\quad dp^i = F^i\,d\tau,$$

$$(2.6.43)\quad ds^{ij} = M^{ij}\,d\tau + V^i p^j - V^j p^i,$$

the isomorphism of which, with (40) and (41), is still more obvious.

The detailed correspondence is given in Table II. So this is an other instance of the 'statification' of a problem by use of the spacetime formalism.

TABLE II

3-*Euclidean space*	4-*Minkowskian spacetime*
Curvilinear abscissa on a filament l	Proper time τ
Tangent unit vector **t**	4-velocity V^i
Tension **T**	Momentum energy p^i
Linear density of applied force **f**	Applied 4-force F^i
Angular tension **S**	6-component angular momentum $S^{[ij]}$
Linear density of applied torque **g**	6-component applied torque $M^{[ij]}$

2.6.12. BARYCENTER AND 6-COMPONENT ANGULAR MOMENTUM AROUND THE BARYCENTER. THE RELATIVISTIC 'GENERAL THEOREMS'

We consider a cloud of spinless point particles, at point-instants x^i, with timelike momentum energies p^i. The total momentum energy, defined as

$$P^i = \sum p^i,$$

is timelike. The 'total mass' of the system is defined as $M = -(i/c)P^4$.

A system of 'sliding vectors' such as p^i is said to constitute a 'wrench' with 'resultant' P^i. Finding its 'axis' is a classical problem, which we solve by writing

$$X^i P^j - X^j P^i + S^{ij} = \sum (x^i p^j - x^j p^i),$$

$$S^{ij} P_j = 0,$$

a system of 6 + 4 equations in the 4 + 6 unknowns P^i and S^{ij}, one of which is free, because $S^{ij} P_i P_j \equiv 0$.

By working in the 'overall rest mass frame' where $\mathbf{P} = 0$, the system is easily solved (Costa de Beauregard, 1966, p. 82) and seen to fix uniquely the 'wrench'. Along its timelike 'axis', parallel to P^i, the 'barycenter' slides, with coordinates X^i. The 6-component 'angular momentum around the barycenter' has three 'spacelike components' $X^u P^v - X^v P^u$ corresponding to the angular momentum proper, and three timelike components $icM(X^u - V^u T)$, where $P^u \equiv MV^u$, generalizing the classical expressions MX^u.

If, now, these point particles interact via a field, we must, in the relativity theory, take account for the momentum energy (or momentum mass) propagated through the field. This we do by introducing a momentum-energy density $T^{ij}(x^k)$, analogous to the Maxwell–Minkowski 'elastic tensor' which is symmetric in the absence of spin. If there is spin, however, T^{ij} must be asymmetric, and a 'spin density' $\sigma^{ijk}(x^l)$ must be considered. Then we also attribute spins s^{ij} to the point particles, as in Equation (43), and we end up with the system of equations

$$P^i = \sum p^i + \iiint_{\mathscr{C}} T^{ij} \, \mathrm{d}u_j = \text{const}$$

$$M^{ij} = X^i P^j - X^j P^i + S^{ij} = \sum [x^i p^j - x^j p^i + s^{ij}] +$$

$$+ \iiint_{\mathscr{C}} [x^i T^{jk} - x^j T^{ik} + \sigma^{ijk}] \, \mathrm{d}u_k = \text{const},$$

expressing, in relativistic form, the 'general theorems': the barycenter of the overall system has a rectilinear and uniform motion; the total momentum energy P^i and 6-component angular momentum M^{ij} are conserved.

2.6.13. ANALYTICAL DYNAMICS OF AN ELECTRICALLY CHARGED POINT PARTICLE

The general idea of analytical dynamics, as proposed by Lagrange, Hamilton, and their followers, is to 'vary' the trajectory of a point particle, thus introducing the concept of spacelike displacements δx^i, to which correspond other variations such as, say, $\delta P^i \equiv \partial^j P^i \, \delta x_j$. Thus, the 'actual' timelike trajectory is viewed as embedded, as a current line, inside a 'fictitious fluid' of 'virtual trajectories'.

It is now well known (Abraham and Marsden, 1978) that such a formalism belongs to the so-called 'symplectic formalism', which I shall not define here, merely showing the pleasant appearance of its formulas. This I do in the case of an electrically charged point particle, the basic dynamical equations of which are (15) together with (8). These are easily recast as

$$(2.6.44)\ \mathrm{d}P^i = Q\, \partial^i A^j \, \mathrm{d}x_j, \qquad P^i \equiv p^i + QA^i,$$

QA^i denoting a 'potential momentum energy' and P^i a 'combined momentum energy'. From these one gets (1966, pp. 92—98)

$$[\partial^j P^i - \partial^i P^j]\, \mathrm{d}x_j = 0$$

and, then a large number of alternative expressions of the fundamental equation of motion, namely (δ denoting a Lagrangian variation)

$$\mathrm{d}P^i \, \delta x_i - \delta P^i \, \mathrm{d}x_i = 0,$$

$$\oint P^i \, \delta x_i = \mathrm{const},$$

$$\frac{1}{2}\iint [\partial^i P^j - \partial^j \mathrm{P}^i]\, [\mathrm{d}x_i \, \mathrm{d}x_j] = \mathrm{const},$$

$$\iint [\delta_1 P^i \, \delta_2 x_i - \delta_2 P^i \, \delta_1 x_i] = 0.$$

All these are action integrals, and covariant generalizations of formulas which, in the Bohr—Sommerfeld 'old quantum theory', are quantized in multiples of Planck's constant h.

The most important equation of the relativistic analytical dynamics is

$$\delta \int_1^2 P^i \, \mathrm{d}x_i = \int_1^2 [\delta P^i \, \mathrm{d}x_i - \mathrm{d}P^i \, \delta x_i] = 0$$

where 1 and 2 denote a fixed 'initial' and a fixed 'final' point-instant. Therefore: *The combined action is stationary along the actual trajectory* (Hamilton—Jacobi theorem).

A significant generalization is obtained by requiring that

$$(P^i \, \delta x_i)_1 = (P^i \, \delta x_i)_2 = 0$$

whence: *If the congruence of the combined momentum-energy lines admits a hypersurface orthogonal trajectory, it admits an infinity of them* (Hamilton—Jacobi surfaces, or 'wave surfaces').

The latter two statements are the relativistic expressions of significant remarks made by Hamilton (1931), Klein (1890; 1901), and Vessiot (1909), pointing to an *isomorphism between geometrical optics and analytical mechanics.* The missing cornerstone for an actual identification was relativistic covariance, *and* Louis de Broglie's group velocity theorem, as explained in Section 15.

2.6.14. WHEELER—FEYNMAN ELECTRODYNAMICS

Wheeler—Feynman (1949) electrodynamics is a clever relativistic scheme for a system of point particles interacting directly at a distance two by two via a lightlike or, as remarked by Katz (1967), via an arbitrary spacelike interaction. Thus it obviously generalizes a familiar Newtonian scheme and, together with a parent scheme due to Katz, it provides, in the realm of relativistic dynamics of interacting point particles (where various schemes have been proposed), an elegant paradigm which happens to be isomorphic to the classical statics of interacting filaments.

The very concept of a spacelike interaction implies that any point along a timelike trajectory interacts with any point along another one, very much as in the Ampère—Laplace statics of current-carrying wires, where any point along a wire interacts with any point along another wire. This remark leads straight back to the contents of Section 11, emphasizing an isomorphism between the relativistic dynamics of a point particle and the classical statics of a filament. As the Wheeler–Feynman particles are spinless, the corresponding filaments are perfectly flexible. At this point, however, a significant difference between the two cases shows up: whereas, in the Ampère–Laplace (AL) scheme, each point of a wire interacts with each point of the same wire, so that there is self-induction and self-energy, in the Wheeler–Feynman (WF) scheme there is no self-action or self-energy, as the interaction is spacelike (or lightlike) and the trajectories are timelike.

The WF interaction essentially rests on a certain bracket, or else on a certain parenthesis, implying three vectors: the infinitesimal displacements $\mathrm{d}a^i$ and $\mathrm{d}b^i$ along two interacting trajectories a and b, (the proper times of which we denote by α and β) and the spacelike or lightlike vector $r^i_{ab} \equiv r^i$ joining them. As will be shown, the very form of this bracket, or parenthesis, follows necessarily from a few general requirements. This bracket, or parenthesis, is identical to the one present in the AL scheme of interacting current-carrying wires. Therefore, what will be displayed is an *isomorphism between the WF relativistic dynamics of spinless point particles and the AL statics of flexible current carrying wires.* It may be added that the rest masses of the WF particles are constant, and that so are the scalar tensions along the AL flexible wires.

Now we proceed, and denote by $\varphi'(r^2)$ the arbitrary spacelike (or lightlike) interaction between the line elements $\mathrm{d}a^i$ and $\mathrm{d}b^i$. We denote it $\varphi'(r^2)$ because it goes into $\varphi(r^2)$ when the field strengths are replaced by the potentials.

Requiring that the length of the momentum energy $p^i_a \equiv mV^i_a$ be preserved is equivalent to requiring that the force $Q_a B^{ij}_{(a)}$ created at a^i by b^i be skew symmetric; therefore, as it is assumed to depend only on the charges Q_a and Q_b and the vectors r^i and $\mathrm{d}b^i$, we must have

$$\mathrm{d}^2 p^i_a = Q_a Q_b \varphi'(r^2)\,[r^i_{ab}\,\mathrm{d}b^j\,\mathrm{d}a_j - r^j_{ab}\,\mathrm{d}b^i\,\mathrm{d}a_j],$$

$\mathrm{d}p^i_a$ being obtained by an integration $\int_{-\infty}^{+\infty} \mathrm{d}b$ of this expression. We remark that by substituting Euclid's space $\mathscr{E}$ for Minkowski's spacetime $\mathscr{M}$, Ampère's currents i for the charges Q, and the tension T_a for the momentum energy p^i_a, this expression is identical to the Laplace one for currents.

Requiring that the above formula leads to an action-and-reaction statement means that the term $-r^j_{ab}\, \mathrm{d}a^i\, \mathrm{d}b_j$ be added in the bracket, and that this term must vanish in the integration $\int_{-\infty}^{+\infty} \mathrm{d}b$. This indeed is the case, as $2r^j\, \mathrm{d}b_j = \mathrm{d}_b[(r^j_{ab})^2]$, $\varphi'(r^2)\, \mathrm{d}r^2 \equiv \varphi(r^2)$, and, by hypothesis, $\varphi(r^2) = 0$ if r^2 is timelike. Finally the 'WF bracket' comes out as

$$(2.6.45)\ \mathrm{d}^2 p^i_a = -\mathrm{d}^2 p^i_b = Q_a Q_b \varphi'(r^2)\,[r^i\, \mathrm{d}a^j\, \mathrm{d}b_j - - r^j(\mathrm{d}b^i\, \mathrm{d}a_j + \mathrm{d}a^i\, \mathrm{d}b_j)].$$

Strictly corresponding to it there is an AL bracket, also entailing an action-and-reaction statement; then the nullity of the integral $\int \mathrm{d}b$ stems from the fact that it is a closed integral, $\oint \mathrm{d}b$.

$A^i_{(a)}$ denoting the potential generated at a by the line element $\mathrm{d}b^i$ ((see formula (8)) we remark that $-Q_b \varphi' r^j_{ab}\, \mathrm{d}a_j\, \mathrm{d}b^i \equiv \frac{1}{2} Q_b\, \mathrm{d}_a \varphi\, \mathrm{d}b^i$ is its variation at a; therefore, setting

$$A^i_{(a)} \equiv Q_b \int_{-\infty}^{+\infty} \varphi(r^2)\, \mathrm{d}b^i$$

(whence follows $\partial_i A^i_{(a)} = 0$), and using formula (44) of the 'combined momentum energy', we can rewrite (45) as

$$(2.6.46)\ \mathrm{d}^2 P^i_a = -\mathrm{d}^2 P^i_b = Q_a Q_b \varphi'(r^2)\,(r^i\, \mathrm{d}a^j\, \mathrm{d}b_j)$$

and call the last parenthesis the WF one. Incidentally, the right-hand side of formula (46) is of a central force type. Similar statements hold in the AL scheme.

From the expressions (45) and (46) Wheeler and Feynman have derived the twin momentum-energy conservation theorems

$$p^i_a + p^i_b + p^i_{ab} = P^i_a + P^i_b + P^i_{ab} = \text{const},$$

where the 'interaction momentum energies' p^i_{ab} and P^i_{ab} are obtained by applying to the 'bracket' or the 'parenthesis', respectively, the operator

$$\int_{\alpha}^{+\infty} \int_{-\infty}^{\beta} - \int_{-\infty}^{\alpha} \int_{\beta}^{+\infty} .$$

No similar statement holds in the AL scheme.

Clearly the whole argument, as presented here for just two interacting elements, can be extended directly to the general case of n elements interacting two by two.

Finally the whole set of the preceding formulas is easily derived (Wheeler and Feynman, 1949; Costa de Beauregard, 1966, p. 98) from an extremum principle: the Fokker (1929) action principle $\delta\mathscr{A} = 0$ applied to the expression

$$\mathscr{A} \equiv -\sum_a \int p_a^i \, \mathrm{d}a_i + \sum_{a \neq b} \iint \varphi(r^2) \, \mathrm{d}a^i \, \mathrm{d}b_i$$

in the WF scheme, or the energy principle $\delta\mathscr{W} = 0$ applied to the expression

$$\mathscr{W} \equiv -\sum_a \oint \mathbf{T}_a \cdot \mathbf{da} + \sum_{a \,\&\, b} \iint \varphi(r^2) \, \mathbf{da} \cdot \mathbf{db}$$

in the AL scheme. The variations $\delta\mathscr{A}$ and $\delta\mathscr{W}$ are performed keeping respectively the Q's and i's fixed (the latter prescription being mathematically clear but physically unrealistic).

Finally, restriction to electromagnetism *stricto sensu* is obtained by setting $\varphi(r^2) = \delta(r^2)$ in the WF case and $\varphi(r^2) = 1/r$ in the AL case.

Table III, supplementing the correspondence table given on p. 83, summarizes the whole affair.

TABLE III

AL statics of filaments	*WF dynamics of charges*
Laplace field	WF field
$\mathbf{dB} = -\mathbf{i}\varphi'(r^2)\mathbf{r} \times \mathbf{db}$	$\mathrm{d}B^{ij} = Q\varphi'(r^2)\,[r^i \, \mathrm{d}b^j - r^j \, \mathrm{d}b^i]$
Laplace potential	WF potential
$\mathbf{dA} = \mathbf{i}\varphi(r^2)\,\mathbf{db}$	$\mathrm{d}A^i = Q\varphi(r^2)\,\mathrm{d}b^i$
Distance dependence	Distance dependence
$\varphi(r^2) = 1/r$	$\varphi(r^2) = \delta(r^2)$

REMARK. In the above formula for $\mathscr{W}$, with $\varphi = 1/r$, the double integral comprises the well-known expressions of the mutual and self-energies of the current loops.

In the simple integral for $\mathscr{W}$, 'minus' the tension of a filament appears as the linear density of a potential energy. That this is physically correct is seen from the two following examples:

1. If to a rope hanging from one extremity in a constant gravity field g, we fix a mass m successively at two different heights differing by l, we realize two configurations differing by the potential energy $mgl \equiv Tl$, which certainly is localized in the rope. Therefore $-T$ is the (negative) linear density of potential energy contained in the rope.

2. If two pulleys of equal radius R connected by a belt rotate at an angular velocity Ω, the one acting as a motor and the other as a receiver, so that torques $+C$ and $-C$ are respectively applied to them, the tensions $T_0 + T$ and $T_0 - T$ of the two straight segments of the belt are such that $C = 2RT$. So, with V denoting the linear velocity of the belt, the power flowing from motor to receiver is $P = C\Omega = 2TV$. Therefore, as it flows with the velocity V along both segments, the linear energy density is $-T$ along the stressed, and $+T$ along the 'dis'-stressed, segment.

3. Dettman and Schild (1954) have proposed a generalization of the WF scheme including a timelike interaction and thus, the existence of self-actions.

2.6.15. DE BROGLIE'S WAVE MECHANICS

Almost a quarter century elapsed after Planck's 1900 discovery of the quantum of action, h, until the proper quantum mechanical formalism could be set up. A key element in this was Louis de Broglie's 1924 'wave mechanics', in which a deep physical intuition was expressed in mathematics much less sophisticated than that put forward shortly afterwards by Heisenberg and coworkers, by Schrödinger and by Dirac. Let us recall that from time to time a great breakthrough in physics is expressed in terms of very simple mathematics; in fact, no mathematics at all are formalized in Carnot's seminal discovery of the 'second law'!

The reason why I choose to present de Broglie's work here rather than in Part 4, devoted to quantum mechanics, is that it fits quite nicely within the general topics of explicit Minkowskian covariance, in the straight line of Einstein's 1905 and 1910 proposal of the 'photon' concept, and in the straight line also of Hamilton's, Klein's and Vessiot's isomorphism of geometrical optics and analytical dynamics (whose arguments were unknown to de Broglie in 1924).

In 1905 Einstein had extended Planck's quantum hypothesis by assuming that the energy quanta $W = h\nu$ apply to the electromagnetic radiation. In 1910 he completed this hypothesis by attributing to these 'light quanta', as he called them, a momentum $\mathbf{p} = h\boldsymbol{\kappa}$, when associated with a plane monochromatic wave of spatial frequency $\boldsymbol{\kappa}$. On the whole he thus produced a proportionality formula of ratio h between the momentum energy of a freely moving 'photon' (as the light quantum is now called) and the 4-frequency of a plane monochromatic wave. Today the 4-frequency vector is denoted by $2\pi k^i$ and a modified Planck constant $\hbar = h/2\pi$ is introduced, so that Einstein's proportionality reads as

$$(2.6.47)\ p^i = \hbar k^i,$$

with, of course, $k_i k^i = 0$ and $p_i p^i = 0$.

In 1923 Louis de Broglie (1925), using a symmetry argument, decided that any elementary particle, such as the electron or the proton, should be accompanied by a sister 'matter wave', the connection between particle and wave being expressed by Einstein's formula. The 'zero spacetime length condition' was relaxed, since de Broglie's 4-vectors p^i and k^i had to be timelike.

The reasons behind de Broglie's idea were as follows. First, the 'discrete quantum conditions' of Planck, Bohr and Sommerfeld reminded him of the resonance conditions in classical acoustics and optics — Newton's rings, for example. Whence the idea of 'matter waves'.

Second, there was the isomorphism between geometrical optics and analytical dynamics, between wave fronts and Hamilton—Jacobi surfaces, between light rays and trajectories, with an unsolved riddle, however: the wave's phase velocity shows up in the denominator of Fermat's extremum principle, while the particle's velocity shows up in the numerator of Hamilton's extremum principle.

The solution to the riddle essentially belongs to the relativity theory, and here de Broglie put down his second trump card: the group velocity v of matter waves is related to their phase velocity u by the formula

$$uv = c^2,$$

so that v comes up in the numerator of Fermat's theorem.

Interpreted in terms of four-dimensional geometry, de Broglie's group velocity theorem is a specification of the classical theory of envelopes as already used for similar purposes, in optics. Fresnel, for example, in an argument combining geometrical and physical optics, thus explained how the illumination at R from a monochromatic point source at S is due mainly to the pencil of light rays surrounding the vector $\mathbf{S} - \mathbf{R} = \mathbf{r}$ and to the associated plane waves of 3-frequency $\mathbf{k}$. The spherical wave reaching S can be considered as the envelope of plane waves of same phase $\mathbf{k} \cdot \mathbf{r}$; those waves adjoining the one that touches the sphere at S are precisely those associated with the pencil. Moreover, he added, except for this tightly packed bunch, there is destructive interference.

Going from three- to four-dimensional geometry, replacing the frequency of Fresnel's source by the proper frequency of de Broglie's matter wave, and Fresnel's sphere by Minkowski's two-sheeted hyperboloïd, one sees that the energy is carried mainly by the pencil of rays surrounding the 4-vector r^i joining S to R, that is also by the tightly packed bunch of associated plane waves. In other words, the timelike 4-vector r^i/r should be interpreted as the group 4-velocity of the matter waves. Moreover, Fresnel's argument of destructive interference can also be transposed, and is in fact included in the formalism of Fourier integrals.

There is no difficulty in expressing the preceding wording in compact mathematics belonging to the calculus of variations, the formulas being the same in both cases. Since $k = |\mathbf{k}|$ is fixed, $\mathbf{k} \cdot \delta\mathbf{k} = 0$. And since $\mathbf{r}$ is fixed, $\delta(\mathbf{k} \cdot \mathbf{r}) = 0$ requires that $\mathbf{r} \cdot \delta\mathbf{k} = 0$. Since both conditions must be identically satisfied, $\mathbf{r}$ and $\mathbf{k}$ must be collinear. Q.E.D.

Following Klein (1890) and Vessiot (1909) we can add the following remarks. According to the geometrical theory of duality, introduced by Michel Chasles and much used in projective geometry, a curve in the two-dimensional plane can be viewed either 'punctually' as comprising its points, or 'tangentially' as the envelope of its tangents, and then expressed by either a 'punctual' or a 'tangential' equation. The link between the two representations is the '1-form' $\mathbf{k} \cdot \mathbf{r} = \varphi$, with φ denoting, in optics, the phase. We can thus view $\mathbf{r}$ as belonging to the punctual and $\mathbf{k}$ to the tangential representation of the curve.

Mutatis mutandis all this can be extended to higher dimensional metric spaces, and, for example, to Minkowski's spacetime. Then de Broglie's *wave-particle dualism* is expressible in terms of geometrical *point-tangent duality*, and inertial motion is viewed either, *à la* Galileo, in terms of straight rays or, *à la* de Broglie, in terms of plane waves.

Regrettably, when de Broglie wrote his 1925 epoch-making thesis, he had read neither Klein nor Vessiot, so that he did not spell out all this.

On receiving from Langevin a copy of de Broglie's 1924 doctoral thesis together with a letter asking his opinion, Einstein remarked that "a corner of the great veil has been uplifted".

It is only fair to mention that these brilliant ideas had ripened inside the environment of Maurice de Broglie's X-ray laboratory. Maurice was adamant that both the wave and the particle aspects of X-rays were of paramount importance, and this may well have inspired Louis's thinking. Bruce Wheaton (1983, pp. 263—301) comments upon this.

The implication of this work was that, to the transition from first-order 'geometrical optics' to second-order 'wave optics', there should correspond one from Newtonian to a new 'wave mechanics' — which indeed was to come very soon in the form of Schrödinger's equation. In the discussion following his PhD Thesis, L. de Broglie did state that the phenomenon of electron diffraction should be observable. In 1923 Davisson and Kunsman unwittingly hit upon this beautiful (and then completely unexpected) phenomenon, soon followed by Davisson—Germer, G. P. Thompson, Kikuchi, Ponte and others.

Thus the true meaning of Planck's 'action constant' h was unveiled, and *unveiled in a fashion emphasizing a fundamental compatibility between the quantum and the relativity theories.*

2.6.16. WHAT WAS SO SPECIAL WITH LIGHT, AFTER ALL?

What is so special with light is that, *in the limiting case where the rest mass of the associated particle is zero, the waves actually travel at the limiting velocity c.*

Had the diffraction of electrons by crystals been discovered before the accomplishment of Michelson's experiment, wave mechanics might have been enunciated much earlier, and then the relativity theory would have followed as a consequence; but the reasoning would have been much less straightforward, as everything is so much simpler in the limiting case.

Moreover, the photon, the Ariel light messenger flying at the ultimate velocity c, also is a very obliging messenger, interacting as it does with the electric charges pervading matter. Therefore, the photon is in some sense unique for gauging the depths of space and of time — from the very large cosmological ones to the very minute elementary particle ones. Also light yields, as we have seen in Chapter 4, the most perfect (primary) time standard and (secondary) length standard.

Finally, light does establish a 'zero spacetime distance' between emitter and receiver, an aspect of things emphasized by Fokker (1965) and others. It may well be that this has deeper implications than those already known, as will be discussed in Part 4 of this book.

2.6.17. CONCLUDING THIS CHAPTER, AND THE SECOND PART OF THE BOOK

My hope is that the various examples presented in this chapter — be they quite central, or more peripheral — have conveyed to the reader the *clarity, brevity and efficiency of the four-dimensional approach to physical problems*. True, there are more practically minded people who tend to liken an approach of this sort to some sort of mathematical pun. Such is the price for going from modelism to formalism.

One thing cannot be disputed, however: covariant reasoning has repeatedly proved itself as not only fast, but also extremely accurate for providing the right answers in delicate matters.

As a theory displaying "the absolute hidden inside the deceitful multifarious apparences", Einstein's so-called 'special relativity theory', together with Minkowski's formalization of it, certainly makes a major 'scientific revolution'.

Among physical mysteries, the most mysterious ones were perhaps those hidden inside the very brilliance of light. Three centuries of hard experimental work and theoretical thinking were needed before the answer was found. It would be unfair to overlook the labors of the early explorers because of the dazzling success of the final discoverers.

As for the fact that the relativistic matter does have 'duration', in Newton's and in Bergson's words (and has it symmetrically toward both the past and the future, so that there is indeed a big question as to how 'process' and 'becoming' should be understood), this discussion is deferred to Part 4 of this book, dealing with interpretative problems of relativistic quantum mechanics.

Concluding Part 2, it can be said that Einstein's 1905 discovery and formulation of the universality of Galileo's 'restricted relativity principle', and Minkowski's 1908 expansion of Poincaré's spacetime concept, make one of the most imposing and far-reaching conceptual discoveries in the history of physics.

PART 3

LAWLIKE TIME SYMMETRY AND FACTLIKE IRREVERSIBILITY

CHAPTER 3.1

OVERVIEW

3.1.1. OLD WISDOM AND DEEPER INSIGHTS

"If only young people knew, and if only old people could" is a French saying. The fact that it is neither possible to 'see into the future' nor to 'act in the past' is by far the most obvious and widely recognized statement of 'physical irreversibility'. And it goes right to the heart of the problem, as the subsequent analyses will show. It is also quite significant that it refers directly to the two faces of Aristotle's information concept: 1°, gain in knowledge, 2°, organizing power.

Between Aristotle and the advent of cybernetics, which has rediscovered that Information is a two-faced coin, information-as-organization was the 'hidden face', an esoteric concept discussed only by a handful of philosophers, such as Thomas Aquinas (who stresses the symmetry much more explicitly than Aristotle[1]) or Schopenhauer (whose title *The World as Will and as Idea* (1883) speaks for itself). The other face, however, was obvious to everybody: the man in the street buys a newspaper for a few cents, in order to procure 'information'.

The reason why the one face is so obvious and the other so hidden is *physical irreversibility* — "factlike, not lawlike physical irreversibility", as Mehlberg (1961, p. 105) puts it. Conversely, (factlike) preponderance of information-as-knowledge over information-as-organization is an expression of the 'Second Law' — and the most profound of all expressions.

Causality will come to the fore in our discussions, and in two different forms: 1° An *elementary*, or *microscopic, intrinsically time symmetric causality concept*, insensitive to the exchange of 'cause' and 'effect', and of 'prediction' and 'retrodiction'; 2° A *macroscopic causality concept*, with a *factlike time asymmetry* built into it. The connection between these two aspects of causality can be displayed quite clearly, and is found in Laplace's 1774 memoir "On the Probability of Causes". It will turn out that the bond between 'lawlike time symmetry' and 'factlike time asymmetry' in both the *information* and the *causality* concepts is so tight that *conditional probability and causality are synonymous, either in the formalization or in the conceptualization.*

At this point an important warning is imperative. The reasoning in this third part of our book rests entirely on *one single and central hinge: lawlike symmetry and factlike asymmetry between prediction and retrodiction.* For instance, such trivial statements as "Physical irreversibility is easily explained by an appeal to probability, because (say) dilution of an inkdrop deposited by a pipette in a glassful of water is far more probable than its spontaneous concentration" will *not* be found in this book, because *they beg the question.* Great care is needed on this very point because the natural trend of habits inherited from macroscopic experience is to fall back into the old rut. Notwithstanding the clearcut warnings of Loschmidt (1876) and Zermelo (1896), loosely worded statements of this sort are very often encountered even today. What the mathematics does say, however, in its transparent language, is that the retrodictive probability that a concentrated inkdrop existing at time zero has condensed spontaneously in the immediate past is exactly the same as the predictive probability that it will dissolve in the future. Bringing the pipette into the picture merely enlarges the problem without changing its nature — and, in fact, one is thus led step by step, via the consideration of 'branch systems', to the very level of cosmology.

So, digging for the root of physical irreversibility is far from an easy task. *One must be very careful to avoid getting caught in an implicit appeal to retarded causality. Causality is the concept to be elucidated, not the assumption to be used.* At each step this hidden trap is wide open, as retarded causality is built into the very tenses of the verbs we are using!

Quotations from some very eminent scholars will be produced, showing how they were caught in the trap, and thus allowed the 'explanation of irreversibility' they were after to fly away unseen.

Merely for brevity in discourse, all through Chapters 3.1 to 3.4 inclusive, we shall speak of time in the traditional fashion, as if it were 'flowing' and carrying 'everything' in its 'present instant' t. Things will be settled in Chapter 3.5 concluding this part of the book, and the proper perspective then displayed.

3.1.2. MATHEMATIZATION OF GAMBLING

Cardano, the author of *De Ludo Aleae* ('concerning the game of dice') composed around 1520, is said to have been an inveterate gambler; Pascal, in the wordly period of his life, made friends with gamblers, and corresponded with Fermat to elucidate the matter of chance games. These mathematicians have thus founded the calculus of probabilities, in the form of combinatorial analysis.

Problems in games of chance provide an excellent approach to the conceptual problems regarding probability and information. Today we see clearly that Cardano, Pascal and Fermat were dealing with information theory — very much as Molière's 'Monsieur Jourdain' was speaking in prose without knowing it.

Problems there are free of many of the intricacies discovered later. For one thing, a playing card is an 'object' endowed with its own 'properties', the whole thing 'existing' whether or not it is turned upside down — whether or not it is looked at. Therefore *lack of knowledge* and *incomplete control* are here exact synonyms for *occurrence of chance*. Dynamics hardly enters the picture in card games (while it does in dice games, thus bringing in extra problems).

As defined by Shannon (1948) for telecommunication problems, where N symbols must be handled ($N = 26$, in the English alphabet), *information* I is the additive quantity $\ln N$.

In a logarithmic basis b other than e, $I \equiv \ln N = (\ln b)(\log_b N)$: I is basis independent; what changes is the number of digits needed for expressing it. 'Binary digits', or 'bits' ($b = 2$), are the most natural system for theoretical matters, as they are directly adapted to 'yes or no' questions.

Consider a deck of 32 cards. It so happens that $32 = 2^5$, so that looking at a card yields 5 bits of information. It is true that John Napier was almost contemporary to Pascal and Fermat. However, three centuries had to elapse before it became clear that the two concepts of *probability* and *information* are germane to each other.

An implicit assumption has been made: that any two of the 32 cards look identical when turned upside-down, so that the 'degree of ignorance' — and of 'control' — is the same for each card. All probability problems must start by conferring 'equal *a priori* probabilities' on certain 'outcomes' (belonging either to the discrete or to the continuum). For now we satisfy ourselves by saying that such an assignment is part of the hypotheses to be tested, so that if, by repeating the test many times, we find that the observed frequencies approach our probabilities, we consider that our *a priori* assignments were correct.

A sharp distinction exists between the two concepts of probability and of frequency. The frequency is the expectation value of the probability; it is not 'the probability'.

Following the 16th- and 17th-century pioneers, Jacob Bernouilli (1713), Bayes (1763) and Laplace (1774) developed the probability theory along this line. Then Ellis (1842), Boole (1854), Venn (1866) and finally von Mises (1919) launched a counterproposal where the two concepts of probability and of frequency were equated to each other, thus leading to undue confusion, as explained by Jaynes (1983, p. 218) in a thought-provoking essay. As Jaynes puts it, "in the interests of . . . clear exposition . . . , we ought to use the word 'probability' in the original sense of Bernouilli and Laplace; and if we mean something else, call it something else"; and as "the historical priority belongs clearly to Bernouilli and Laplace" the wording "frequency theory of probability is a pure incongruity . . .".

By this quotation I make clear that my philosophy definitely is the so-called Bayesian one, recently revived and extended by Keynes (1921), Jeffreys (1961), Cox (1946; 1961) Jaynes (1983), and a few others — including Wald (1950) who, having started as a frequentist, ended as a Bayesian.

The hypothesis implied in the assignment of equal *a priori* probabilities has been termed by Bernouilli "principle of insufficient reason". It is the negative form of Leibniz's "principle of sufficient reason", and consists of this: *if no obvious reason is seen for deciding otherwise, assign equal a priori probabilities to the possible outcomes.*

Then if, in a test, the observed frequencies conform to the probabilities (as computed from the *a priori* ones), the hypothesis is confirmed — and this is the end of the story. If, however, a discrepancy turns out, this is proof that some 'sufficient reason' has been overlooked, and this use of the probability theory affords a very good clue as to which 'reason' was overlooked. As emphasized by Jaynes, a historical example is Planck's discovery of the quantum, in his statistical study of thermal radiation, with Einstein's confirmation, in his statistical study of the heat capacity of solids. This sort of game is called 'hypothesis testing', or 'extracting signal from noise'.

Keynes (1921) helped avoid semantic discussions by renaming 'principle of indifference' the principle of sufficient (or insufficient) reason. *If the hypothesis is experimentally verified, this means that the 'indifference' exists in both the theorist's mind and the facts in Nature.* Any accord of this sort is considered proof that 'the theory conforms to the facts'. In this particular case, however, it is generally felt that there is something mysterious in such a 'conspiracy'; Landé (1965, p. 29), for example, is adamant in this. This feeling, in my opinion, proceeds from the fact that *nowhere else in physics is the interaction between mind and matter so directly expressed.* There have been endless discussions to decide if the probability concept should be thought of as objective or as subjective. My contention is that *the probability and the information concepts should be said to be neither objective nor subjective because they are inherently both.* Probability, or Information, is the very hinge around which mind and matter are interacting, holding the gate through which information as knowledge flows out of, and information as control flows into, our so-called material world.

3.1.3. PROBABILITY AS DATA DEPENDENT

A famous and often discussed problem of Joseph Bertrand (1888, pp. 4—5) is finding the probability that the length of a chord in a circle is greater than that of the inscribed equilateral triangle. Bertrand considers three natural procedures for drawing such a chord:

1. One extremity, A, being fixed, the direction of the chord is chosen at random. Then the probability obviously is 1/3.
2. The direction of the chord being fixed, and the diameter BC perpendicular to that direction being drawn, the middle D of the chord is randomly chosen on BC. Then the probability is 1/2.
3. The middle D of the chord is chosen at random inside the area of the circle. Then the probability is 1/4.

Bertrand, and many of his followers, pondering upon this, concluded that probabilities on the continuum are ill defined. Borel (1914, p. 84), however, remarked that all three conclusions are correctly drawn, but, as they refer to different procedures, they differ from each other. Therefore, *probabilities are data dependent, or 'relative'* — somewhat as the 'projection' of an object depends on the view point. This settles the matter.

Jaynes (1983, pp. 133—148), however, has something interesting to say concerning the application of the principle of indifference to continuous probabilities: appeal to a group invariant property.

Rotational invariance in Bertrand's problem has been stressed by many authors, but, as Jaynes remarks, this is not sufficient. If we require that the solution be independent of the size of the circle and of its position in the plane, what will turn out? Let it be made clear that these requirements refer to a fourth procedure for drawing the chord, such as 'dropping long straws from high above the plane'. Jaynes, using elementary formalizations of what could be handled more abstractly by using a 4-parameter transformation group, ends up with solution 3: probability 1/4. This could have been intuitively guessed.

Quoting Jaynes (p. 134):

> The point at issue is far more important than merely resolving a geometrical puzzle; . . . applications of probability theory to physical experiments usually lead to problems of just this type. . . . Yet physicists have made definite choices, guided by the principle of indifference, and they have [produced] correct and non-trivial predictions.

And again (p. 143):

> Over a century ago, without benefit of any frequency data on positions and velocities of molecules, . . . Maxwell was able to predict all these quantities by a 'pure thought' probability analysis. In the case of viscosity the predicted dependence on density appeared at first sight to contradict common sense, casting doubt on [this] analysis. But . . . the experiments . . . confirmed [his] predictions, leading to the first great triumph of kinetic theory.

3.1.4. THE SHANNON—JAYNES PRINCIPLE OF ENTROPY MAXIMIZATION, OR 'MAXENT'

To optimize the channel capacity in communications technology Shannon demonstrated in 1948 that the function H of a discrete set of probabilities p_i which is required to be continuous, monotonic, increasing with n when all p_i's equal $1/n$, and invariant when the p_i's are arbitrarily clustered, is, up to a multiplicative factor,

$$(3.1.1)\quad H = -\sum p_i \ln p_i .$$

This is identical to Gibbs's expression for the entropy of a dynamical system at thermal equilibrium.

Shannon's prescription, raised by Jaynes to a general principle for optimizing the use of all information available is: "maximize H subject to all known constraints". For example, in the classical statistical mechanics of thermal equilibrium ('thermostatics', as it is now called) the constraints are, with W denoting the energy,

$$(3.1.2) \quad \sum p_i = 1, \qquad \sum p_i W^i = \langle W \rangle.$$

This prescription was precisely the one used by Gibbs (1902) who, as it seems, arrived at it by considerations akin to Shannon's (1948, p. 143).

Thus, by an appeal to information theory, Jaynes has reduced the previous lengthy reasonings of statistical mechanics to their bare skeleton. This falls well in line with Gibbs's own philosophy, as can be read in his writings (1902, p. 17).

It seems to me that the Shannon—Jaynes MAXENT principle essentially is a (very powerful) extension of the 'principle of indifference'. In statistical mechanics, expectation values of quantities other than the energy can be used as constraints – for example, angular momentum in the theory of magnetism, or momentum energy in problems where Lorentz invariance is needed. The energy W can contain a potential contribution, and Jaynes has finally found that the MAXENT method (including Liouville's invariance of six N-dimensional phase cells) is extendable straightaway to irreversible processes (1983, pp. 83—85). Therefore, MAXENT today is, in some sense, the nec plus ultra with discrete probabilities.

3.1.5. 'HOW SUBJECTIVE IS ENTROPY?'

The following dialogue between Jaynes (1983, p. 85) and Denbigh (1981) will help set the stage. Under the subtitle 'The "anthropomorphic" nature of entropy' Jaynes writes:

> It may come as a shock to realize that . . . thermodynamics knows of no such thing as the 'entropy of a physical system'. Thermodynamics does have the concept of the entropy of a *thermodynamic* system; but a given physical system corresponds to many different thermodynamic systems.

This is the transposition of Bertrand's problem on a much larger scale.

> It is . . . meaningless to ask 'What is the entropy of [a system]?' unless we . . . specify the set of parameters which define its thermodynamic state . . . [Entropy] is a property not

of the physical system, but of the particular experiments you or I choose to perform on it.

This is why Hobson (1971) proposes to speak of the entropy not of the system, but 'of the data'. So it turns out that, by definition, *thermodynamics does not deal with physical systems per se, but with what we know and can do with systems* (of which we decide what we do not want to know or control). A mischievous quotation from Poincaré (1906a, Pt 1, Ch. 4) is significant here:

You ask me to foretell future phenomena. If by ill-luck I knew their laws I would be led to intractable calculations, and I could not answer. But as, by good luck, I dont know them, I will answer you right-away.

With this we are back to a successful use of the 'principle of indifference', and to the fact that *indifference is present in both*[2] *the mind (as 'ignorance') and in the facts (as 'indetermination')*. As Poincaré (1906a, Pt 1, Ch. 4) puts it: "Chance is not *only* [italics mine] the name we give to our ignorance." In other words, *successful use of probabilities is merely a clever shortcut.*

Jaynes continues thus:

Engineers have their 'steam tables' which give measured values of the entropy of superheated steam at various temperatures and pressures. But the H_2O molecule has a large electric dipole moment; and so the entropy of steam depends . . . on the electric field strength present. It must be understood . . . that this extra . . . degree of freedom is not tampered with during the experiments . . . which means . . . that the electric field was not inadvertently varied.

To this Denbigh (1981) objects as follows:

There is no need to bring in 'you or I' for Jaynes's sentence could be worded '[Entropy] is a property of the variables required to specify the physical system under the conditions of the particular experiment'.

A little before he had written that Grad (1967)

has remarked that a change in the estimate of an entropy can occur 'when some relevant facet of the problem has changed, even if only in the mind of the observer' . . . But the fact that an entropy value is changed, or . . . not . . . , according to whether it is known, or . . . not . . . that such factors are operative is surely no more an indication of subjectivity than would be the change in the value of almost any other physical property when new factors become known. After all we do not regard the age of a rock stratum as being subjective simply because the estimate of its age is subject to revision! To be sure the case of entropy is a little more involved since, as Grad points out, we may wish to use different measures for different purposes — for example, we may wish to neglect the difference between oxygen and nitrogen in matters relating to the flow of air in a wind tunnel.

All this is very clearly stated. The question then boils down to something akin to this: Is projective geometry dealing with the 'true shape' (if any) of an object viewed in perspective, or rather with consistent representations *per se*? The latter is the case. A corroborative remark may be added: in the statistical theory of thermal equilibrium the entropy enters on exactly the same footing as the expectation value $\langle W \rangle$ of the energy. Therefore the key question is tantamount to this: Does the expectation value of the energy of a thermodynamic system at equilibrium belong to the system *per se*, or rather to the system as we choose to know and control it? The correct answer is the latter — by definition: *of course* $\langle W \rangle$ does not belong to the system! If it were argued that the mean value $\langle W \rangle$ 'belongs' to the Gibbs's ensemble representative of the system, the rejoinder is that this ensemble 'exists' only in our minds and that, with fairly common thermodynamical systems, the whole world would not be large enough to contain the N members of the ensemble.

So, as Lewis put it in 1930: "Gain in entropy always means loss of information, and nothing more." He went too far, however, when he added "It is a subjective concept", because, as we have said, 'indifference' is *both* in the mind and in the things — in a way essentially different, however, from the one accepted in deterministic physics: *one implying an active coupling between mind and matter.*

We have been discussing entropy without yet making the essential point that, mathematically speaking, 'minus the entropy', or negentropy, and information, have exactly the same expression. For Jaynes, Léon Brillouin (1902), Rothstein (1958), Katz (1967) and others, physical negentropy *is* information; and we have made their thesis ours. As for the question of sign, a significant remark is in order: a probability being defined as the ratio n/N of the 'favorable' to the 'possible' cases, in classical statistics one reasoned with N constant and n variable, while in information theory and, most usually, in quantal statistical mechanics, one reasons with n constant and N variable. Then the signs of $\mathrm{d}P/\mathrm{d}n$ and $\mathrm{d}P/\mathrm{d}N$, and also those of $\mathrm{d}\ln P/\mathrm{d}n$ and $\mathrm{d}\ln P/\mathrm{d}N$, are opposite.

3.1.6. LOSCHMIDT-LIKE AND ZERMELO-LIKE BEHAVIOR IN CARD SHUFFLING

If we have at time $t = 0$ an 'ordered' deck of cards (that is, the permutation of which belongs to some specified sparsely populated subensemble of all the permutations) the probability calculus, when used predictively, explains quite well how this order is destroyed by shuffling the deck. Nobody relies, however, on shuffling for ordering the deck at will. There is a paradox in this, because *no time arrow is built into combinational analysis*; therefore, using 'blindly' (Watanabe, 1955) the probability calculus in retrodiction (that is, symmetrically to what is done in prediction) brings the result that the ordered deck existing at time $t = 0$ *has emerged* from the series of shufflings performed before.

In other words, *it cannot be stated that the probability calculus yields per se a time arrow.* Lewis (1930) among others was adamant on this. *The time arrow*, as we know it, *emerges if a boundary condition is added*, namely, that *blind statistical prediction is physical, while blind statistical retrodiction is unphysical.* Therefore, in macrophysics blind statistical prediction is accepted, while blind statistical retrodiction is rejected. And so *the time arrow is not statistically deduced, because it is postulated.*

This is the (1876) Loschmidt-like behavior of card shuffling. Now we come to the (1896) Zermelo-like behavior.

Shuffling the deck again and again will probably bring back, from time to time, any permutation that existed at time 0 — just as it had before, if the shuffling was already going on. The expected recurrence period is $N!$ shufflings for a deck of N cards — an enormous number if N is not small. Nevertheless the phenomenon exists. But the point is that *we can neither foretell when a given permutation will turn up, nor produce it at will.*

In other words, using the second sentence in Section 1, "we can neither see into the future nor act in the past". At least, we cannot do so 'normally'. Whether or not we can do it 'paranormally' is a question to be discussed later.

Two remarks need be made here.

First, the very symmetry of the mathematics (together with the existence of statistical fluctuations) does imply that *the factlike taboo just stated is not a lawlike prohibition.*

Second, the folklore (even in our Western culture) repeatedly mentions rare, but striking occurrences of 'paranormal' precognitions and the ability to influence random outcomes by strength of will. Sir Francis Bacon, the Founding Father of experimental methodology, in his books *The Advancement of Learning* and *Sylva Sylvarum*, did not exclude the 'paranormal' from scientific study and proposed "deliberate investigation of telepathic dreams . . . and of the influence of imagination on the casting of dice" (Jahn, 1982, p. 137).

It would be flatly antiscientific to reject *a priori* the possibilities opened by a mathematical symmetry, and to deny the widespread existence of 'paranormal' reports.

3.1.7. LAPLACE, THE FIRST, AND PROFOUND THEORIST OF LAWLIKE REVERSIBILITY AND FACTLIKE IRREVERSIBILITY

Laplace, the champion of universal determinism, is also an important name in the history of the calculus of probabilities. It is worth mentioning that his approach in this was informational, not frequentist.

Expressing very far-reaching thinking in quite unpretentious wording, he produced results some of which were far ahead of his time (see Jaynes, 1983). More than a century before Loschmidt and Boltzmann, in 1774 he formalized, at the basic level of the calculus of probabilities *per se*, the statistical lawlike reversibility and factlike irreversibility which is none else than 'The Second Principle of the Science of Time'. This is found in his 1774 'Memoir on the Probability of Causes from the Events' (and in subsequent writings, as he repeatedly came back to the subject of probabilities). Therefore, Boltzmann's (1964) excellent exposition of the matter in terms of statistical mechanics is in fact a specification of Laplace's earlier and much more general one. A van der Waals' article in 1911 stresses that a time asymmetric use of Bayes's conditional probability formula is implied in the derivation of Boltzmann's '*H*-theorem' – and of course this same formula is at the core of Laplace's 1774 discussion.

Although Laplace's discourse is very much time laden, tacitly assuming that 'cause precedes effect', his mathematics is not. It turns out that logical, not time ordering, is what he is really considering. Therefore, he is discussing essentially lawlike reversibility and (possible) factlike irreversibility of the causality concept *per se*. The time aspect of this concept is secondary to the logical one.

All this is very relevant to our forthcoming discussions pertaining to relativistic quantum mechanics because, *mutatis mutandis*, this whole Laplacean problematic survives the '1926 revolution', in which Born (1926) and Jordan (1926) defined a radically new 'wavelike probability calculus' tailored so as to suit the Einstein and de Broglie wave–particle dualism.

3.1.8. TIMELESS CAUSALITY AND TIMELESS PROBABILITY

Logical implication is a form of causality, and it is timeless.

Let it be emphasized that timing is not essentially implied in the probability concept either.

The *joint probability* $|H) \cdot (W| = |W) \cdot (H|$ that a male U.S. citizen has a height H and a weight W is a number which may or may not depend on time; that is not the question.

In the absence of prior probabilities of H and W alone, this number is also the *conditional probability* $(H|W)$ that a U.S. citizen has height H *if* he has weight W, and the *conditional probability* $(W|H)$ that he has weight W *if* he has height H.

It may be that prior probabilities are relevant. For example, the *prior probabilities* $|H) = (H|$ of H and $|W) = (W|$ of W must be considered if, say, the subclasses of basketball players or of weightlifters are at stake. Then, using the symmetric concept of an *intrinsic conditional probability*

$$(3.1.3)\quad (A|B) = (B|A)$$

we write the joint probability of A and B in the form

$$(3.1.4)\quad |A) \cdot (B| \equiv |B) \cdot (A| = |A)(A|B)(B| = |B)(B|A)(A|.$$

Another interpretation of the $(A|B) = (B|A)$ concept is often relevant: the *transition probability* between two *representations* of the system. In this jargon we should speak here of the 'height representation' and the 'weight representation' of the male U.S. citizen.

3.1.9. FACTLIKE IRREVERSIBILITY ACCORDING TO LAPLACE, BOLTZMANN AND GIBBS

Laplace and his followers used a definition of *converse conditional probabilities*[3] slightly different from the one above, namely

$$(3.1.5)\quad |A|B) = |A)(A|B), \qquad |B|A) = |B)(B|A),$$

and used it for discussing factlike irreversibility.

Obviously,

$$(3.1.6)\quad |B|A) \neq |A|B) \qquad \textit{iff} \qquad |B) \neq |A).$$

Maximal irreversibility shows up if, say, all $|B)$'s are equal to each other, so that they need not be mentioned. This is what both Laplace and Boltzmann assumed for their 'effects'. Laplace's motivation was that 'one has no sufficient reason' for ascribing different $(B|$ values. Boltzmann's (1964) motivation was, in Watanabe's (1955) wording, that blind statistical prediction is physical while blind statistical retrodiction is not. This conforms to an often quoted sentence by Willard Gibbs (1902, p. 150):

> It should not be forgotten, when our ensembles are chosen to illustrate the probabilities of events in the real world, that while the probabilities of subsequent events may often be determined from the probabilities or prior events, it is rarely the case that probabilities of prior events can be determined from those of subsequent events, for we are rarely justified in excluding the consideration of the antecedent probability of the prior events.

3.1.10. LAWLIKE REVERSIBILITY

Laplace's notation (5) has two drawbacks. It somewhat conceals the intrinsic cause—effect symmetry underlying the whole problem (a symmetry he had assumed explicitly in 1774), and it unduly mixes the extrinsic or prior probabilities $|A)$ and $|B)$ with the intrinsic transition probability $(A|B)$. Therefore, from now on, we shall forget the (widely used) Laplacean conceptualization in favor of the symmetric one expressed in formulas (3) and (4).

Let us verify by a few examples that it is very relevant, and that cases definitely exist where there is not only 'sufficient reason', but even compelling reason, for ascribing prior probabilities $(B|$ to the 'effects'.

Table IV displays a few disciplines where the *prior probabilities* $|A)$ and $(B|$, the *intrinsic conditional or transition probability* $(A|B) = (B|A)$, and the *joint probability* $|A) \cdot (B| = |B) \cdot (A|$, taken in this order, have relevance, and receive specific denominations.[4]

TABLE IV

Discipline	$\vert A)\ \&\ (B \vert$	$(A \vert B) = (B \vert A)$	$\vert A) \cdot (B \vert$
Physical Chemistry	Concentrations	Mutual Affinity	Rate of Reaction
Interconnected Phone Cabins	Occupation Rates	Channel Capacity	Traffic Rate on a Line
$\wedge$ $\vee$ Statistical Mechanics	Occupation Numbers (O.N.)	Mutual Cross Section	Predictive or Retrodictive Collision Probability
$<$	Initial O.N. of Initial State and Final O.N. of Final State	'Naked' Transition Probability	'Dressed' Transition Probability

Statistical mechanics, the fourth discipline considered, has paramount significance for us, not only because of its importance in the history and in the very corpus of physics, but also because of its 'correspondance' (in Bohr's sense) with the *S*-matrix formalization of relativistic quantum mechanics, to be discussed later. Three spacetime (or else momentum energy) shapes of a collision between two particles are considered: the $\wedge$ shape, associated with the prediction of a collision, the $\vee$ shape, associated with the retrodiction of a collision, and the $<$ or, should we say, the *C* shape, associated with the transition of a specified particle between a state before and a state after the collision. Not much comment is needed for the $\wedge$ or the $\vee$ shapes: obviously, the occupation numbers at stake then are the initial occupation numbers of the initial states and the final occupation numbers of the final states, respectively. But a comment is needed in the $<$ or *C* case, in which neither Laplace nor Boltzmann felt obliged to multiply the 'naked' transition probability by the prior probability $(B|$ of the final state — while of course both of them did multiply it by $|A)$.

The point is that, in this, their behavior was 'intrinsically illogical' for the following reason: multiplication by $|A)$ does imply statistical indistinguishability of particles of given species. Therefore, there are $(B|$ ways in which the particle undergoing transition can reach the final state B, and therefore multiplication by $(B|$ is compulsory, as a corollary of multiplication by $|A)$.

What does physics teach in this respect? It teaches that particles of a given species indeed are statistically indistinguishable, and that in fact they are either 'bosons', which are such that $|A), (B| = 0, 1, 2, \ldots$, or 'fermions', which are such that $|A), (B| = 0, 1$.

So, the teaching of physics, via the Bose or the Fermi statistics, is that *we must compute the dressed transition probability assuming that the initial and the final occupation numbers are actually present in spacetime* (or in the momentum-energy space), the ones in the initial state, the others in the final state. This is very Minkowskian indeed, and may come as a shock, because *a priori* one would rather feel that a time-extended four-dimensional picture is not compatible with a calculus of probabilities!

Let us postpone the discussion of this intricate matter to the point where we can resume it in the still more puzzling context of the quantum mechanical wavelike probability calculus.

For the present, let us simply emphasize that there is an isomorphism between this spacetime (or momentum-energy) problem and the one of interconnected phone cabins, also considered in Table IV.

Finally I find it useful to bring in one more example illustrating the present thesis: the well-known balls-in-boxes game, discussed in elementary textbooks.

There are balls differing at most by their colors, $(B|$ balls of color B, all enclosed in boxes differing at most by their colors, $|A)$ boxes of color A. The rule of the game is that all A boxes contain the same number $(A|B)(B|$ of B balls. Then, picking blindfolded one ball from one box, we see that the joint number of chances for hitting an A box and a B ball obviously has the expression (4). As in fact one hits first the box and then the ball, this is either the predictive probability of picking, or the retrodictive probability of having picked, a B ball from an A box. Cause—effect symmetry is obvious. The 'intrinsic conditional or transition probability' $(A|B)$ can be thought of as connecting a 'ball representation' and a 'box representation' of the system.

3.1.11. MATRIX CONCEPTUALIZATION OF CONDITIONAL OR TRANSITION PROBABILITIES

Speaking of 'representations of a system' implies, for consistency, that two different elements of a representation be mutually exclusive. This we express by the condition[5]

(3.1.7) $(A|A') = \delta(A, A')$,

with, of course, $\delta = 0$ if $A \neq A'$, $\delta = 1$ if $A = A'$.

Then, a set of conditional or transition probabilities is naturally thought of as consisting of the elements of a matrix, and a generally asymmetric one. The meaning of the Laplacean symmetry condition (3) is thus matrix transposition.

3.1.12. LAPLACEAN REVERSAL AND TIME REVERSAL

Among the many possible specifications of chance occurrences A and B we have the 'initial' and 'final states' of an 'evolving system' which, borrowing a term from quantum mechanics, we can call 'preparations' and 'measurements'.

Such paired events are exchanged by a Loschmidt (1876) motion reversal, which is an 'active time reversal'. In this way, *a Loschmidt time reversal does induce a Laplacean reversal* (3), and this is exactly how logical and chronological lawlike reversibility and factlike irreversibility are tied together.

3.1.13. MARKOV CHAINS IN GENERAL

The H and W symbols have helped us in understanding an aspect of probability by means of a simple little fable. They will help again, with a different meaning however.

Suppose that, in a U.S. National Park, we ask what is the joint probability of finding a male bear at H and a female bear at W, these being mates (H now stands for 'husband' and W for 'wife'). The HW 4-vector can be either spacelike, or future timelike, or past timelike, which, from the standpoint of probabilities *per se*, makes no difference.

That H and W are mated means that an interaction exists between them — say, that they meet at some hidden spacetime point I. Therefore, the transition probability $(H|W) = (W|H)$ between the 'H representation' and the 'W representation' of the couple is expressible as a sum over the I states, and we can consider three cases.

If the HW vector is spacelike, the meeting place I can be either in the common past or in the common future of H and W. Then the HIW

zigzag respectively assumes a $\vee$ or a $\wedge$ shape. If the HW vector is, say, future timelike, the meeting place can be in the future of H and in the past of W, in which case the HIW zigzag assumes a $<$ or C shape.

The present discussion is confined to cases where the expression of a conditional or transition probability $(A|C) = (C|A)$ *as a sum*

$$(3.1.8) \quad (A|C) = \sum (A|B)(B|C)$$

has topological invariance with respect to distortions of an ABC zigzag into a $\vee$ *or a* $\wedge$ *or a C shape.* This class is indeed a very large one, with much relevance in physics.

For example, the statistical mechanics of colliding molecules belongs to this class. Suppose our molecules are spherical. The 'hidden state' existing while the molecules are in contact can be characterized by the line of their two centers – and, indeed, the common tangent plane is a reflecting plane for the overall motion.

Formula (8) is the generating formula of the so-called 'Markov chains'; and indeed, for colliding molecules, according to what has been said in Section 10, formula (8) has this topological invariance. Therefore, *a Markov chain consisting of a series of links* $(A|B)$ *can zigzag arbitrarily* throughout spacetime (or throughout the momentum-energy space) *regardless of the macroscopic time or energy arrow.*

Intermediate summations such as $|B)(B|$ are thought of as being over 'real hidden states' – but this, as we shall see, is no longer true in the 1926 Born—Jordan wavelike probability calculus.

End prior probabilities in formula (4) are truly parts of conditional probabilities $(E|A)$ and $(C|E')$ linking the system to its environment. In the Bayesian wording E is the 'available evidence'. The implication is, of course, that $(E|A)$ and $(C|E')$ are the values of the prior probabilities $|A)$ and $(C|$.

3.1.14. FACTLIKE IRREVERSIBILITY AS BLIND STATISTICAL RETRODICTION FORBIDDEN

As previously said, neither Laplace nor Boltzmann was willing to consider the prior probabilities, or occupation numbers, of their final states. However, we have shown that, either in physics *stricto sensu*, or in such familiar examples as the balls-in-boxes game, there are cases where this rejection cannot be maintained.

Here I shall show that, from the strict standpoint of logic, there are cases where the Laplace—Boltzmann prescription should be used in reverse.

Consider, in the Darwinian descent towards the horse, the 'eohippus'. Given the eohippus can we predict the horse? Of course not. But we can retrodict a 'primeval molecular soup' correctly, by merely ignoring the $|A)$'s. This is an instance of problems where, using a well-known philosophical term, 'finality' prevails over 'causality'.[6]

This being said, we can summarize the whole question of factlike irreversibility, in its temporal aspect, by formalizing it, in Watanabe's (1955) words, and in conformity with Gibbs's (1902; p. 150) solemn sentence, as 'blind statistical retrodiction forbidden'.

3.1.15. CAUSALITY IDENTIFIED WITH CONDITIONAL OR TRANSITION PROBABILITY

In the direct line, I believe, of Laplace's (1774) thinking, and in the light of the previous considerations, I submit that *the physical causality concept be merely identified with that of the conditional, or equivalently, with that of the transition probability concept.* This confers clarity and definiteness on an otherwise rather shaky concept.

Thus defined, causality is endowed with the Laplacean intrinsic cause—effect symmetry expressed by Equation (3), meaning matrix transposition, and with the possible factlike asymmetry displayed in the expression (4) of a joint probability, where the prior probabilities $|A)$ and $(B|$ need not be equal.

These are essentially timeless definitions. However, whenever two such occurrences A and B are, say, a 'preparation' and a subsequent 'measurement' performed upon a 'system', the Loschmidt motion reversal induces the Laplacean symmetry (3). This is exactly how the time aspect of lawlike reversibility versus factlike irreversibility of physical causality stems from its more basic purely logical aspect.

3.1.16. CONCLUDING THE CHAPTER: A SPACETIME COVARIANT, ARROWLESS CALCULUS OF PROBABILITIES

A widely circulated statement is that the special relativity theory, in its geometrical, four-dimensional presentation, is essentially a superdeterministic theory, where everything is written down once and for ever. There is good reason, however, to question this statement. Special relativity provides quite adequate descriptions of, say, fluid motions, or of interacting electric currents. But, physically, the momentum-energy density of a fluid, or the electric 4-current density, has statistical meaning only, so that their Minkowskian formalization does not, strictly speaking, represent something that really exists. The implication is that, underlying this apparently deterministic picture, there must exist a covariant way of handling the probabilities that are subsumed. The corresponding 'hidden events' need not be describable in spacetime; we require only that what is observable — the preparations and measurements of physical evolutions, in the quantal jargon, together with the conditional or transition probabilities linking them — be expressible in four-dimensional geometric style.

This is exactly what we have found to follow quite easily from an adequate handling of the conditional or transition probabilities defined by Bayes and by Laplace. And so it turns out that *the Poincaré–Minkowski four-dimensional paradigm is not constrained to determinism, but can very well handle the calculus of transition probabilities.*

Emphasis, in this context, is upon the Bayesian concept of probability, as there is only one world history. However repetitions of a test do make sense, up to those parameters the knowledge and control of which are neglected.

So let us disregard these for a while, and undertake a guided tour through the museum of classical thermodynamics and statistical mechanics. *These we will examine, keeping always in mind that there is an intrinsic symmetry, and only a factlike asymmetry, between prediction and retrodiction* — the latter expressed as a time asymmetric boundary condition.

What biases almost all basic analyses of probabilistic problems in physics is that *the adjective 'predictive' is almost always tacitly assumed, thus fraudulently introducing what is later said to be disclosed: irreversibility.* From this one should guard oneself as from touching an electric line. For example, such assertions as 'transitions from lowly populated to highly populated subensembles are by far the most probable' express no more than *factlike* irreversibility.

An important specification of this class of problems consists of lawlike reversibility and factlike irreversibility of the *negentropy* $\rightleftharpoons$ *information*, or $N \rightleftharpoons I$ *transition*, exhibiting the twin faces of the information concept, the *patent one*, $N \rightarrow I$, *gain in knowledge*, and the *hidden one*, $I \rightarrow N$, *organizing power*. Intrinsic reversibility is expressed by the two arrows, and factlike irreversibility (the 'second law') by the preponderance of the upper over the lower one.

3.1.17. APPENDIX: COMPARISON BETWEEN MY THESIS AND THOSE OF OTHER AUTHORS WHO HAVE DISCUSSED THE FUNDAMENTALS OF IRREVERSIBILITY

Due tribute being paid to him, Boltzmann (1964, pp. 446—448), in brief but extremely penetrating remarks, has by far anticipated the essentials of all subsequent studies of irreversibility. He pointed out that the mathematical expression of irreversibility lies not in the evolution equations, but in the retained solutions; that irreversibility is a cosmological phenomenon; that the time arrow does not belong to physics, but to biology, in the sense that, for some reason, life has to go up, and not down, the entropy curve. It is odd indeed that the numerous authors (including myself) who later delved into the problem did not notice Boltzmann's indisputable priority.

The well-known papers by the Ehrenfests (1911) and Smoluchowski (1912) stressed lawlike time symmetry in the Loschmidt and the Zermelo behaviors, together with intrinsic symmetry between prediction and retrodiction. In this same year, 1911, J. D. van der Waals, in a short and penetrating paper, pointed out that the statistical derivation of the '*H*-theorem' must start with a temporally asymmetric use of 'Bayes's formula', that is, by using Laplace's 'principle of probability of causes'.

In his 1930 article entitled 'The Symmetry of Time in Physics', G. N. Lewis stated as a new principle that "throughout the sciences of physics and chemistry, symmetrical or two-way time everywhere suffices", so that "we can no longer regard effect as subsequent to cause" and that "if we think of the present as pushed into existence by the past, we must in . . . the same sense think of it as pulled into existence by the future." According to Lewis, if "our common idea of time is unidirectional . . . this is largely due to the phenomena of consciousness and memory". Discussing the entropy of mixing in terms akin to Szilard's, he defined entropy as missing information, and stated that "gain in entropy always means loss of information and nothing more. It is a subjective concept."

All this is very interesting, but falls short of a complete treatment of irreversibility. Is there really *nothing* objective in the phenomenon of radiating stars? Is information not *also* organizing power?

Then Lewis went on sketching, but in words only, what came out later as Wheeler and Feynman's time symmetric electrodynamics.

Finally he discussed transition probabilities, not noticing that the very approach he used would have led to replacing Boltzmann's statistics by either the Bose—Einstein or the Fermi—Dirac one.

In 1939 C. von Weiszäcker carefully discussed what Mehlberg (1961) later called 'Lawlike time symmetry versus factlike time asymmetry'. So also did Schrödinger in 1950, and, later, Watanabe (1951), Ludwig (1954) and I (1964).

And then, starting in 1955 with Watanabe's important studies 'Symmetry of Physical Laws, I, Symmetry in Space-Time and Balance Theorems' and 'III, Prediction and Retrodiction', and in 1956 with Reichenbach's celebrated posthumous book *The Direction of Time*, there came, like a pack of hounds, an avalanche of articles or books all stressing the same fundamentals — without excluding, however, sudden bursts of harsh back-biting concerning peculiar points.

One of the differences between Reichenbach's thesis and the one presented here is that Reichenbach sees matter as 'becoming', whereas Grünbaum and I see it as 'time extended'. For Reichenbach, consciousness is a mere epiphenomenon, whereas Hugo Bergmann (1929, pp. 27—28) and I see it as the sole selector of the 'present instant'.

Popper (1956—1957), in his series of papers entitled 'The Arrow of Time', discussed scatterings of waves and of particles, without considering their connection via quantum mechanics. He excludes *a priori* the very possibility of a statistical discussion by using an infinite Euclidean space. He excludes advanced waves by stating that, to produce them causally a miraculous conspiracy of causalities would be required; but this begs the question, as these waves are defined by a unique final condition.

Büchel (1960) discussed 'The *H*-theorem and its Reversal'. Mehlberg, in his 1961 paper 'Physical Laws and the Time Arrow', went so far as to write that 'Time has no arrow' (meaning no 'lawlike', but only a 'factlike' arrow in physics).

Grünbaum (1962; 1963, pp. 209—329) has repeatedly expressed ideas extremely close to mine (except on the point that the information concept is peripheral to him). The following titles speak for themselves: 'Temporally Asymmetric Principles, Parity between Explanation and Prediction, and Mechanism versus Teleology.'

Then there is a bunch of 1960 to 1962 articles by physicists all discussing lawlike time symmetry versus factlike time asymmetry: Terletsky (1960), 'The Principle of Causality and the Second Principle of Thermodynamics' (the title of which speaks for itself); Adams (1960), 'Irreversible Processes in Isolated Systems'; McLennan (1960), 'Statistical Mechanics of Transport in Fluids', stressing the connection between increasing entropy and retarded waves; Wu and Rivier (1961), 'On the Time Arrow and the Theory of Irreversible Precesses'; Penrose and Percival (1962), 'The Direction of Time'.

In 1975 appeared Belinfante's book *Measurements and Time Reversal in Objective Quantum Theory.*

Davies (1974), *The Physics of Time Asymmetry* and Gal-Or (1981), *Cosmology, Physics and Philosophy* are cosmology oriented, and emphasize that gravitation is the ruling field in the large.

CHAPTER 3.2

PHENOMENOLOGICAL IRREVERSIBILITY

3.2.1. CLASSICAL THERMODYNAMICS

Sadi Carnot's 1824 'Réflexions sur la Puissance Motrice des Machines à Feu' is an extraordinary piece of scientific literature: a very short one, containing not a single mathematical formula. It turns an easily observed twin-faced fact into a far-reaching principle yielding, even today, an understanding and intercorrelation of innumerable phenomena. It seems that it was perhaps intended as a brochure for launching a company building steam engines.

The two aspects of 'Carnot's principle', the 'second law' of thermodynamics, are:

1. Heat flows spontaneously from high- to low-temperature regions, not the converse.
2. In an environment of uniform temperature, work can be converted directly into heat (think of a brake), but not the reverse.

These are two time asymmetric statements, mutually reducible to each other (an exercise in thermodynamical reasoning). Clausius has used the first one, and Planck the second one, in their presentations of thermodynamics.

The 'first law', stating equivalence between heat and work, was clearly understood only later. When writing his paper Carnot ignored it, and used instead the 'conserved caloric' concept. However, he discovered it very soon afterwards, as evidenced in his personal notes, published posthumously, in 1927. The 'first law' was established by Joule's 1843 experiments; Robert Mayer had previously arrived at it by a clever reasoning using the two heat capacities of a perfect gas, at constant pressure or volume, respectively. Even earlier, Rumford had understood it while conducting work on cannon-boring experiments. That heat — or should we rather say 'internal energy' — is nothing else than a 'hidden kinetic energy' was the idea of Rumford, Daniel Bernouilli and others. In a remarkable memoir of 1786, Lavoisier and Laplace discuss the matter.

Firmly resting on the two pillars of the 'first' and the 'second' laws, the discipline of classical thermodynamics was established by Clausius and by J. J. Thomson (Lord Kelvin). The concept and name of entropy were discovered, and coined, by Clausius. The absolute temperature T scale, now denoted '°K', was defined by Kelvin and him. As is well known, the Gay-Lussac 'perfect gases temperature scale' coincides with the Kelvin scale; so, N denoting Avogrado's number and k Boltzmann's constant, the state function of a perfect gas assumes the simple form $\varpi v = NkT$.

Clausius's entropy S is the 'extensive' magnitude associated with the 'intensive' magnitude T, as the volume v is associated with the pressure ϖ in, say, the exact differential of the 'internal energy'

$$\mathrm{d}U = \varpi\,\mathrm{d}v + T\,\mathrm{d}S,$$

in units such that Joule's conversion coefficient J is 1. Classical thermodynamics uses 'Pfaffians', such as $\varpi\,\mathrm{d}v$ or $T\,\mathrm{d}S$. It can use also the reversed style differentials $v\,\mathrm{d}\varpi$ or $S\,\mathrm{d}T$, writing, for example, the 'enthalpy' as

$$\mathrm{d}(\varpi v + TS - U) = v\,\mathrm{d}\varpi + S\,\mathrm{d}T.$$

Clausius's discovery, upon which Caratheodory elaborated, is that $\mathrm{d}S$ is (like $\mathrm{d}v$) an 'exact differential', so that T^{-1} is (like ϖ^{-1}) an 'integrating factor'. In other words, the entropy is (like the volume) a 'state function' of a 'thermodynamic system' studied in terms of 'thermodynamic variables'.

3.2.2. FACTLIKE THERMODYNAMIC IRREVERSIBILITY AND ITS RELEVANCE TO CAUSALITY AND INFORMATION

Poincaré (1906a, Pt 1, Ch. 4) asked how things would look if Carnot's time arrow were reversed. We would find ourselves, says he, inside a catastrophic sort of world, where foresight and action would be impossible. As Grünbaum (1963, p. 227) puts it, it would be dangerous to get into a lukewarm bathtub, because one would never know which end is going to boil and which to freeze. Also, viscosity would be an accelerating instead of a damping force, and would set resting bodies into unpredictable states of motion. The most carefully considered of our actions would have enormous and unforeseeable consequences; it would be dangerous to play at bowls; and a fly inside a quarry would be far more dangerous in that world than a bull in a china shop in ours.

We are accustomed to hearing the complaints of all sorts of engineers about the tribute they must pay (and which they try by all means to minimize) to friction, viscosity, resistance, drag and all sorts of 'energy degradation'. Truly, this is a fortunate tribute, by which the Universe efficiently takes care of our security, guaranteeing the return to calmness after any sort of tempest — even one caused by us! In Poincaré's paradoxical world, forecasting estimation (even the intuitive sense used by luggage carriers or bowlers) would be impossible. . . . Poincaré says of this sort of Universe that it would be 'lawless'. The truth is, however, that *it would be a world obeying final rather than causal laws*; final instead of initial conditions would be needed when formalizing problems.

How the two laws of increasing entropy and of retarded causality are coupled to each other is clarified by statistical mechanics. But, already, this sort of 'thermodynamical philosophy' makes clear that the time arrow (or, should we rather say, that *our biological adaptation to the time arrow*) is such that we are able to 'see in the past and to act in the future' — thus directing, so to speak, our psyche's eye along the direction where there is security. Thus we can continuously enlarge the sphere of what we know, and control our action step by step. Thus we can already guess that some quite crucial tenet of Natural Philosophy may well lie in the connection between our subjective time arrow and the second law's time arrow.

3.2.3. ENTROPY INCREASE AND WAVE RETARDATION

Observationally speaking, the mechanical phenomenon mediating entropy increase is very often a retarded wave of one kind or an other.

Consider, for example, a meteorite entering the Earth's atmosphere. While friction slows it down, a shock wave and an electromagnetic retarded wave are emitted; and, of course, the entropy of the whole system increases. Finally the meteorite is consumed. The reversed procedure, as displayed in a film run backwards, would appear simply miraculous.

As another example consider a stone thrown into a pond. A diverging wave is seen to propagate on the surface, while an unseen sound wave is emitted inside the water. Finally the stone rests at the bottom; all waves subside; and the temperature of the pond increases slightly. Running a film backwards will show a converging surface wave building up, then being annihilated at its center, while the whole energy that has been concentrated ejects the stone, which flies nicely into the opening hand of a person walking backwards along the pond side. Again a miracle.

Examples of this sort can be multiplied *ad infinitum*. All display a coupling between a localized 'irreversible' event and the whole environment — a coupling accepted as quite natural when seen in the usual way but looking 'ridiculous' when seen in reverse.

Briefly speaking, the environment is a sink of retarded waves of all sorts; it is in this sink that negentropy finally disappears. Imagining that it may reappear from there in the reversed fashion would look like a fantastic 'conspiracy of causalities'.

As Gold (1966, p. 316) puts it, "could a plane fly backward?" Just think of the air current lines around the wings, and in and out the jet engines. . . .

Basically, the question of an essential connection between the two (factlike) principles of entropy increase and of wave retardation came up in the context of developing quantum mechanics. It was present in Planck's thinking, and in a heated 1906—1909 discussion between Ritz and Einstein. Quantum mechanics settled the matter, as explained in Sections 4.1.1 and 4.2.7.

3.2.4. LIGHT WAVES

While it is quite easy to produce a diverging, or retarded, light wave by lighting a bulb, 'there is no hope' of summoning from the past a converging or advanced light wave by picking up a piece of charcoal from inside a box. There is no more hope of this than of extracting mechanical power by applying a brake to a wheel.

However, in the case of light waves, there exists a fairly well-known recipe for mimicking an advanced wave by using a 'conspiracy of causalities', the image generation in optical systems. A burning source can, in this way, generate at a distance a burning sink, and Archimedes knew that. The secret here consists in preservation of phase coherence. To this we shall return.

As an other example consider (Figure 7(a)) a linear grating which, from some incident plane wave labeled i, can generate a finite set of g well-defined outgoing plane waves labelled j. Any of the outgoing waves j can be generated by a well-defined set of g incident plane

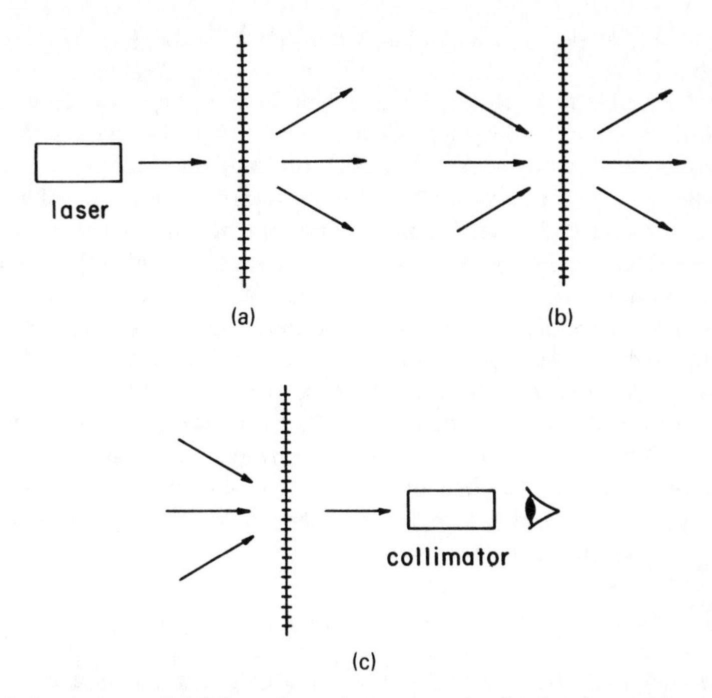

Fig. 7. Increasing probabilities and retarded waves go hand in hand: a grating receiving a plane monochromatic wave generates g such waves, upon which the incident corpuscles are statistically distributed, with probabilities equal to the intensities (Figure 7a). Figure 7(c) pictures the ideally reversed procedure: 'blind statistical retrodiction' as reasoned when receiving from a grating a plane monochromatic wave (and knowing nothing more). Figure 7(b) pictures 'intrinsic reversibility' of the elementary phenomenon: a 'transition $g \times g$ matrix' connects any one incident to any one outgoing plane wave.

waves, one of which was the one considered first (Figure 7c). The intensities of these waves can be arranged in a square matrix of rank g; Figure 7(b) displays the 'intrinsically symmetric' elementary procedure out of which the two asymmetric ones, pictured in Figure 7(a) and (b), are extracted.

It is, of course, quite easy to excite, by means of a laser, one of the i-waves, and to observe the g phase coherent j-waves thus generated (Figure 7(a)). But, by merely placing a collimator to receive one of the j-waves, there is no hope of generating the set of g phase coherent i-waves which could build the j-wave. This is feasible, however, in the form of a 'conspiracy of causalities', and M. von Laue (1907a) did it in his PhD Thesis.

Nevertheless the 'retrodictive problem' does make sense: receiving a plane wave j in a collimator (Figure 7(c)) and knowing only that the grating is there, we can ask in which form the light entered the grating. One thing is sure: the light has come on one of the i plane waves that can generate our j-wave. Nothing more can be said. To surmise that it may have come in g phase-coherent i-waves would imply that the observation was performed in M. von Laue's laboratory.

Quantum mechanically speaking, the intensities of our plane waves are fluxes of photons. The elements of the transition matrix $(i \rightleftharpoons j)$ are *transition probabilities*, which can be used either predictively, from i to j, or retrodictively, from j to i. Therefore the predictive and the retrodictive calculation use, respectively, wave retardation and wave advance, with the complete time symmetry illustrated in Figure 7(b). Rejection of advanced waves is equivalent to rejection of 'blind' statistical retrodiction. To this we shall return.

3.2.5. WAVES AND INFORMATION THEORY

Fourier analysis is used in the problems of image-building in optical instruments, in holography, and in transmitting messages. Thus the well-known cells associated with reciprocal Fourier transforms, $\delta x. \delta k_x \simeq 1$ in their 'space' aspect, or $\delta t. \delta \nu \simeq 1$ in their 'time' aspect, are considered, and 'bits' of information are distributed among them.

According to 'wave mechanics', as discussed in Chapter 4.1, these cells are nothing else than those of dimension h (or h^2, or h^3) appearing in statistical quantum mechanics, where the information consists of integer occupation numbers distributed among them.

We shall not delve here into these matters.

3.2.6. LAWLIKE TIME SYMMETRY AND FACTLIKE TIME ASYMMETRY IN THE WHEELER—FEYNMAN ELECTRODYNAMICS

The very clever Wheeler—Feynman relativistically covariant and time symmetric electrodynamics we have summarized in Section 2.6.14 lends itself to an interesting discussion of "lawlike time symmetry versus factlike time asymmetry" (in Mehlberg's 1961 wording).

To this end Wheeler and Feynman (1949) ideally enclose their system inside a perfectly absorbing box, thus treating it as an isolated system — a procedure familiar in thermodynamics or statistical mechanics. Then their formulas show that the use of full wave retardation and the well-known Lorentz damping force, as is done classically, is strictly equivalent to the use of full wave advance and an anti-Lorentz force, of half retardation and half advance of waves and no Lorentz force. This shows clearly the intrinsic time symmetry of classical mechanics and electromagnetism.

Physically speaking, however, it is well known that such a symmetry is broken by the coupling of electromagnetic radiation with something else. If a physicist arbitrarily displaces an electric charge, he experiences the Lorentz damping force and observes an emitted retarded wave; also, in a broadcasting station, a retarded wave diverges from the antenna while power is consumed at the station. The reversed sort of procedure occurs at the receivers, on a smaller scale, but an overall diverging wave rushes into the depths of space, completely washing out all convergences at the receivers.

So, the facts are such that *a very strong time asymmetry must be attributed to Wheeler and Feynman's adiabatic wall at infinity* — and this they have implicitly recognized, because an *ad hoc* asymmetry is slipped into their calculation.

What is asymmetric is the very concept of 'perfect absorption', the opposite of which is perfect emission. Symmetry of absorption and emission by the walls of a cavity leads to the concept of thermal radiation discussed in the next section. Wheeler and Feynman's cavity is of a different sort: it is a *perfect sink*; so *where does the energy go?* Their assumption of a perfect sink, rather than of a perfect source, is the assumption of a time asymmetry — and an extremely radical one.

On the whole, in order to fit the facts, the clever Wheeler—Feynman theory (which, I believe, is intrinsically true) must be supplemented by the following 'second statement': If the system of interacting charges is open, either to the depths of space, or by a coupling to another system, then electromagnetic energy radiates out, and non-electromagnetic energy is consumed.

This very much resembles Carnot's statements, although Wheeler and Feynman are not doing thermodynamics, as their radiation has an arbitrary spectrum. There is in this, however, a very strong hint in favor of a close connection between the two principles of entropy increase and of wave retardation — a connection established by Planck's quantal hypothesis and by his definition of the entropy of radiation.

3.2.7. THERMAL EQUILIBRIUM RADIATION

Among the numerous successful applications of classical thermodynamics the most far-reaching certainly is the one pertaining to the problem of thermal equilibrium radiation.

Inside a closed oven at absolute temperature T the radiation is at equilibrium; that is, isotropic, with a uniform energy density w or, per frequency element $\mathrm{d}\nu$, $\mathrm{d}w(\nu)$. Nothing can be seen, as uniform brightness or darkness prevails.

The enclosed equilibrium radiation can be considered as a Fourier superposition of plane standing waves, so that the formula $3\varpi = w$ holds between the pressure ϖ on the cavity wall and w. From this Boltzmann, writing the differentials $\mathrm{d}U$ of the internal energy and $\mathrm{d}S$ of the entropy of the radiation in terms of $\partial w/\partial T$, and requiring that $\mathrm{d}S$ be an exact differential, found that

$$(3.2.1) \quad w = aT^4$$

and that, in a reversible adiabatic transformation,

$$(3.2.2) \quad vT^3 = \text{const.}$$

The latter is known as Stefan's law.

The next step was aimed at improving the expression for $w(\nu)$. Wien, reasoning with a perfectly reflecting wall and a moving piston within it, used the Doppler shift formula to arrive at the highly significant formula

$$(3.2.3) \quad \nu^{-3}w(\nu) = F(T^{-1}\nu)$$

where $\nu^{-1}w(\nu)$ is an 'adiabatic invariant'.

All these deductions were perfectly well confirmed by the observations. But this was as far as classical thermodynamics was able to go. An unexpected Niagara Fall into the unknown lay ahead of all this 'normal science' paddling on a peaceful lake.

It all began with counting the normal standing modes in the cavity. This number, $n(\nu)$ per volume $\mathrm{d}v$ and frequency $\mathrm{d}\nu$ elements, is analysis independent, so that the simplest way to obtain it is to think of a perfectly parallelepipedic oven. Then, as shown in 1900 by Lord

Rayleigh, one gets $n(\nu) = 4\pi\nu^2/c^3$ or rather, as there are two independent polarization states per plane standing wave, $n(\nu) = 8\pi\nu^2/c^3$; c denotes, of course, the speed of light in the vacuum.

Thinking in terms of statistical mechanics and assuming the equipartition of energy in the form $w(\nu) = kT$ (k denoting Boltzmann's constant), Lord Rayleigh was led to

$$F(\nu/T) = kT/\nu,$$

a formula fitting the experimental curve when $\nu/T \to 0$. However, the integral $\int_0^\infty w\,\mathrm{d}\nu$ diverges, leading to the so-called ultraviolet catastrophe. Wien, on the other hand, found that he could fit the curve for $T/\nu \to 0$ by writing down

$$F(\nu/T) = \exp(-h\nu/kT),$$

with h denoting some universal constant.

How Planck, in 1900, straightened the matter by setting

$$F(\nu/T) = [(\exp(h\nu/kT) - 1]^{-1}$$

is a story to be told in Chapter 4.2.1.

3.2.8. IRREVERSIBILITY AND THE COSMOLOGICAL COOL OVEN

What role does irreversibility play in all this? Equilibrium thermostatics, by definition, describes some sort of an eternal inferno either hot or cold, but with no thermostatic adjustment in any case.

So let us consider an oven, the contour of which is arbitrarily complicated, and pierce a little hole in its wall. In fact, this is exactly how the spectrum of the thermal radiation is studied experimentally.

The radiation will leak out to infinity and the oven's temperature drop to 0 °K. This we understand well under the assumption of wave retardation.

Blind statistical retrodiction, using the assumption of advanced waves, would have it that our oven at T °K was obtained by just closing the aperture at the time $t = 0$. As Eddington put it: "Do not ask me if I really believe that this will happen."

In 1965 Penzias and Wilson found that we are living inside an oven of cosmological size at 2.7 °K, the radiation of which provides a cosmological preferred spatial reference frame. So the vacuum around us is not at 0 °K after all. However, we are far from thermal equilibrium, because we can observe numerous very hot stars, and we know that our own modest Sun makes our Earth agreeably warm.

Then, how is it that the cosmic oven is not at the temperature of the stars, if the stars are uniformly distributed throughout an infinite Euclidean space? This question is termed the 'Olbers Paradox', to be discussed in Section 3.4.5.

CHAPTER 3.3

RETARDED CAUSALITY AS A STATISTICAL CONCEPT. ARROWLESS MICROCAUSALITY

3.3.1. POINCARÉ'S DISCUSSION OF THE LITTLE PLANETS' RING

In *Science and Hypothesis* Poincaré (1906a, Pt 4, Ch. 11) remarks that if the angular velocities a and the initial longitudes b of the little planets were related by the formula $ap + b = q$ (p and q denoting two constants, $p > 0$), a 'miracle' would occur at time $t = p$: all the little planets would coalesce at the longitude q (perhaps welding themselves into a big planet). By what argument, asks Poincaré, should we exclude such a 'ridiculous' assumption? On the argument of 'sufficient reason:

> It seems to us that there is no sufficient reason by which the unknown cause which has generated [the little planets] has acted according to a law so delicately adjusted that it seems it was chosen on a purpose.

All right. But, if $p < 0$, the 'miracle' has occurred at $t = -p$, where it *could* have been the explosion of a big planet. Therefore it is not the formal structure of the law that is paradoxical (all important physical laws have a delicately adjusted structure) but the fact that *in one case* ($p > 0$) *the improbable structure follows the initial condition, whereas in the other one* ($p < 0$) *it precedes it.* So what Poincaré is really saying is that 'blind statistical prediction' is accepted (in this case as in many others) while 'blind statistical retrodiction' is rejected as being 'unphysical'. So the truth is that Poincaré has attached to his "principle of sufficient reason" a hidden label saying: 'retarded causality', or 'one way sufficient reason'. This amounts to 'causality accepted, finality rejected' — as his very words "on a purpose" do say. Finally, *far from deducing irreversibility, Poincaré has assumed it at the start.*

Poincaré then expands his discussion. Replacing the swarm of little planets by a continuous fluid, he introduces the so-called 'characteristic function'[1]

$$C(t) = \iint \exp[i(at + b)] f(a, b) \, \mathrm{d}a \, \mathrm{d}b$$

and remarks that, provided simply that f is continuous in a, $C(t) \to 0$ when $t \to +\infty$. Therefore the final distribution will be quite homogeneous. All right. But it was so at $t = -\infty$. Therefore the magic spell has not been exorcised.

This discussion by Poincaré provides a vivid sketch of a much more general one inherent in the Maxwell—Boltzmann—Gibbs statistical mechanics, to which we come now.

3.3.2. BOLTZMANN, GIBBS AND THERMODYNAMICAL ENTROPIES

Jaynes (1983, p. 77), in his important and penetrating article entitled 'Gibbs versus Boltzmann Entropies', essentially discusses the respective relevances of Gibbs's negentropy

$$(3.3.1)\quad I_G = \int \mathscr{W}_n \ln \mathscr{W}_n \, d\tau,$$

where $\mathscr{W}_n$ $(x_i \ldots ; p_i \ldots ; t)$ denotes Liouville's $6N + 1$ variables function, and of Boltzmann's negentropy no. 1.

$$(3.3.2)\quad I_{B1} = -\ln \mathscr{W}$$

where $\mathscr{W}$ measures, in some sense, the phase volume of 'reasonably probable microstates'. He also discusses Boltzmann's negentropy no. 2.

$$(3.3.3)\quad I_{B2} = N \int w_i \ln w_i \, d\tau_i,$$

where $w_i(x_i, p_i, t)$ is the single particle probability density

$$(3.3.4)\quad w_i = \int \mathscr{W}_N \, d\tau_{-i}$$

with $d\tau_{-i} \equiv d\tau/dx_i\, dp_i$. He shows that $I_{B2} \leqslant I_G$, with equality if, and only if, $\mathscr{W}_N$ factors 'almost everywhere' as $\Pi(w_i)$.

Applying this to Gibbs's canonical ensemble governed by the Hamiltonian $H = \Sigma(2m)^{-1}p_i^2 + U(x_i - x_j)$ he demonstrates that, along a reversible path,

(3.3.5) $\mathrm{d}I_G = -\mathrm{d}Q/kT,$

T and Q denoting the thermodynamical temperature and heat as functions of the pressure ϖ and volume v; as for I_{B1} one has

$$\mathrm{d}(I_{B1} - I_G) = (kT)^{-1}[\mathrm{d}\langle U \rangle + (\langle \varpi \rangle - \varpi_0)\,\mathrm{d}v],$$

ϖ_0 denoting the pressure of an ideal gas with the same temperature and density. So, what are neglected in I_{B1} are the interparticle forces.

Jaynes expresses the 'Second Law' by comparing the behaviors of I_G and I_{B1}. Starting at time $t = 0$ with a gas in thermal equilibrium where [according to formula (5)] the 'thermodynamical negentropy' I_T is taken equal to I_G, we perform an adiabatic change by, say, moving a piston, or applying a magnetic field (representing this by a perturbative, time dependent term, in the Hamiltonian), suppress at time t the perturbation, and let the system relax into a new equilibrium. Then, as a consequence of the previous remarks, $I_T(t) \leqslant I_G(t) \leqslant I_G(0)$. *This is the 'Second Law'.*

This is a brilliant demonstration — and one not found elsewhere, as it seems. However, in the light of my own perspective, there is one important point needing to be stressed. Jaynes writes that, according to his demonstration, the thermodynamic negentropy I_T decreases "in any 'reproducible' experiment". The point is that, once again, *by this sentence, irreversibility is not deduced, but postulated.* What the mathematics does say is that $I_T(t) \leqslant I_G(0)$ *whenever* $t \neq 0$, *be t positive or negative . . .*

Jaynes's final remark is also quite interesting. Let it be recalled that W, in the expression of I_{B1}, is the volume in phase space. Jaynes considers a 'high probability region' R in phase space, and shows that, in the previous thought experiment, $R(t) \geqslant R(0)$. This is exactly 'lawlike reversibility and factlike irreversibility' as expressed in terms of *conservation and deformation of Liouville's phase volume element.* It is the $6N$-dimensional analog of the diffusion of an ink drop. Liouville's $6N$-dimensional volume is constant, but 'in the course of time' it twists and elongates itself; if it were compact at $t = 0$, at any other time a compact volume containing it is larger than at $t = 0$. Everybody knows how a clockwork's spiral spring goes into an intractable tenia-like shape when it escapes from its holder, and how difficult it is to put it back properly there! But, again, *this is not a demonstration, it is an authentification of irreversibility.*

Throughout this Jaynes reasoning no 'coarse graining' has been used. Or, rather, *one 'coarse graining' that has operational meaning has been used, consisting of those macroscopic thermodynamic variables 'you or I' decide to use.*

3.3.3. LOSCHMIDT'S OBJECTION AND BOLTZMANN'S FIRST INAPPROPRIATE ANSWER. RECURRENCE OF THIS SORT OF PARALOGISM

In 1876 Loschmidt argued that, according to Newton's equations of motion, to every mechanical evolution of a system there corresponds a time-reversed one. Therefore, if the entropy increases in the first evolution, it must decrease in the second one.

Boltzmann's first reaction to this is worthy of an exact quotation (1964, p. 117) being one of these paralogisms so easily generated by relying carelessly on macroscopic thinking habits:

> This in no way contradicts what was proved [before]; the assumption made there that the statistical distribution is molecular-disorded is not fulfilled here, since after exact reversal of all velocities each molecule does not collide with others according to the laws of probability, but rather it must collide in the previously calculated manner.

Marcel Brillouin (1902, p. 196) was not taken in by this specious reasoning, his comment being: "I cannot accept that if the direct motion was 'molecular ungeordnet' the reversed one is 'molecular geordnet' merely because we have learned the time ordering of the collision."

In order to show into what sort of conceptual mess some otherwise brilliant minds have been led, I quote the following lines from Borel's *Classical Statistical Mechanics*, under the subtitle 'The Irreversibility Paradox' (1925, pp. 59–61):

> We have just shown that, in the natural evolution of a gas taken in any state, it is highly probable that [the negentropy] H will constantly decrease. This conclusion may seem unacceptable, because if at some time t one exactly reverses the speeds of the molecules (which does not change the . . . expression of H) the gas will go through its previous states in reversed order; thus H will go up if it previously went down. Thus it is impossible that $[H]$ decreases whichever was the initial states, and it seems that there are as many chances that it goes up or down.

This of course is Loschmidt's argument. But Borel continues thus:

This only is the semblance of a contradiction, because, while our reasoning is valid in the case of any natural motion, it is no longer such in the case of the reversed motion, which incidentally is physically impossible.

However, Borel brings in a constructive remark:

Because of the tremendous scattering produced by the collisions, the positions of the molecules are no more dependent on the state of the gas a few collisions before than on anything else in the universe exerting an influence . . . For the [reversed] motion thus initiated [*à la* Loschmidt] to continue . . . the universe's whole motion should also be rigorously reversed; even the minute displacement of a molecule on Sirius would almost immediately prevent our gas from going back through its previous states. Therefore the reversal of a natural evolution is practically impossible.

In other words, isolation of the system under study is not a valid assumption when discussing irreversibility. The interaction between the molecules of our gas and those on Sirius is gravitational, propagated by waves. Thus, reversal of mechanical motions implies also reversal of wave motions. Also, on the astronomical scale, gravity is the important physical interaction. Thus Borel, well aware that the enormous values of numbers in Gibbsian ensembles are of a cosmological magnitude, does bring cosmology into the irreversibility problem, implicitly tying the question of entropy increase to that of wave retardation, and mentioning gravitation as the important field in cosmology. These are exceedingly important points Borel makes *en passant*, and almost unwittingly. . . .

In order to show how the central point at issue is missed by falling back into the old, muddy track of routine thinking, let me quote a recent example, chosen almost at random.

In an otherwise very interesting article entitled 'Thermodynamics as a Science of Symmetry', Callen (1974, p. 437) writes:

An example is provided by the spontaneous $2p \rightarrow 1s$ transition in a hydrogen atom, with the emission of a photon. The photon is radiated in all directions in space, with relative phases depending on the initial direction of the angular momentum in the $2p$ electronic state. This simple atomic process is reversible in principle as are all fundamentally quantum mechanical processes. But how is it operationally reversible? Simply sending in a photon from some particular direction to be absorbed in the $1s \rightarrow 2p$ transition is not sufficient, for it will result at best in a $2p$ state with some accidental direction of orbital angular momentum, unrelated to the original direction. We must *prepare* [italics mine] an incoming spherical wave with the proper phase relations over the entire wavefront. This is so utterly impractical that for all *reasonable* [italics mine] purposes the initial radiation can be thought of as operationally irreversible.

In other words, an 'incredible conspiracy of causalities' would be needed to mimic the *unique* finality requirement formalized as 'blind statistical retrodiction'. *This is the point, and it must be stated explicitly.*

Incidentally, Callen is discussing here a problem in the '1926 Born wavelike probability calculus', where phase relations are significant. This immediately raises the level of 'factlike irreversibility' far above what it is in the classical probability calculus. To this we shall return.

3.3.4. RETARDED CAUSALITY AS IDENTICAL TO PROBABILITY INCREASE. CAUSALITY AS ARROWLESS AT THE MICROLEVEL

The quintessence of physical irreversibility is this: *retardation of causality is not an elementary, but a statistical — and a factlike — law, strictly identical to the 'second law'. At the elementary level causality is arrowless.*

This is the essential kernel of Loschmidt's reversibility argument. At a general and abstract level, it is also the essence of Laplace's discussion of 'probability of causes', as expressed in formulas (3.1.4) and (3.1.5). *It is from there that the discussion must start.*

Why is it that, in macrophysical applications of the probability calculus, we are allowed to compute 'blindly' (Watanabe, 1955) in prediction, but not in retrodiction? In prediction we just use the intrinsic transition probabilities of the system, but *in fact* this will not do in retrodiction. Instead, Laplace enjoins us to use Bayes's conditional probability formula, including a set of extrinsic transition probabilities, correlating the system with its environment; these express at best what we know — or think we know — concerning the insertion of the system into its surroundings.

Then the question is: *Why is it that we must consider the interaction of the system with its environment in retrodictive but not in predictive problems?*

Let us first discuss this by using an example.

Suppose that, between times t_1 and t_2, a physicist moves a piston in the wall of a vessel containing a gas in equilibrium, returning it at t_2 where it was at t_1. The *fact* is that Maxwell's velocity distribution is altered after time t_2, not before time t_1. Also, the *fact* is that the disturbance is propagated as a retarded wave emitted from the piston, not as an advanced wave absorbed by it. On the whole, work has been done, and, when equilibrium re-established, the gas is hotter than it was.

This example is one among many displaying a *one-to-one binding between the two (macroscopic) laws of wave retardation and of entropy increase.*

Moreover, beyond this technical point, such examples yield a statement of tremendous generality. *An interaction limited in time between two initially and finally separated systems produces on each of them its effect after it has ceased, not before it has begun.* So, in ceasing, it leaves each system in an improbable state to be taken as the 'initial state' for a predictive statistical calculation — not the converse. This is a paraphrase of Laplace's 'principle of probability of causes': to say that we are required to use Bayes's extrinsic probabilities in retrodictive, not in predictive, problems amounts to saying that the two concepts of probability increase and of retarded causality are merely two different statements of one and the same *physical concept.*[2] *As a physical concept, retarded causality is statistical in its nature.* It is completely erroneous to think otherwise. This we will have to remember when we discuss the Einstein—Podolsky—Rosen correlations in Chapter 4.6.

As previously said, this central issue has repeatedly escaped eminent scholars in the past. Even today, it is very frequently overlooked in the form of an *undue identification of probability with predictive probability.* Among those physicists who have explicitly made this point there are, of course, Terletsky (1960), Adams (1960), McLennan (1960), Penrose and Percival (1962) and, to some extent, van der Waals (1911) previously quoted. Owing to the paramount importance of the issue, I now quote very strong and explicit statements by these two eminent philosophers of science, Reichenbach and Grünbaum.

Reichenbach (1956, pp. 154—155) writes:

> The convention of defining positive time through growing entropy is inseparable from accepting causality as the general method of explanation. . . . The word 'produces' is a statistical concept.

In a similar vein Grünbaum (1963, p. 288) writes:

> We saw earlier how reliance on entropy enables us to ascertain which one of two causally connected events is the cause of the other. . . .

Not surprisingly, both authors draw their conclusion in connection with a discussion of branch systems.

I find it extremely surprising that so important a conclusion, explicitly drawn by these two philosophers of science and so clearly implied in the writings of those physicists previously mentioned, remains today very largely ignored. Retarded causality and/or blind statistical prediction are like two sister deities being worshipped blindly by a superstitious congregation.

The connection, and the difference, between lawlike, or intrinsic, time symmetry and factlike time asymmetry of the causality concept are, of course, expressed in the two Laplace formulas (3.1.4) and (3.1.5).

Application of these remarks to various problems is straightforward. For example, the little planets' ring, in Poincaré's thought experiment, has been generated by the explosion of a big planet under some cause either external (a disturbing celestial body) or internal (a volcano); a cause operating statistically, of course, very much like those weak and continuously acting astronomical influences, which do prevent the little planets from coalescing anew.

Let it be mentioned, however, that there seems to be, inside our overall causal world, a small contribution from the anti-Carnot, or final world. Evolutionary phyla, mentioned near the end of Section 3.1.14, illustrate this point.

3.3.5. RETARDED CAUSALITY AND REGISTRATION

If retarded causality can be defined by the fact that an interaction limited in time between two initially and finally separated systems produces, on each of them, its effect after it has ceased, and not before it has begun, this means that past, not future events can be registered. Both Reichenbach and Grünbaum explicitly make this connection.

Reichenbach (1957, p. 155) writes:

> The statement that although the past can be recorded, the future cannot, is translatable into the statistical statement: Isolated states of order are always postinteraction states, never preinteraction states.

Grünbaum (1963, p. 283) writes:

> Reliable indicators in interacting systems permit only retrodictive inferences concerning the interactions for which they vouch but no predictive inferences pertaining to corresponding later interactions.

A persistent registration consists of some structure remaining (and the longer the better) in a state of metastable equilibrium. And, of course, the optimal recording devices are those initially having the highest possible entropy.

How can we reconcile the possibility of registration with the principle that blind retrodiction is forbidden? First of all, by noting that registration is *not* blind statistical retrodiction.

If I see human footprints on a beach (Schlick, 1932) I conclude that someone has gone by there, not that someone will come, walking backwards, and erasing the footprints. Registration makes sense only to the extent that one is able to interpret it, having sufficient *a priori* knowledge. If I see a wake on a lake I conclude that a boat has passed, but I can validly imagine what sort of boat only to the extent that I know what sorts of boats exist on the lake. If an archaeologist is shown a harpoon, he recognizes it as such only if he knows what a harpoon is for. Many amusing stories exist concerning the discovery of the true use of an archaeological artifact.

This is to say that retrodiction is an art, and an art such that when memory or intuition that can decipher the meaning of registrations or 'fossils' is lacking, creative inspiration is needed to invent a correct theory. As we can see only one planetary system — ours — the problem of its origin is much further away from us than that of the origin of the stars. Problems in the 'probability of causes' are the most difficult of all.

3.3.6. ZERMELO'S RECURRENCE OBJECTION, AND THE PHENOMENON OF SPIN ECHOES

Poincaré's 1890 'recurrence theorem' states that any dynamical system with a finite number of degrees of freedom, obeying Newton's equations, has passed, and will pass, an infinite number of times arbitrarily near any configuration that it once goes through. In 1959 J. A. McLennan completed the theorem by showing that Poincaré's pseudoperiods grow longer as the state considered is further from equilibrium — which, of course, was expected.

Relying on Poincaré's theorem, Zermelo (1896) argued that an isolated Boltzmannian gas has gone, and will go, infinitely many times through any state it has once gone through — including low entropy states.

Of course, the Zermelo sort of behavior is terribly sensitive to the fact that there is no such thing as an isolated system. Any physico-chemical system does feel the cosmological environment — if only via neutrinos or low-frequency electromagnetic radiation. Also, observation of the system inevitably brings in an interaction.

Notwithstanding all this, a very striking Zermelo sort of behavior has been observed, by Hahn (1950), in the form of 'spin echoes'.

3.3.7. OTHER INSTANCES OF LAWLIKE SYMMETRY AND FACTLIKE ASYMMETRY BETWEEN BLIND STATISTICAL PREDICTION AND RETRODICTION

Why is it that the decay of a radionuclide, or of an elementary particle, into two or more particles is 'far more probable' than the converse syntheses? And why is it that, if the initial and final numbers of particles are the same, the transition from heavier to lighter particles is 'far more probable' than the converse reaction? As an example of the latter we have the *electron* + *positron* → 2 *photons* transition.

These reactions are intrinsically reversible, of course. The point is that the more probable (coarse grained) state is the one in which the 3-component momenta (in the center of mass frame) are the larger, so that the accessible phase cells are the more numerous. Therefore a 'blind statistical predictive reasoning' yields the observed irreversibility just stated, while a 'blind statistical retrodictive reasoning' would yield the converse 'miraculous' synthesis.

As an other example consider a gas (not necessarily in equilibrium) inside a container, and suppose that at time 0 we open an aperture in the wall. The (factlike) increasing probability principle has it that the gas will escape, leaving the container empty — and this is what happens in the vacuum. If, in order to meet an objection by Popper (1956–1957), we want the calculation to be feasible, we merely include our container inside an extremely large evacuated one. Thus, the most probable state is no longer a pure vacuum, but rather a gas of exceedingly low density.

And what of the reversed evolution in which all molecules rush back into the small container (which then we would just have to close)? Zermelo has it that this will happen (and has happened) time and time again. 'Blind statistical retrodiction' precisely describes this. In wave theory the analogous problem exists, if it is a standing wave rather than a swarm of particles that was initially enclosed in the box. In quantum mechanics the two problems really are only one, because of wave-particle duality. Then the cosmologist will say that this is really the sort of problem raised by the twin observations that the sky is dark at night and that there exists a 3 °K background radiation. Finally, the information-theoretician will ask if it is or is not possible to foretell exactly, or to select at will, the precise moment when the door should be closed, because all the molecules have come inside.

Of course the same sort of game can be played with radionuclides, or with Bohrian atoms emitting or absorbing photons, all enclosed within a tight container. Fluctuations will be occurring, even large ones over a long period. But, while we can place at will an excited atom in the box and let it decay, we cannot pick one out at will in the excited state. In the limit where the container is very large, blind statistical prediction and retrodiction yield, respectively, the symmetrical decay or build-up laws in $\exp(\mp at)$, the latter being rejected as anti-physical.

3.3.8. STATISTICAL MECHANICS: FROM MAXWELL'S THREE-DIMENSIONAL BILLIARD-BALLS GAME TO SHANNON'S INFORMATION CONCEPT

Maxwell, in 1859 (p. 713), started his statistical analysis of a gas made of spherical molecules thus:

> In order that a collision may take place, the line of motion of one of the balls must pass the center of the other at a distance less than the sum of their radii; that is, it must pass through a circle whose center is that of the other ball, and radius the sum of the radii. ... Within this circle every position is equally probable. ...

Having thus chosen his prior continuous probabilities, Maxwell successfully predicted

> ... such things as the equation of state, velocity distribution law, diffusion coefficient, viscosity, and thermal conductivity. ... The case of viscosity was particularly interesting because Maxwell's ... prediction that viscosity is independent of density ... seemed to contradict common sense. ... (Jaynes, 1983, p. 223).

The next step was taken by Boltzmann. He considered a swarm of N molecules with energies $E = \frac{1}{2}mv^2 + U(x)$ enclosed in a rigid and impermeable container, and asked: Given N, E and $U(x)$, what is the best prediction of the numbers N_k of molecules inside the phase space (position momentum space)? Treating the molecules as indistinguishable, he found the number of ways this can be done to be $W(N_k) = N!/\Pi(N_i!)$. This must be maximized under the constraints $E = \Sigma N_k E_K$

and $N = \Sigma N_k$, yielding a calculation found in all textbooks, which formally is the same as that associated with the MAXENT principle. We must say 'formally' because the MAXENT calculation plays with $\langle E \rangle$ and $\langle N \rangle$ rather than E and N.

From this Boltzmann got, among other things, the distribution law of molecules inside, say, the gravitational field, and the Maxwell velocity distribution law, with the unexpected consequence that the latter also holds inside a force field.

Jaynes (1983, p. 226) writes: "This is only [one example] where it appears that we are getting 'something from nothing', the answer coming too easy to believe." After alluding to Poincaré's epigraph (previously quoted) that we learn more from ignorance than we do from too much knowledge, Jaynes adds: "In fact Boltzmann's argument does take the dynamics into account, but in a very efficient manner . . . (1) the conservation of energy; and (2) . . . Liouville's theorem."

That was all, but it was enough.

Then Gibbs stepped in with a generalized use of the ensemble concept. This language he used

> rather as a concession to established custom. . . . We can detect a hint of cynicism . . when he states: 'It is in fact customary in the discussion of probabilities to describe anything which is imperfectly known as . . . taken at random from a great number of things which are completely described. (Jaynes, 1983, p. 228).

It is thus quite clear that Gibbs was *not* a 'frequentist'!

In 1948 'Enters Shannon' obtaining from his information concept exactly Gibbs's entropy formula. Whose information, or whose degree of ignorance, is it, asks Jaynes? *That of the communications engineer.*

To make a long story short, this is how the two concepts of negentropy and of information came to be unified, as meaning *something that is neither objective nor subjective because it is indissolubly both.*

G. N. Lewis (1930), L. Szilard (1929), J. von Neumann (1932), W. Elsasser (1937), L. Brillouin (1967), A. Katz (1969), and some others, are the *dramatis personae.*

3.3.9. BOLTZMANN'S SECOND THOUGHTS CONCERNING THE LOSCHMIDT OBJECTION

Boltzmann's first reaction to Loschmidt was way off the mark, as we have seen in Section 3.3.2. Later in his book (1964, pp. 446–448), under the subtitle 'Application to the Universe', he returned to the question, this time giving the right answer. The lines here quoted are significant in more than one respect.

> Is the apparent unidirectionality of time consistent with the infinite extent or cyclic nature of time? He who tries to answer these questions . . . must use . . . equations in which the positive and negative directions of time are equivalent, and by means of which the appearance of irreversibility . . . is explicable by some special assumption. . . . In the universe, which is in thermal equilibrium throughout . . [there are fluctuations of extensions in space and time comparable to those of our galaxy]. For the universe, the two directions of time are indistinguishable, just as in space there is no up or down. However just as at a . . . place on . . . earth we call 'down' the direction toward the center . . . , so will a living being . . . distinguish the time direction toward the less probable state [and call it the past]. By virtue of this terminology . . . small regions in the universe will always find themselves 'initially' in an improbable state. . . . That the transition from a probable to an improbable state does not take place as often as the converse can be explained by assuming a very improbable initial state of [that part of the universe] surrounding us . . .

Let us summarize. *Actual time extendedness* is implied in Boltzmann's argument (which would not make sense otherwise). *Irreversibility is considered to have indissolubly a cosmological and a biological aspect.* The time arrow is (locally) relative to the perceptions of the living beings there. *Mathematical expression of the time arrow is found not in the equations (which are reversible) but in the solutions selected as significant* 'here and now'.

These are quite decisive steps forward; one wonders why they had to be rediscovered again and again, in an avalanche of articles or books (which often make interesting reading), all agreeing in the essentials — even if perhaps disagreeing sharply in one important point or an other.

Additional remarks by Boltzmann are also worth reading, such as this one:

> This method seems to me the only way in which one can understand the second law.

and

> No one would consider such speculations as important discoveries or even — as did the ancient philosophers — as the highest purpose of science. However it is doubtful that one should despise them as completely idle. Who knows whether they may not broaden the horizon of our circle of ideas, and by stimulating thought, advance the understanding of the facts of experience?

CHAPTER 3.4

IRREVERSIBILITY AS A COSMIC PHENOMENON

3.4.1. LIMINAL ADVICE

It is more and more widely recognized that all the mutually consistent, locally observable aspects of physical irreversibility — entropy or probability increase, wave retardation, preponderance of 'causality' over 'finality', of statistical prediction over statistical retrodiction, of information-as-knowledge over information-as-organization, darkness of the sky at night, Hubble redshift — are not only tied together, but are also all tied with the cosmological phenomenon of world expansion.

While the astrophysicists — Bondi, Gold, Narlikar together with Gal-Or and others — make the connection from above, I shall tie it upwards starting from Laplace's formula for the 'probability of causes', and the 'theory of branch systems' as discussed independently by Reichenbach and others.

3.4.2. BRANCH SYSTEMS. THE 'STATISTICAL BIG BANG'

That 'isolation' of a system is an unwarranted idealization in problems of statistical irreversibility has been widely recognized. Becoming aware of this immediately ties local irreversibility to cosmological irreversibility. In Section 3.3.3 a characteristic quotation from Borel has been produced, quite astonishing in its mixture of insight and blindness: insight as to what are the significant elements, blindness as to how they should be neatly put together.

Rather than upon 'weak continuous interactions' we shall now concentrate upon 'strong interactions limited in time', as a way of going directly up from local to cosmological irreversibility. This is known as the 'theory of branch systems', discussed by Reichenbach (1956), Popper (1956—1957), Grünbaum (1963) and myself (1964).

When asking how statistical mechanics really formalizes irreversibility we have only found a terse decree stating *blind retrodiction forbidden*, formalized in Laplace's formula (3.1.6) for the 'probability of causes'.

When asking what reasons are adduced in support of this decree, we found that Bayes's extrinsic probabilities (as present in the Bayes–Laplace formula) do represent the initial interaction that has generated the system. And then, when asking why such an interaction is retarded, not advanced, we just found that, if it were advanced, the entropy of the combined system would go down. So we were back at the starting point.

Are we thus stuck in a circle, like in the well-known egg-and-hen problem, the question being 'which came first?'

Not exactly. At each step, we had to include our system in a larger one. So there is no end to this query till the whole Universe is brought in. The whole of its spatial extension, because the smallest of the evolutions considered (be it that of a radionuclide bathed in the neutrino flux) is in fact coupled to that of the Universe, and, through it, to all the other partial evolutions exhibiting the same arrow. And the whole of its temporal evolution is replicated, because all partial evolutions repeat in mass production the similar process having generated each of them, like the branch repeats the tree and the twig the branch. Whether they are limited in time but strong, or weak but continuous, these interactions couple any partial system to the whole universe. That is the reason why all local evolutions do exhibit the same 'entropic' or 'retarded' time arrow, which is the Universal Time Arrow. Here we have a topical case where explanation rests on an 'action of the whole upon its parts'.

The Universe is thus portrayed as a huge 'branch system', an enormous tree; or else a Niagara cascade, the gigantic flow of which again and again divides into rivulets. It is the Universal Negentropy cascading, another name of which is 'cosmological causality' — that is, *factlike retarded causality*. At some time ahead of us the prodigious fireworks (to take still another example) must have flashed. If one pauses to wonder, the conclusion is inevitable: *it must all have started as a big explosion*. This is implied in both the twin concepts of *retarded causality* and of *entropy increase*. And evidence of such a momentous fact is quite trivial and close at hand: we just have to pick it up, like a pebble on a beach. Its name is *one-way local irreversibility*, that is *retarded causality*.

The present state of the Universe consists of a tremendous collection of improbable situations: radiating stars, radionuclides, pairs of systems evolving separately, until, suddenly, their interaction goes from very weak to very strong. All these are nothing else than vestiges of improbability, remains from the towering Negentropy given some time ago. They are fossils and witnesses of a 'Big Bang' given in a nutshell.

That the statistical Big Bang just considered must be the same one as the cosmological Big Bang derived from general relativity is a natural assumption that has occurred to many.

If this is so, should we speculate that gravitation also is a statistical phenomenon? Borel wrote in 1914 that "the statistical explanation of Newton's law would be regarded as progress". Today, as is well known, much thinking is being devoted towards producing a quantal theory of gravitation which, of course, would just be that, with gravitons as the particle side of gravitational waves. However, even before this is achieved, it must be definitely stated that a strong connection exists between the phenomena of the world expansion, of entropy or probability increase, and of wave retardation. As Davies (1981, p. 109) puts it: "The origin of all thermodynamic irreversibility . . . depends ultimately on gravitation."

3.4.3. UNUSUAL STATISTICS OF SELF-GRAVITATING SYSTEMS

As emphasized by Tolman (1974), the statistical mechanics of gravitating systems has unique features. Quoting Davies (1974, p. 97):

> The relevance of gravitation to the subject of entropy increase is twofold. Firstly, the growth of density perturbations by gravitational concentration of the cosmological fluid is responsible for the existence of the inhomogeneous structures such as stars [leading to] the production of entropy by starlight. Secondly, gravitational condensation and collapse [in] itself [entails] an increase in entropy.

This is because, like electromagnetism, gravitation is a long-range field, and because, unlike electromagnetism where positive and negative charges cancel each other in the large, the monopole contribution of the gravity field is felt everywhere, as is ominously expressed in the cosmological formula $GM/c^2R \simeq 1$ which holds for static or permanent regime models of the Universe.

Gravitational collapse exists, of course, in the Newtonian theory, but not with the dramatic traits it assumes in the general relativity theory, where contraction of a spherical source below the 'Schwarzschild radius' $r = 2Gm/c^2$ produces the startling object named a 'black hole' — an unsaturable 'cold sink', in thermodynamical terminology.

Coming back to Davies we read:

The application of thermodynamics to a self-gravitating system is a highly non-trivial matter . . . The long range nature of gravitational forces introduces novel features into . . . equilibrium properties . . . A non-gravitating system subject to only short range forces can be maintained in equilibrium by . . . control over boundary conditions [whereas] a self-gravitating system can only be constrained . . in equilibrium by internal forces, so that these systems . . . tend to collapse . . . Formally, the entropy of a gravitating system has no maximum . . . ; complete collapse is permitted . . . Internal forces usually prevent this fate, allowing local entropy maxima. These metastable states are of several varieties for example, stars [of various types].

This being so, what can be said of the Loschmidt-like and Zermelo-like behaviors of the Universe as a whole?

Many authors have discussed the Zermelo-like behavior of oscillating cosmological models, but *a priori* this cannot make much sense. According to general relativity, the cusps existing at both ends of an arch of the curve figuring, in terms of the 'cosmological time', the variation of the 'Universe's radius' $R(t)$, are singular points; so how could 'the Universe go through them'? All this is very much idle speculation. The tremendous depths of space and time into which our telescopes and radiotelescopes plunge show nothing else than expansion. Even if observed estimations of the so-called 'deceleration parameter' did suggest that our Universe is 'closed' and 'oscillating', large extrapolation would still be very hazardous.

This is to say that Davies's (1974, pp. 91 and 97) considerations regarding the loose coupling between various relaxation times — for example, those of viscosity or of retarded light emission as compared to the 'Hubble time constant' — are not very compelling to me. Instead I would stress that his strong desire to 'preserve causality' (that is, *retarded* causality) hardly fits the very concept of an oscillating universe, where time symmetric action-at-a-distance interactions obviously are the adequate concept.[1] But, again, discussing the Zermelo-like behavior of the whole Universe does not make much sense.

3.4.4. LOSCHMIDT-LIKE BEHAVIOR OF THE UNIVERSE: BIG BANG AND TIME REVERSAL

Let us recall Loschmidt's 1876 recipe: At *time t reverse exactly all velocities of the system*. How do we do that with the Universe?

Like Newton's, Einstein's equations for the generation of the gravity field are insensitive to the time arrow, so that any solution of them has a time reversed twin. This exactly is Loschmidt's argument.

On the cosmological scale, a wide class of such solutions is expressed in the well-known Robertson—Walker metric

$$d\tau^2 = dt^2 - R(t)[(1 - ar^2)\, dr^2 + r^2(d\varphi^2 + \sin^2 \varphi\, d\theta^2)]$$

where 'comoving coordinates' are used, with respect to the mean motions of matter and radiation. The constant a can assume the values -1, 0, or $+1$. If $a = -1$, the 3-space is closed and spherical, with R denoting the radius of the hypersphere. If $a = 0$, the 3-space is Euclidean, and the Universe is de Sitter's; if $a = +1$, the space is open and hyperbolic. In both cases $a = 0$ and $a = +1$, R is a scale parameter.

The Loschmidt behavior shows up in the time reversal operation $t \rightleftharpoons -t$ the meaning of which, however, must not be misunderstood. *Relativistically speaking the Universe is a four-dimensional object* so that, as Weyl (1949, p. 116) put it: "the objective world simply is, it does not happen." Therefore when one reads that 'the universe we live in is expanding' one should understand that 'inside our time asymmetric four-dimensional Universe, the subjective time of our consciousness is experiencing $R(t)$ as increasing'.

An explicit expression for $R(t)$ can be obtained by integrating Einstein's equations after inserting in them 'constitutive formulas' (involving mass densities and pressures) for the momentum-energy tensor. And, of course, the time reversal operation must be understood as affecting both the randomized equations we were speaking of, and all the elementary evolution equations subsumed in them. *That* is the essence of the Loschmidt argument.

What follows then? All radioactive disintegrations are turned into syntheses; all radiating stars are turned into active sinks similar to the one we see when emptying our bath; all determinations by arbitrary initial conditions are turned into ones implying final conditions; retarded causality becomes advanced causality, and viscosity, antiviscosity; information as gain in knowledge is changed into information as organizing power; the Hubble redshift becomes a blueshift which, however, is not seen, because it is 'anti-seen'. Finality rather than causality becomes the prevailing concept. Thus all the various, self-consistent, time arrows, as they are locally experienced, turn out to be multifarious aspects of one and the same cosmological time arrow.

And this time arrow ultimately consists of this: in our subjective past, some 10^{12} years ago, there 'was' an exceedingly hot negentropy source;

and in our subjective future there is a universal, intensely black and cold energy sink, called 'expanding space'.

This is what Bondi, Gold, and others have been saying — and, as it seems, saying correctly.

The prominent role in this of the gravity field has been emphasized by Davies and others. Gal-Or (1981) proposes a name and a doctrine: "gravitism", or even "dialectical gravitism". I must say, however, that the dialectics of my own gravitism widely differ from his, as will be explained in Chapter 3.5.

3.4.5. THE OLBERS PARADOX

On a beach many pebbles can be picked up. Also everybody sees the sea. The darkness of the night sky is another quite ordinary proof of cosmological irreversibility.

The argument was given first in 1720 by Halley in a cryptic form; then, in 1744, quite explicitly by the Swiss amateur astronomer Loys de Chéseaux; and finally in 1823 by Olbers, a medical doctor by profession, and astronomer by dedication.

Anyone who has been within a blizzard up in the mountains, where the ground is white also, knows that everything beyond a very short distance is completely hidden, as if by a uniform white blanket. Only the near objects are visible, only if dark, brilliant, or coloured. Everything else is nothing but white light. This is because every light ray reaching the eye has been scattered many times, with no frequency change.

Similarly, reasoned Halley, de Chéseaux and Olbers, if the brilliant stars we see are of brightness similar to our Sun, and are distributed regularly throughout the infinite Euclidean space, the night sky (and, for that matter, the day sky also) would have the same uniform brilliance. Everyone having felt the warmth of the sunshine in Arizona can imagine easily how one would then feel on Earth . . .

Lambert's thoughts on photometry were published in France in 1784, and Kirchhoff's thoughts, which started the whole subject of thermal radiation, in Germany, in 1859. Olbers was well aware that his working hypothesis implies the existence of an isotropic, uniform radiation, as brilliant as our Sun's.[2] S. L. Jaki (1969, p. 135) writes:

The fact that every point in the sky is not an emitter of sunlight gave Olbers no small confort. He was not concerned though about the consequences of scorching heat and blinding brilliance resulting from a uniformly brilliant sky. Almighty God, he noted with piety, would have easily adapted the human organism . . . What alone seems to have worried Olbers . . . was the fate of astronomy . . .: 'We would not know anything about the fixed stars; we would detect our own sun only through its spots and not without troublesome efforts; and we would distinguish the moon and the planets . . . only as darker disks. . . .' One may understand [continues Jaki] Olbers's relief about the fact that astronomers have . . . a dark night sky at their disposal. A man who trained himself to four-hour sleep that he may turn a long-day medical practice into an observer of the sky at night had . . . to be gratified with the actual darkness of the night sky.

So the fact is that, instead of a very bright thermal radiation, constant in space and in time, we experience, always and everywhere, only retarded electromagnetic radiation, together with the damping Lorentz force. Therefore there lies in the future of any electromagnetic radiation a general absorber, so nearly black that it prevents any one of our local black absorbers from capturing into itself an advanced wave from the past. In the wording of thermodynamics, there lies in the future of any hot source of radiation a universal cold sink, nearing absolute zero.

Today we know that this universal black and cold sink is the expansion of space.

Let us note that the Halley—de Chéseaux—Olbers thinking had to consider gravitation. If the stellar and nebular system were thought of as finite inside an infinite Euclidean space, this distribution could not last very long: any sort of friction would bring together the slowly moving bodies, while the fastly moving ones, including photons, would escape at infinity. Therefore the distribution of the stars had to be postulated as uniform in space and time, whence necessarily followed the 'Olbers paradox'.

3.4.6. THE 2.7 °K COSMOLOGICAL RADIATION

Today the night sky is not absolute black; the oven we are living in is not yet completely cold. This is Penzias's and Wilson's 1965 discovery. The primeval fireball has left a 'universal' remnant: thermal radiation at 2.7 °K, providing a 'preferred rest frame', with respect to which our Galaxy is moving at a velocity of some 600 km/s.

3.4.7. BUILDING ORDER BY FEEDING ON THE UNIVERSAL NEGENTROPY CASCADE

This is what all living beings are doing — and even what Carnot's heat engine was doing in a rudimentary form. Bergson, among other philosophers or philosopher-biologists, has elaborated upon this in his *Creative Evolution*. Prigogine and his school (Nicolis and Prigogine, 1977) have produced an elaborate formalism of the non-equilibrium thermodynamics appropriate for handling this matter, which turns out to be surprisingly large, extending from self-organizing Bénard vortices, to intricate oscillating physico-chemical reactions, and (as they claim) to the very organization of living organisms.*

3.4.8. CONCLUDING THE CHAPTER

At cosmic time zero there has been given in a nutshell (according to our human parlance) an intensely hot 'source', a tremendous negentropy or information package, the 'explosion' and later expansion of which 'has generated' and 'is generating' the Universe, yielding as by-products all sorts of 'order generating' evolutions which feed upon it.

In the words of Mehlberg (1961) this 'universal irreversibility' is 'factlike, not lawlike', as it shows up not in the equations, but merely in the convention chosen for directing the time axis with respect to the asymmetry existing in the solution retained.

World expansion, entropy or probability increase, wave retardation, cause—effect succession, information-as-gain-in-knowledge, merely are various aspects of the 'universal irreversibility' *where two essential ingredients enter: a cosmological one* just studied, and *a psychological one* to be discussed in the next chapter, according to which 'we learn by living inside our cosmos'.

* See 'Added in Proof' p. 319 ff.

CHAPTER 3.5

LAWLIKE REVERSIBILITY AND FACTLIKE IRREVERSIBILITY IN THE NEGENTROPY-INFORMATION TRANSITION

3.5.1. PRELIMINARY CONSIDERATIONS

For the sake of brevity the prerelativistic wording according to which time is conceived as 'flowing forward' has been used until now. The nature of the problems facing us now requires that we change both our wording and point of view. Rather than observers on a bridge gazing at the rushing river 'time', we should think of ourselves as being immersed in the river, swimming 'upstream'. As Ronsard the poet put it: "Time is going by, going by, Madam! Alas, not time: we are going by."

Our awareness of time is such that 'we', at our conscious 'now', are exploring 'forward' our personal time-extended history, as it exists all at once — 'at once', of course, *not* being synonymous with 'at the same time'! This is the way the relativity theory sees things. And we have seen in Section 3.3.9 that Boltzmann had come to a similar view, by pondering the question of lawlike symmetry versus factlike asymmetry in time.

For one thing, how is it then that all of us, living beings, exchanging 'information' (mainly by various sorts of waves) feel that we are swimming together along the river?

The gigantic Universe, as we observe and understand it presently, differs from the one envisaged by Boltzmann nearly a century ago, in that *it displays one and the same time arrow throughout*. Therefore no fundamental difficulty associated with an exchange of information (a problem implicitly felt by Boltzmann (1964, pp. 446—448) and explicitly discussed by Wiener (1958, pp. 44—45)) can arise. Disregarding problems raised by the immense transit times involved, citizens of the world's galaxies could in principle understand each other.

This being said, the reason why we earthlings feel that (for all practical purposes) we are living 'at the same time' is that the space and time units we find 'practical' (say, the meter and the second) make the speed of light exceedingly large. We can speak easily to each other on the telephone from continent to continent, and even, via microwaves, from a spacecraft to Earth and vice versa. The transit time is really not felt. This must be an adaptative trait of earthly living beings, a trait facilitating life in society.

By immersing living beings within the relativistic four-dimensional spacetime, we are *ipso facto* rephrasing the 'lawlike reversibility versus factlike irreversibility' problem in the form: Why is it that we are all, so to speak, rushing together, like a school of dolphins, through spacetime in the direction where we experience waves as retarded, entropies as increasing, and the world as expanding?

Information theory can teach us something here. Among many other examples consider the case of telecommunications, where a message, encoded as a 'signal' endowed with a Shannon negentropy N (a 'structural negentropy', in the wording of Brillouin (1967)) runs along some channel. When 'received' it is 'decoded', and (we hope) 'understood', yielding an information-as-cognizance I_2. When 'emitted' it has been 'coded', that, is 'expressed', after having been 'conceived' in the form of an information-as-organizing-power I_1. Symbolically

$$I_1 \rightarrow N \rightarrow I_2.$$

Lawlike reversibility and factlike irreversibility are synthetically expressed in the form

$$I_1 \geqslant N \geqslant I_2.$$

The macroscopic 'facts' are that there is noise along the channel and that errors creep in both at emission and at reception (some of them attributable to lack of attention), whence the inequality signs. If everything were perfect, equalities would hold.

Let us emphasize, before we proceed, that the sender and the receiver may be the same person, as when one uses a computer. Thus any quantitative measurement, even with more social than scientific meaning (such as, say, the value of the inch in centimeters); any social communication, or even any private deliberation requiring writing or recording; any thinking, even one using only the brain; all these need physical agents to allow the mind to enter the material world. Therefore factlike irreversibility creeps in, the material world retaining a tax upon any message it handles. This is one aspect of the very far-reaching discovery of an 'equivalence' between negentropy N and information I.

In the words of Brillouin (1967), the inequality signs express a "generalized Carnot principle" according to which, in the *de jure* reversible transition

$$N \rightleftharpoons I,$$

the upper arrow prevails in fact.

It seems, therefore, that, once the existence of the cosmological time asymmetry is recognized, we can restate the 'factlike irreversibility law' in the form that *information as knowledge increases in living beings.* We can learn things by reading (and understanding) a book from page 1 to page ω; but we cannot erase our previous knowledge in some field by 'anti-reading' a book from page ω to page 1! It seems that perhaps the World, as parametrized by the spacelike surfaces $\mathscr{E}$ considered in Part 2, is like a big book we are perusing, by the very fact that we are 'living through it'.

An objection that has been levelled against the lawlike equality $N = I$ is that a lecturer does not lose the information he conveys to his audience, although in principle the N listeners can each gain this information I. Brillouin's answer is, first, that one must think of the process as comprising N channels going from the lecturer to the listeners; and the fact is that (in the absence of a microphone) the speaker must speak almost N times louder than for just one listener (the 'almost' taking care of absorption of sound by the walls, or by empty space). If the lecturer uses a microphone, this device feeds upon a negentropy source — as does in fact the lecturer himself. The information he conveys is not lost to him, because his organism is so built that it just borrows the 'broadcasting negentropy' from its biological reserves; he must have had a restful night, and perhaps a good breakfast.

One more discussion is needed for settling commonsense matters before moving to deeper ones.

Innumerable authors have discussed the problem of free action by comparing it to, say, the problem of driving an automobile. The automobile's tank contains fossil fuel (the quality of which has been improved by the gasoline industry). By turning the contact-key the driver starts an exothermic reaction, which he controls by means of the accelerator pedal, and also with the wheel. What we do have here is control of very large energies by means of very small ones — all the more so if we were piloting a large jetliner! But this is not the solution of the problem of free action — any more than Poincaré's discussion of small causes ending in large effects radically solves the problem of 'chance'! Hiding the large behind the small is playing the ostrich, or sweeping the dust under the rug. The dust will inevitably keep trickling out; and this, no less inevitably, will make the audience — and the lecturer also — cough from time to time.

The problem here is essentially the same in the discussions both of free action and in that of chance. *It is a problem in information*: information-as-organizing-power in one case, information-as-lack-of-knowledge in the other.

This leads us straight to a discussion of the matter known as 'Maxwell's demon'.

Maxwell's demon is a minute goblin who can play with molecules as a soccer player does with a football. Maxwell sees him as standing near a small aperture in the wall separating two gas containers, and being able to move at will a little rigid, impenetrable, and massless door. Thus he can, for example, separate fast molecules on one side and slow ones on the other, or separate two mixed gases such as oxygen and nitrogen. So this demon can make the Clausius entropy and the Gibbs mixing entropy decrease. In this he is more clever than the airplane pilot, who only borrows negentropy from the universal cascade.

However, there are objections to this, the first being that the little door is a miraculous sort of door: both massless and impenetrable! A door is made of matter. If it is in thermal equilibrium with the gas it will fluctuate, as Smoluchowski (1912) and Demers (1944; 1945) have remarked. One has then to face the dilemma of either high precision in position — that is, a strong hold — or else weak binding, whence imprecision in position. So finally the demon, no less than the airliner's pilot, has to tap at will some energy source.

But that is not all. If the demon wants to see the molecules, he cannot just sit inside thermal radiation at the temperature T. So Brillouin (1967, pp. 164—166) gives him a little torch feeding on some negentropy source. But then, discussing the overall information and negentropy balance, Brillouin ends up with the conclusion that the sum $N + I$ has to go down, as required by the "generalized Carnot principle". Then he triumphantly writes that "the demon has been exorcised".

For one thing, a demon defined as so incarnate that, like any one of us, he depends on the "generalized Carnot principle", really need not be exorcised! He could just as well be a robot. And in fact, throughout the large realm of experimental physics and of technology, thousands of robotized Maxwell demons are harnessed to various tasks. Of course, there is then the question of conceiving and constructing these robots; so, once again, we end up sweeping the dust under the rug — and, this time, an enormous rug.

With this we have practically ended the preliminary tour we had to take before going into the more fundamental issues facing us now, the two main ones being:

1. How should we understand the connection between mind and matter, if matter is conceived as space-and-time extended?
2. How should we fundamentally understand the lawlike reversibility between negentropy and information, $N \rightleftharpoons I$?

In both cases we shall put at work the fundamental lancet we have assumed, from the start, to be at hand: *intrinsic symmetry between blind statistical prediction and retrodiction.*

As we shall see, use of this lancet will *necessarily* involve confirmation of 'the claims of the paranormal' — precognition and psychokinesis.

3.5.2. IS THE SUBCONSCIOUS MIND TIME-EXTENDED, AS MATTER IS?

If we must conceive of the connection between mind and matter in terms of the four-dimensional spacetime geometry, there is something akward in the psychological 'now' with its absence of temporal thickness. Bergson feels that this must be an idealization, and that, for instance, the perception we have of the colors of the spectra might well imply that 'trillions of oscillations' are treated together — a Fourier analysis we would say mathematically. Thus Bergson felt that it is only 'pure consciousness', or, as he called it also, 'attention to life', that tends to focus sharply within a very narrow sliver of time thickness. The subconscious mind, he thought, relaxing from sharp 'attention to life', should rather be time extended — but only, said he, towards the past, because he considered the future to be pure 'potentia'.

Today, however, we know that relativistic covariance requires such a time extendedness to penetrate *both* the past and the future. At first sight this may seem to contradict the existence of free will. This is not so, however, as will be explained in Section 4.7.8 which aims at resolving the central 'paradox' inherent in relativistic quantum mechanics.

Postponing this discussion, we can presently think of spacetime as a sort of book, the pages of which are the layered spacelike surfaces $\mathscr{E}$ used throughout Chapter 2.5, which our conscious 'now', or "attention to life", is 'reading', each in turn, in the 'right' order. There may well be, in this comparison, a hint as to the type of connection that exists between temporally arrowed causality and logical causality, as discussed in Section 3.1.12.

Suppose now that we take cognizance for the first time of a logically articulated treatise on a subject we have much interest in. Then, perhaps, before undertaking a thorough study, line after line, page after page, we do not resist glancing through the pages. As a general rule, we shall not really understand what we see. However, if we are strongly motivated, we may sometimes get an intuitive feeling of what it is all about – and a feeling that was most apposite, as we recognize when we carry out our study. Even so we cannot place our intuition in any context; it floats in the air like those mirages displaying something which exists behind the horizon.

This is exactly how 'precognitions' (of which there is abundant circumstantial evidence) look. It would be irrational to deny their existence, first because (as just said) their possible existence is quite rational indeed, and, second, because there is such an abundant record of them.

Then, if precognition is a normal attribute of the subconscious mind, geared as it is to the material world, why is it that its emergence at the conscious level is so unusual? It need not be recalled that repression of dangerous subconscious knowledge below the conscious level is a psychological rule; and it is almost self-evident that extensive conscious foreknowledge might be extremely destructive to spiritually undeveloped minds.

Concluding this section, and without implying that the authors I am now quoting do agree with what has just been said, I produce a few texts drawing very well the distinction between the extended time of the Universe and the unextended time of consciousness.

3.5.3. LAWLIKE REVERSIBILITY BETWEEN NEGENTROPY AND INFORMATION

Descartes (and this, for more than one reason, may come as a surprise) has never varied in his conviction that our mind directly moves our body 'at will'; that is, definitely *not* as a ball moves another ball when hitting it. In a letter to Arnauld dated 29 July 1648 this is what he writes (1971, p. 219, Letter 302): "That our mind, which is incorporated, can move our body, is shown not by some ratiocination or comparison taken from something else, but by the most certain and evident everyday experience." The argument is exactly the same as in the "cogito" — direct internal evidence — but, here, it is applied to willing awareness rather than to cognitive awareness. In other words, it is applied to the *information* → *negentropy* rather to the *negentropy* → *information* transition. In an other letter, written to Elisabeth, he writes (1974, p. 663, Letter 525): "The principal cause of our mistakes resides in trying to apply notions in explanations not belonging to them, as when one tries to conceive how the soul moves the body in the same manner as a body moves an other body."

As implied in Brillouin's "generalized Carnot principle", there is a one-to-one connection between (retarded) causality and information-as-gain-in-knowledge, just as there is a *de jure* symmetric one between (advanced) finality and information-as-organizing-power. One often wonders why finality (which intuitively seems so obvious in many biological phenomena) is so difficult to *demonstrate* by an argumentation. The reason for this is quite clear to me: by virtue of their very natures, *the evidence of causality belongs to cognitive awareness just as the evidence of finality belongs to willing awareness* — as is cryptically summarized in the $N \rightleftharpoons I$ formula. As Descartes says, it is a mistake to try explanations not fitting their object . . .

An eminent physicist who, also arguing on the basis of intrisic formal symmetries, came to the conclusion that there must exist a direct action of mind upon matter (symmetric to the direct action of matter upon mind named cognizance) is Eugene Wigner (1967, pp. 171—184). To a few technical arguments, belonging mainly to quantum mechanics, he adds the general principle that "to every action there corresponds a reaction". Then he adds (tongue in cheek, as it seems) "Every phenomenon is unexpected and most unlikely until it has been discovered — and some of them remain unreasonable for a long time after they have been discovered."

The symmetry we are speaking of is the one basically inherent in information theory, in the form of the reciprocal $N \rightleftharpoons I$ transition. In this we must understand both the lawlike symmetry just mentioned, and the factlike asymmetry consisting of an extremely large proponderance of the $N \rightarrow I$ over the $I \rightarrow N$ transition.

Very much as we can understand how the Einstein 'equivalence' between time and space is largely hidden from us 'in fact', merely by remarking that, in the formula $d\mathbf{r}^2 - c^2\, dt^2 = 0$, c, as expressed in 'practical' units, is 'very large', similarly we can clarify the present point by resorting to 'practical' units. The conversion coefficient between a negentropy N, as expressed in thermodynamic units, and an information I, as expressed in binary units, is proportional to Boltzmann's constant k:

$$N = k \ln 2I;$$

$k \ln 2$ is an exceedingly small constant,[1] with the value 0.956×10^{-16}.

As Gabor, quoted by Brillouin (1967, p. 168) puts it, the cybernetical discovery is that "we cannot get anything from nothing, not even an observation". In other words, cybernetics asks consciousness-the-spectator to pay her admission — at an extremely low price. But this alone confers existence to consciousness-the-actor — this time, of course, at very high wages, because the change rate plays the other way round.

What happens if we let k go to the limit zero? This is a traditional physical game (also played with the relativistic $1/c$ and the quantal h). By letting k go to zero we fall back on a once fashionable theory named "epiphenomenal consciousness", where gain in knowledge was rigorously costless, and free action rigorously impossible. Cybernetics says that now this time is over.

Cybernetics also makes quite clear why information-as-knowledge is so trivial, and information-as-organization so esoteric a concept: *it is a direct consequence of the (factlike) second law.* The man in the street buys a newspaper for a few cents in order to find 'information' in it, and many advertisements go straight to the waste basket. But what is costly in a man-made or a 'manufactured' object is the skill of the craftsman or engineer; even when the matter used is costly, this is because it is difficult to find or to refine it: again we have to 'pay the price' when producing negentropy.

By stating quite explicitly, as do Descartes, Wigner, and possibly others, that the key of free action resides inside the recesses of our nervous systems (as far as men and animals are concerned), in the form of a *direct* $I \rightarrow N$ transition, I am definitely not denying that amplifying devices, analogous to those used by pilots of large airliners, or to Bergson's (1907, Ch. 2) stored 'vital explosives', do exist inside living organisms. Most certainly they do exist. What I am saying is that the ultimate and quite hidden key to free action is of a radically different kind.

Then the question is: Can experiments be devised demonstrating a direct action of mind upon matter, occurring outside our animated bodies? In fact this sort of phenomenon, the name of which is psychokinesis, must be just as rare as the other 'paranormal' phenomenon, precognition. Here also, however, there exists an oral tradition of people believing they can influence the fall of dice or the stopping of a roulette ball.

Serious experimental work has been done in the field by J. B. Rhine (1964), H. Schmidt (1982) and others. Here I adduce as evidence, with his written permission, graphs of experiments conducted by R. Jahn (see Jahn *et al.*, 1986) at the Engineering Faculty of Princeton University (see Figure 11).

Concluding this section I quote Léon Brillouin (1967, p. 294) (without asserting, however, that he would have endorsed the preceding statements):

> Relativity theory seemed, at the beginning, to yield only very small corrections to classical mechanics. New applications to nuclear energy now prove the importance of the mass—energy relation. We may hope that the entropy-information connection will, sooner or later, come into the foreground, and that we will discover how to use it to its full value.

If and when this happens, let us hope that wisdom prevails!

3.5.4. 'SEEING IN THE FUTURE AND ACTING IN THE PAST'

Repression, not suppression is the 'Second Law's' statement.

Lawlike symmetry between the $I \rightarrow N$ and the $N \rightarrow I$ transitions implies that the converse of 'physical irreversibility'; that is, 'seeing in the future and acting in the past' is not strictly forbidden, but is in fact very much repressed. Somewhat similarly, antiparticles are much rarer than particles; their tracking or production require sophisticated experimental approaches; but their search has opened a scientific El Dorado.

Let it be made clear that psychokinesis *essentially* is retropsychokinesis. If, looking at a screen displaying the recording of yes-or-no outcomes as produced by a random outcome generator, an agent is able to influence the recording, as in Schmidt's (1982) or Jahn's (Jahn *et al.*, 1986) experiments, he most certainly acts upon the random event before its amplification — that is, in the past. Also, if influencing falling dice, he *must* do so ahead of their coming to a standstill.

I need not recall that, according to the views presented here, psychokinesis is (together with precognition) the key element inherent in finality: the hidden 'event' triggering the conspiracy of causalities as displayed in biological ontogenesis for example. To say that finality is a 'causality acting backwards in time' may be dangerously misleading. Rather, one should speak of finality as 'sucking from the future'. A permanent regime of hydrodynamic flux governed by sources and sinks is a better image of causality (the sources) and of finality (the sinks). Therefore, the worn out objection that 'causality as acting backwards in time is self-contradictory, because it could be used to cancel an already written history' completely misses the point. Being materially written as spacetime events, *a history is what it is*; in the future, it cannot turn out as other than what it will have been. But although, in some sense, everything is written down, nothing that 'happens' does 'happen' without some cognitive information flowing out of, and some organizing information flowing into, the cosmos. That a relativistically covariant formalization of a material world where real chance is an ingredient is possible, will be explained in Chapter 4.7. What is mathematically formalizable also is conceptualizable. But, right now, some essential elements, belonging to quantum mechanics, are missing.

Summarizing this section, precognition and psychokinesis are two closely associated phenomena, and indeed really two faces of the one (rare, or 'paradoxical') $I \to N$ transition. Very rare in the open, this phenomenon, however, is quasi-omnipresent (though very hidden) inside those extremely numerous chains through which order is extracted from the universal negentropy cascade.

The game of theoretical physics largely consists of extending the domain described by equalities, by defining concepts adequately filling some remaining inequalities. Thus it is that 'potential energy' has been defined, Poincaré (1906a, Pt 3, Ch. 8) insisting that the 'first law' largely consists of *definitions* such that 'something' is conserved: energy. And, of course, Fermi's neutrino has been introduced just in this fashion, preserving energy and angular momentum conservation — and then, later, lepton number conservation. Thermodynamicists often introduce an 'uncompensated entropy' just filling the hole left open by the inequality sign in the 'second law'.

After all, should not Brillouin's 'generalized Carnot principle' be written as $N + I = \text{const}$? Is there really a universal negentropy hemorrhage, as that of water leaking into a sand desert? Or else is not the seemingly lost information perhaps constantly coming back in other forms? Can the desert not bloom after all? This I leave as an open question.

3.5.5. A PROPOSED EXPERIMENT IN PSYCHOKINESIS

A very simple experimental scheme is here proposed, using a random event generator (REG), which, if successful, would clearly display psychokinesis as 'retropsychokinesis' or alternatively display backwards-in-time and faster-than-light telegraphing, without violating relativistic covariance.

Suppose we have at C one of those REGs now widely used for experiments in psychokinesis; they may be monitored by radioactive decay or (with just the same efficiency, and much more ease) by an electronic setup. From C two channels, CL and CN, diverge, transmitting identical messages, one of them being observed by an agent at L, the other one simply registered at N.

The agent influences the outcome as he sees it at L, and thus identically the other outcome registered at N. This occurs through *psychokinesis before amplification*, at C, so that the signal carrying a negentropy N (because random outcome is biased in it) 'travels' from L to N, via C, along two timelike vectors, once towards the past, once towards the future.

If the two lines CL and CN are identical, this is 'telegraphing outside the light cone', but not directly of course — somewhat like a sailboat which, by 'tacking about', can go up wind. If a delay is inserted in the CL line, the recording at N can be in the past of that at L; this is 'telegraphing backwards in time' — and a much more direct proof of retropsychokinesis than those previously produced.

This experiment is really not more difficult than those already successfully performed; I do hope that some laboratory undertakes it.

Against this sort of 'telegraphing faster than light' (which does not violate Einstein's prohibition against direct signalling outside the light cone, but does trespass against his taboo of 'signalling backwards in time'). P. Davies (1981, p. 124) has made the following objection:

Autocidal machines . . . are programmed to self-destruct at two o'clock if they receive a

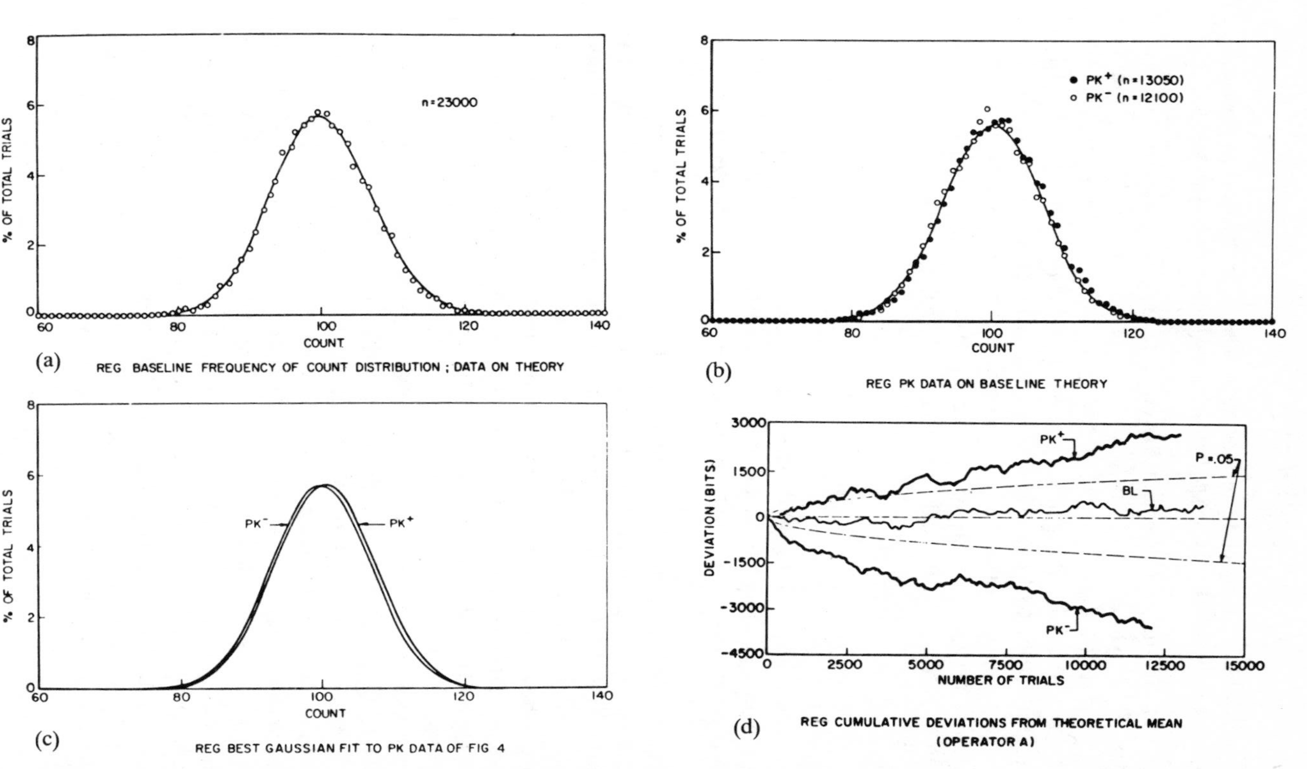

Fig. 8. Four graphs pertaining to a series of experiments in psychokinesis performed by R. Jahn's group in Princeton: (a) 'natural' Gaussian distribution produced by a 'random event generator' (REG); (b) series of outcomes as biased by psychokinesis towards one side (PK^+) or the other (PK^-); (c) Gaussian fits to biased data; (d) the corresponding cumulative deviations.

signal at one o'clock transmitted by them at three o'clock . . . This obvious contradiction seems to rule out reversed time signalling and hence faster than light messages.

There is in this, as it seems to me, a misunderstanding of the question. As previously said, history, as recorded, is one. 'Telegraphing backwards in time' does *not* mean reshaping history, it means *shaping* it — as when the fully fledged bird is said to be the 'final cause' of its development from the egg. Direct proof of this sort of phenomenon essentially requires repeated observation of biasing 'at will' an independently assigned 'blind predictive probability distribution. Again, the suction by hydrodynamical sinks is a good analogy here.

3.5.6. CONCLUDING THE CHAPTER, AND PART 3 OF THE BOOK

What sort of relation does exist between the cosmological irreversibility and its universal time arrow, as discussed in Chapter 3.4, and the mini-negentropy sources, with their lawlike $N \rightleftharpoons I$ symmetry, as discussed in this chapter?

It seems that the mini-sources are present almost everywhere and at every time, building, in their own ways, order from the negentropy cascade — Bénard whirlpools, oscillating physico-chemical reactions, and many sorts of Prigogine-like sequences; also, cosmological generation of particles, atoms, molecules, organic molecules; and astronomical building of stars, planets, and life on planets, starting as in the Miller experiment; producing phylogenesis, ontogenesis, plant, animal, and man activity.

No sharp barrier has ever been drawn between the living and the non-living: on which side should we put the virus? It is said that the Japanese workers in the automobile industry feel that there is some sort of rudimentary soul inside their robots; they may not be quite wrong. *Operationally* speaking, where is the difference between the 'materialist' believing that 'the evolution of matter' finally produces feeling in animals, reasoning in men, and that cybernetics will some day produce self-programming and sentient machines; and the 'spiritualist' who sees, conversely, matter as a by-product of 'mind' in a broad sense? The discussion is not on facts, it is in interpreting them — which can make a very big difference when pursuing scientific work. The difference is not in the physics, it is in the metaphysics. Then, as both the materialist and the spiritualist admit that no clearcut severance can be drawn between the non-living and the living, it can be said that life is virtually present everywhere, as Bergson says of 'consciousness' that, in plants, it is "rather asleep than absent" (1907, Ch. 2). To those objecting that 'consciousness' (in a broad sense) requires the existence of a nervous system, Bergson answered that the same argument applied to digestion would lead to deny it in organisms having no stomach. But, of course, amoebas do digest!

So, on the whole, cosmological irreversibility and elementary, intrinsic, *negentropy* ⇌ *information* reversibility do go hand in hand. And *at the very root of this line of thinking, ending in lawlike symmetry of the N ⇌ I transition, there is Loschmidt's* 1876 *reversibility argument.* There is also Laplace's 1774 discussion of the 'probability of causes', and Boltzmann's 1898 second thoughts on Loschmidt's argument.

PART 4

RELATIVISTIC QUANTUM MECHANICS AND THE PROBLEM OF BECOMING

CHAPTER 4.1

OVERVIEW

4.1.1. QUANTUM THEORY AS THE CHILD OF WAVE PHYSICS AND OF A PROBABILITY CALCULUS

While a detailed account of the history of quantum mechanics leaves the impression of disconnected advances through a jungle maze, when viewed in retrospect the main lines of the discovering expedition come out as having been fairly straightforward.

As displayed in the second of Planck's two famous 1900 articles, the newly born action quantum h obviously has two progenitors: the physics of electromagnetic waves and a *sui generis* probability calculus.

During its first twenty-five years the 'quantum theory' continued growing as it was born. Einstein's numerous contributions to the building of the theory in 1905, 1909, 1910, 1916, 1924—1925 have all proceeded from his profound knowledge of statistical mechanics and also, by the very nature of things, from a consideration of electromagnetic waves.

Two important issues are then at stake. The use of waves implies that relativistic covariance is an underlying problem. The use of waves *and* of probabilities implies that two 'factlike irreversibilities' are at stake, and that their mutual interplay also is an underlying problem. Let us consider these two points in turn.

Relativistic covariance of the quantum concept arose first with Einstein's 'Lichtquantum', endowed by him in 1905 with an energy, and in 1910 with a momentum. In covariant form Einstein's formula $p^i = \hbar k^i$ holds between the momentum energy p^i of the 'photon' (as Lewis baptized it in 1926) and the 4-frequency $2\pi k^i$ of the plane wave, associated in inertial motion; $\hbar \equiv h/2\pi$ is a 'modified Planck constant' now widely used.

In Einstein's 1910 and 1916 papers relativistic covariance is again an essential element.

Relativistic covariance is also central to Louis de Broglie's 1925 presentation of wave mechanics (his 1924 PhD Thesis). In it a 'matter wave' is associated with every material particle according to Einstein's formula $p^i = \hbar k^i$. However, the restriction to lightlike 4-vectors is of course removed.

As for the two aspects of the irreversibility concept — factlike preponderance of increasing over decreasing probabilities and factlike preponderance of retarded over advanced waves — both figured again and again in Planck's thinking, leading to the discovery of the quantum (Klein, 1970) and afterwards; they blend together in his definition of the entropy of a light beam, and in his demonstration that the entropy increases in phase coherent scattering.

In Einstein's thinking this two-headed beast also appears. From 1906 to 1909 Ritz and Einstein argued vehemently with each other, Ritz insisting that wave retardation was a prerequisite when deriving entropy increase and Einstein maintaining that wave retardation should be derived from probabilistic considerations. It did not occur to them that, far from being antagonistic, their positions were *reciprocal* to each other, because, while Einstein's photon had been identified, de Broglie's matter wave had not. Therefore it was not clear that scattering of waves and of particles do go hand in hand. It is only in the wake of the 1925 'quantum mechanics' that the whole matter was fully clarified.

By its very nature the probability concept mediates between the continuous and the discontinuous aspects of things. In 1926 Max Born turned to it for reconciling the continuity of the wave and the discreteness of the particle, which Einstein and de Broglie had tied together. Quite naturally he assumed that the intensity of the wave expresses the probability that the particle shows up at some point-instant. By doing this, without much fanfare he changed the rules of the probability game, introducing a brand new *wavelike probability calculus*, where partial amplitudes rather than probabilities are added, and independent amplitudes rather than probabilities are multiplied. This is because, as is well known from classical acoustics and optics, in phase coherent phenomena it is the amplitudes, not the intensities, which combine so as to 'interfere' in space or to 'beat' in time. Jordan systematized the whole matter in that same year (1926).

As it 'corresponds' to a classical intensity, Born's probability is (loosely speaking) a squared amplitude, thus comprising diagonal and off-diagonal terms. If alone, the diagonal terms would yield the old, classical rules. But the off-diagonal terms are there — even if rendered invisible by an *ad hoc* 'change in representation'. Via 'interference' or 'beating' they entail the thousand and one 'paradoxes' of quantum mechanics, all very well confirmed experimentally — the thousand and oneth being the 1935 Einstein—Podolsky—Rosen correlation as discussed in Chapter 4.6.

Born's and Jordan's 1926 wavelike probability calculus also provided the answer to the irreversibility enigma. As explained independently by Fock in 1948, by Watanabe in 1955, and others (for example, my article, 1964), retarded waves are used in quantum mechanics for statistical prediction and advanced waves for statistical retrodiction, so that the two macroscopic, 'factlike' statements that blind statistical retrodiction is unphysical and that advanced waves do not appear are *reciprocal* to each other. Also, J. von Neumann's (1932, Ch. 5) quantal version of Boltzmann's 'H-theorem', according to which the entropy of a quantal ensemble increases if identical measurements are made at time t on all its members, is derived from the tacit assumption that the measurement is used for prediction, not for retrodiction, so that the wave equation is integrated forwards, not backwards.

Thus, the quantum theory is essentially a combined wave and probability theory. In its most advanced version it is a 'theory of quantized waves'. As such, it formalizes a *universal spacetime telegraph transmitting information.* So do all telegraphs. But this one has properties quite *sui generis.* It obeys Born's wavelike probability calculus; and it possesses a relativistic invariance stronger than the macroscopic one, named 'Lorentz and *CPT* invariance', to be discussed right now.

4.1.2. MACRORELATIVITY AND MICRORELATIVITY, LORENTZ-AND-*CPT* INVARIANCE

Relativity theory is defined as expressing the invariance of physical laws under changes of inertial frames, which are interpreted as Cartesian tetrapods in a pseudo-Euclidean spacetime. Thus, assuming continuity of the transformation group, a change of inertial frame is figured as a rotation of the tetrapod.

In ordinary, Euclidean three-dimensional space, a rotation of a Cartesian tripod cannot exchange right and left. For this exchange a space symmetry, also called 'parity reversal' and denoted P, is needed; it is a 'discrete symmetry'. As classical physical laws are expressible in 'intrinsic' geometrical terms, their invariance under rotations of the tripod is assumed. Their invariance under a parity reversal may need some discussion. Notwithstanding what a first glance may suggest, the

Ampère—Laplace 'electrodynamics' is parity invariant, because *two* vectorial cross-products intervene: one in the generation of the field, and one in the force acting on a current element. Therefore Coulomb's 'magnetic charge' must be defined as a 'pseudoscalar'.

Loschmidt's 1876 time-reversal invariance essentially states the symmetry of the transition probability P_{ij} in the collision of two molecules. Such a trait appears in a very large class of probabilistic theories, including that used in quantum mechanics, to be discussed now. As we have seen, even if a transition probability P_{ij} were not symmetric in i and j, it could be interpreted either predictively or retrodictively, as explained in Section 3.1.10. If $P_{ij} = P_{ji}$, the 'principle of detailed balance' holds. Loschmidt's statement is that *microphysical laws are T-symmetric*, while 'in fact', as explained in Chapters 3.2 and 3.3, macrophysical laws are not.

Now we come back to the relativity theory, and emphasize that *a sharp distinction must be made between the classical or macrorelativity theory and the quantal or microrelativity theory.*

A continuous rotation of the Poincaré—Minkowski tetrapod can neither exchange right and left in the (xyz) subspace nor reverse the time arrow. Therefore it is called the 'orthochiral' and 'orthochronous' Poincaré—Lorentz transformation. Leaving aside the question of right and left exchange, this is exactly what is needed for expressing the invariance of macrophysical laws, including the irreversibility law in its various aspects. So, *macrorelativity invariance is invariance under the orthochronous Lorentz group.*

But of course there must exist a 'microrelativity theory', where Loschmidt's T invariance finds its covariant generalization. It would seem at first sight that PT invariance is the needed generalization, but Nature has more subtlety; the existence of antiparticles brings in something specific.

After de Broglie's wave mechanics, the next confrontation between the quantum and the relativistic concepts came with Dirac's 1927 electron theory, with its negative energy states interpretable as positive energy states of a positively charged anti-electron. This 'positron' was first observed by Anderson in 1935. Stueckelberg in 1942, and then Feynman in 1949, showed how the positron is elegantly and efficiently interpreted as a particle with a 'momentum energy pointing backwards in time'. More generally, the various 'spinning wave equations' allow the symmetrical existence of twin 'particles' and 'antiparticles', with

4-frequencies respectively 'pointing forwards or backwards in time'. It may happen that particle and antiparticle are physically indistinguishable from each other, which is the case with photons.

Geometrical intuition then has it that, if macrophysical laws are invariant under (hyperbolic) rotations of the Poincaré—Minkowski tetrapod, microphysical laws should be invariant under full reversal of the tetrapod — or rather, in the 'active' interpretation, under "strong reversal" of motions. These we denote $\Pi\Theta$.

Strong reversal has two effects. The one is reversal of the network of collisions *à la* Loschmidt, implying exchange of emission and absorption processes and, in quantum mechanics, of 'preparations' and 'measurements'; this we denote PT and call *covariant motion reversal.* The second effect is reversal of the arrows of 4-vectors from which, according to the Stueckelberg—Feynman description, particle—antiparticle exchange arises, which is denoted C. Therefore

$$(4.1.1)\quad \Pi\Theta = CPT = 1,$$

an important result first obtained by Lüders (1952).

Therefore, we can speak of a 1952 *microrelativity theory*, bearing the same relation to the 1905 *macrorelativity theory* as does Loschmidt's '*lawlike reversibility*' to Boltzmann's '*factlike irreversibility*'. And indeed, as we shall see, CPT invariance is the legal heir of Loschmidt's T invariance.

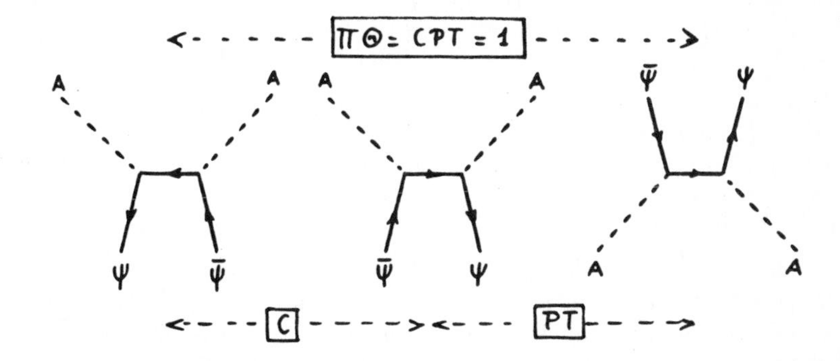

Fig. 9. CPT invariance as exemplified in the electron + positron ⇌ two photons transition: C, particle—antiparticle exchange pictured as reversal of the arrows; PT, covariant motion reversal, or emission—absorption exchange, pictured as reversal of the trajectories; $CPT = \Pi\Theta = 1$ overall reversal, or identity operation.

Using the relativisitic wording, we can say that covariant motion reversal PT — that is also *emission—absorption exchange* — and *particle—antiparticle exchange* C are two 'relative' pictures of essentially the same process. This is a basic assumption in the second quantization scheme, where emission of a particle and absorption of an antiparticle (or vice versa, of course) are assumed to be mathematically equivalent.

Clearly, the $CPT = 1$ statement implies a *principle of detailed balance* written as

$$A + \bar{B} + \cdots \rightleftharpoons C + \bar{D} + \cdots$$

where (beware!) a bar means 'particle' on the left-hand side and 'antiparticle' on the right-hand side — and vice versa of course. This describes a generalized sort of collision between particles, where the CPT invariance exactly is a quantal and relativistic extension of Loschmidt's 1876 T invariance. Incidentally, had Loschmidt imagined his colliding molecules as spinning rifle bullets rather than spherical balls, so that the right and left, fore and aft distinctions make sense, he would have been led to something quite akin to the CPT-invariance concept.

4.1.3. 'CORRESPONDENCE' BETWEEN THE CLASSICAL AND THE QUANTAL, WAVELIKE, PROBABILITY CALCULUS

Introduced in 1926 by Born, and systematized that same year by Jordan, the radically new 'wavelike probability calculus' derives from a *complex transition* (*or conditional*) *amplitude*, endowed with the Hermitean symmetry

$$(4.1.2)\quad \langle A|C\rangle = \langle C|A\rangle^*,$$

the *transition or conditional probability*

$$(4.1.3)\quad (A|C) = |\langle A|C\rangle|^2.$$

Dirac (1928) 'bras' $\langle\,|$ and 'kets' $|\,\rangle$ denote vectors in a (complex, multidimensional) Hilbert space, $\langle A|C\rangle$ being a Hermitean scalar product of two of them; the symmetry property (2) is understood as matrix conjugation. All these are time independent definitions, yielding a scheme 'corresponding' to the Laplacean scheme discussed in Sections 3.1.8 to 3.1.16. The *transition* is between two *representations* of the system, each consisting of an orthonormal set of 'state vectors' such that, by definition,

(4.1.4) $\langle B | B' \rangle = \delta(B, B')$.

The expression (3) of a Born—Jordan transition probability contains diagonal and off-diagonal terms. The diagonal terms alone would reproduce the Laplacean scheme. The off-diagonal terms are responsible for many of the 'paradoxical' aspects of quantum mechanics.

Corresponding to the classical expression (3.1.4) of a joint probability we have that of a *joint amplitude*

(4.1.5) $|A\rangle \cdot \langle C| = |A\rangle\langle A|C\rangle\langle C|$.

Corresponding to the classical expression of a Markow linking (3.1.6) we have that of a Landé linking (1965, pp. 79—84)

(4.1.6) $\langle A|C\rangle = \sum \langle A|B\rangle\langle B|C\rangle$.

The intermediate summation in (6) is over orthonormal state vectors satisfying the condition (4). The very presence of off-diagonal terms in the transition (or conditional) probability (3) does not allow these intermediate states to be thought of as 'real hidden states' [as were those in formula (3.1.8)]: *in the wavelike probability calculus intermediate summations are over virtual states* — a point of the utmost significance in the world view provided by quantum mechanics. Thus, the *similarity* between the Jordanian and the Laplacean schemes is such that *amplitudes* correspond to *probabilities*, the bridge between these schemes being formula (3).

End prior amplitudes such as $|A\rangle$ or $|C\rangle$ are truly parts of conditional amplitudes $\langle E|A\rangle$ or $\langle C|E'\rangle$ linking the system to the environment. In the spacetime picture $\langle E|A\rangle = \langle x|A\rangle$ and in the momentum-energy picture $\langle E|A\rangle = \langle k|A\rangle$; these are the Dirac (1947) and Landé (1965) interpretations of the state vector, to which we shall come back in the following sections.

In both classical or quantal mechanics, space and time as well as momentum and energy must appear. Thus, the Hilbert-space vectors $|\,\rangle$ are often (but not always) functions of space and time or momentum and energy (or sometimes functionals with a spacetime connotation). Quite often the problem treated is that of a system evolving from a *preparation* $|A\rangle$ to a *measurement* $|C\rangle$. Then, a *CPT* symmetry, as discussed in Section 2, interchanges preparation and measurement, and this induces the Hermitean symmetry (2). This is exactly how, in quantum mechanics, the time-independent concepts of lawlike reversibility and possible factlike irreversibility of conditional amplitudes, as respectively expressed in formulas (2) and (5), are turned into their time-dependent conterparts.

Self-consistency of the quantal formalism is quite impressive at this point. As explained by Wigner (1932), bra and ket exchange is required in a consistent definition of time reversal of solutions of a first order in time wave propagation equation. Also, bra and ket exchange is required for consistency of the particle—antiparticle exchange. These points are discussed in our presentation of the Dirac equation, Section 4.4.2.

When transition or conditional amplitudes $\langle A|C\rangle$ are referred to spacetime or to the momentum-energy space, they are termed 'propagators'. These show up in Landé chains, and in their generalization known as Feynman graphs (1949a, b), where more than two 'links' $\langle A|B\rangle$ can be attached to a 'vertex' A. Due to the symmetry property (2) such Landé chains or Feynman concatenations are allowed to zigzag arbitrarily throughout either spacetime or the momentum-energy space, regardless of the macroscopic time or energy arrow. They are of course endowed with the *CPT* symmetry, and of topological invariance *vis-à-vis* $\vee$, or $\wedge$, or C shapes of an ABC zigzag, as already were the classical chains discussed in Section 3.1.10.

A Feynman network visualizes how the overall transition should be computed; that is, the conditional amplitude $\langle A|\langle B|\langle C|\ldots|K\rangle|L\rangle|M\rangle$. The end 'kets' $\langle A|$ or 'bras' $|M\rangle$ can also be thought of as definite projectors $|A\rangle\langle A|$ or $|M\rangle\langle M|$ linking the system to its environment, in which case the $|A\rangle$'s and $\langle M|$'s are interpreted as the *prior amplitudes* of the initial and final partial states. The corresponding prior probabilities respectively are the initial occupation numbers of the initial states and the final occupation numbers of the final states; therefore, what we have here is a spacetime or a momentum-energy scheme 'corresponding' to the classical ones discussed in Section 3.1.10.

4.1.4. TOPOLOGICAL INVARIANCE OF LANDÉ CHAINS AND OF FEYNMAN GRAPHS: WHEELER'S SMOKY DRAGON; EPR CORRELATIONS

'Corresponding' to the topological invariance of Markov chains discussed in Section 3.1.13, we here have topological invariance of formula (5) under distortions, either in spacetime or in the momentum-energy space, of a Landé chain ABC, into a $\vee$, or a $\wedge$, or a C shape.

The $\vee$ and the $\wedge$ shapes illustrate respectively the so-called Einstein—Podolsky—Rosen correlations, proper and reversed. In the EPR correlations proper, measurements are performed at two distant places A and C upon paired particles issuing from a common source B. In the reversed EPR correlations preparations are performed at two distant places A and C upon particles, of which those having the right phase relation are absorbed in a common sink B; these are paired via their future interaction, and an intrinsic symmetry exists between this case and the one above. Both are described by the same transition amplitude; the operational asymmetry between the two cases is merely 'factlike'.

By definition of these correlations, the state of the common source or sink B is a pure state. However, in terms of the Hilbert spaces within which are embedded the two measurements or preparations, $|A\rangle$ and $|C\rangle$, this pure state is expressed as a superposition, a sum of products of orthonormal state vectors (of course, defined up to an arbitrary rotation). In this sense the state of the common source or sink should be said 'virtual', meaning that it *certainly does not* consist of paired 'real hidden states' of the two particles measured, or prepared, at A and C. This is the very essence of the so-called 'EPR paradox', to be discussed in Chapter 4.6.

What then of the $\langle$ or C shape of an ABC zigzag? It illustrates a quite ubiquitous procedure in quantum mechanics: evolution of a system between a preparation $|A\rangle$ and a measurement $|C\rangle$. What is the state of the evolving system between $|A\rangle$ and $|C\rangle$? Is it the retarded state generated by $|A\rangle$, or the advanced state $|C\rangle$ in which it ends? It cannot be both if there is a transition. But, then, why should it be the one rather than the other, since there is the intrinsic symmetry (2)? The truth is that the evolving system is neither in the retarded state generated by the preparation, nor in the advanced state selected by the measurement, because *it is actually transiting from the one to the other* (thus feeling symmetrically, so to speak, the influences of the preparation and of the measurement).

The transition amplitude $\langle A|C\rangle$ is expressible in the form (6), with an intermediate summation over virtual evolving states $|B\rangle$, or $|\psi(t)\rangle$ as they are usually labeled. And these are defined up to an arbitrary rotation of the Hilbert space reference frame. So, *there is really no such thing as an evolving state vector.* The 'evolving state vector' $|\psi(t)\rangle$, which is ubiquitous in textbooks and scientific papers, is no more than a ghost, as worthless as was the 'luminiferous aether'. *What makes sense is the transition or conditional amplitude* $\langle A|C\rangle$, *nothing else.*

The 'evolving state vector' $|\psi(t)\rangle$, or $|B\rangle$ as it is labeled here, enters in the intermediate summation in formula (6). As expressed in terms of the separate representations $|A\rangle$ and $|C\rangle$, it enters as a summation over virtual states. However, it is expressed as a pure state in terms of the composite $A \otimes C$ system. What is then its interpretation? It is the state of a virtual particle, fastened to a vertex B between A and C, exactly wiping out the difference between $|A\rangle$ and $|C\rangle$. And thus, formally speaking, the overall preparing-and-measuring device is equivalent to that particle.

Of the 'evolving system' Wheeler (Miller and Wheeler, 1983, pp. 136—152) says that it is a 'smoky dragon', of which only the 'tail' held as $|A\rangle$, and the 'mouth' biting as $|C\rangle$, have physical presence. He also writes that "no elementary quantum phenomenon is a phenomenon until it is recorded", thus emphasizing the advanced influence of the measurement, drawing down the dragon from the higher world $A \otimes C$ where it was living since its birth as $|A\rangle$, into the $|C\rangle$ world where it bites and dies.

A familiar example may help to clarify the whole matter.

Consider a light beam crossing in succession two birefringent crystals A and C. From classical optics we deduce that, α denoting the angle between the associated linear polarization planes in A and C, the conditional amplitude $\langle A|C\rangle$ that a photon prepared with linear polarization $|A\rangle$ is measured with linear polarization $|C\rangle$ is

$$(4.1.7)\quad \langle A|C\rangle = 2^{-1/2}\begin{cases}\cos\alpha & \text{if } \alpha = A - C\\ \sin\alpha & \text{if } \alpha = A - C + \pi/2\end{cases}.$$

If, now, a third birefringent crystal, B, is inserted between A and C, its length being such that a zero phase shift (modulo $2n\pi$) is introduced, formula (7) is still correct and can be written as

$$(4.1.8)\quad \langle A|C\rangle = 2^{-1/2}\begin{cases}\cos A\cos C + \sin A\sin C & \text{if } \alpha = A - C\\ \sin A\cos C - \cos A\sin C & \text{if } \alpha = A - C + \pi/2\end{cases}$$

the angles A and C being referred to the orthogonal polarization planes in B. So, the sum in formulas (8) really is over the 'virtual' polarization states in B. These states must be said to be 'virtual', because it is impossible to ascribe either the one or the other of them to a travelling photon. Only their phase coherent superposition has physical meaning. Therefore, Equation (8) is clearly the specification of formula (5) appropriate in this case.

The virtual state of a photon travelling from A to C is expressible as a sum of products of two orthogonal states defined up to a rotation of the crystal B. In particular, the travelling photon is neither in the retarded state emitted as $|A\rangle$, nor in the advanced state received as $|C\rangle$; it is actually transiting from $|A\rangle$ to $|C\rangle$, and this is all we can say.

The pure state in the right-hand side of formula (7) or (8) belongs to the composite $|A\rangle|C\rangle$ representation and can be thought of as symbolizing the crystal B.

Formulas (7) and (8) are identical to those valid for correlated linear polarizations prepared or measured on spin-zero photon pairs.

On the whole, invariance of the generating formula (6) of Landé chains *vis-à-vis* $\vee$, $\wedge$, and C shapes of an ABC zigzag has been clearly illustrated.

4.1.5. COVARIANT FOURIER ANALYSIS AND THE SECOND-ORDER KLEIN–GORDON EQUATION

De Broglie matter waves, discussed in Section 2.6.15, in empty space obey the 1926 Klein–Gordon equation

(4.1.9) $(\partial_i^i - k_0^2)\psi(x) = 0,$

where x stands for x, y, z, ict, and ∂_i^i is the 1747 Dalembertian operator (often denoted as $\square$); the 'proper frequency' k_0 is related to the rest mass m of the particle via $k_0 = (2\pi/h)cm_0 \equiv (c/\hbar)m_0$, $\hbar = h/2\pi$ being a widely used 'modified Planck constant'.

The four-frequency representation of this equation reads

(4.1.10) $(k_i k^i + k_0^2)\theta(k) = 0$

and is derived from (9) via the Heisenberg–Schrödinger recipe

(4.1.11) $k_j = -i\underset{\rightarrow}{\partial}_j = +i\underset{\leftarrow}{\partial}_j = -(i/2)[\underset{\rightarrow}{\partial}_j - \underset{\leftarrow}{\partial}_j] = -(i/2)[\partial_j];$

$[\partial_j]$ is the Gordon current operator we have met in Section 2.6.7. $\psi(x)$ and $\theta(k)$ are Fourier associated representations of the wave–particle system. A relativistically covariant formalization of Fourier analysis, as introduced by Marcel Riesz (1946), and pursued by me (1955) can be thus summarized.

We write the 'Hermitean scalar product' of two 'state vectors' $|a\rangle$ and $|b\rangle$ as $\langle a||b\rangle$, the double bar reminding us that they are solutions of a second-order equation. By definition

$$(4.1.12)\quad \langle a||b\rangle = -\frac{i}{2k_0}\iiint_{\mathscr{E}} \psi_a^*\,[\partial_j]\,\psi_b\,\mathrm{d}u^j = \begin{cases} \dfrac{1}{k_0}\displaystyle\iiint_{\mathscr{H}} \theta_a^*\,k_j\,\theta_b\,\varepsilon(k)\,\mathrm{d}\eta^j \\ \displaystyle\iiint_{\mathscr{H}} \theta_a^*\,\theta_b\,\varepsilon(k)\,\mathrm{d}\eta \end{cases}$$

the second equality expressing covariantly the so-called 'Parseval equality'. The $\mathscr{E}$ integral in spacetime is over an arbitrary spacelike surface, with volume element $\mathrm{d}u^i$ (as defined in Section 2.6.4), and is $\mathscr{E}$-independent by virtue of $\partial^i[\partial_i] \equiv 0$. The two $\mathscr{H}$ integrals, which are equivalent by virtue of

$$(4.1.13)\quad k^j\,\mathrm{d}\eta = -k_0\,\mathrm{d}\eta^j, \qquad k^j\,\mathrm{d}\eta_j = k_0\,\mathrm{d}\eta,$$

are over the 4-frequency hyperboloid

$$(4.1.14)\quad \mathscr{H}(k) \equiv k_j k^j + k_0^2 = 0$$

having a positive $\mathscr{H}_+$ and a negative $\mathscr{H}_-$ energy sheets. The discrete function $\varepsilon(k) = +1$ on $\mathscr{H}_+$, -1 on $\mathscr{H}_-$, 0 otherwise, is introduced for elegance of the formulas. Equation (10) says that $\theta(k)$ is arbitrary over $\mathscr{H}$, and zero outside.

The 'Fourier nucleus' is defined covariantly as

$$(4.1.15)\quad \langle x||k\rangle = \langle k||x\rangle^* \equiv \begin{cases} (2\pi)^{-3/2}\exp(ik\cdot x) \text{ if } \mathscr{H}(k) = 0 \\ 0 \text{ otherwise} \end{cases}$$

$k \cdot x$ being a shorthand notation for $k_i x^i$. Using the twin definitions (12) we can write the reciprocal Fourier transforms in compact form as

$$(4.1.16)\quad \langle x||a\rangle = \langle x||k\rangle\langle k||a\rangle, \qquad \langle k||a\rangle = \langle k||x\rangle\langle x||a\rangle$$

with an automatic *intermediate summation*; $\langle x||a\rangle$ and $\langle k||a\rangle$ denote, *à la* Dirac (1928, p. 79), $\psi_a(x)$ and $\theta_a(k)$ in the form of transition amplitudes.

The 1928 Jordan—Pauli (JP) propagator is

$$(4.1.17)\langle x'||x''\rangle \equiv \langle x'||k\rangle\langle k||x''\rangle = \langle x'||x\rangle\langle x||x''\rangle.$$

Being a scalar and odd in *ct*, it is zero outside the light cone, thus expressing orthogonality of two Fourier nuclei with spacelike separation of their x's. Expressed as an x integral, it is a 'convolution product', and displays orthogonality of two JP propagators with spacelike separation of their apexes x' and x''.

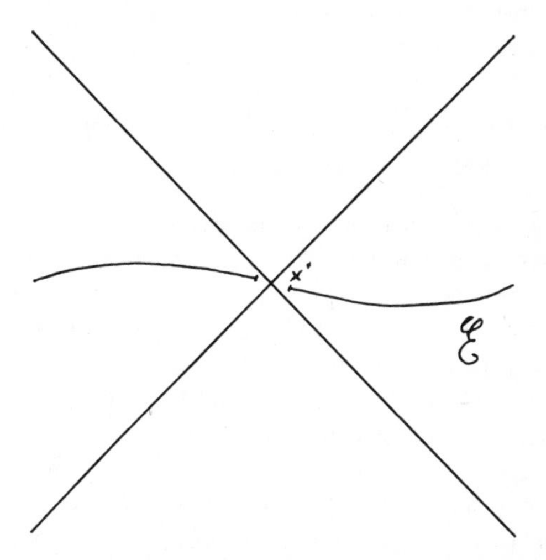

Fig. 10. Covariant position measurement answering the question: does the particle cross an arbitrary spacelike surface at a given point-instant x'? The corresponding eigenfunction is the Jordan—Pauli propagator $D(x - x')$, which is zero outside the light cone, and generalizes covariantly Dirac's $\delta(x - x')$.

Using the JP propagator we can expand the wave function $\langle x||a\rangle$ at any point-instant x' in terms of JP propagators with apexes x' over an arbitrary spacelike surface $\mathscr{E}$, the 'coefficients of the expansion' being the values of $\langle x||a\rangle$ on $\mathscr{E}$. This is said to 'solve the Cauchy problem' for the second-order Klein—Gordon equation; both the wave function and its 'normal derivative' are implied.

This same formula expresses covariantly *and jointly* a *position measurement-and-preparation process*. It says that a particle found crossing the three-dimensional surface $\mathscr{E}$ at x has come inside the past light cone and will leave inside the future light cone. Thus, the JP propagator is the eigenfunction attached to the manifestation of the particle 'at $\mathscr{E}$ at x'. Extending a Schrödinger argument and remarking that, by definition, the JP propagator is the Fourier transform of the Fourier nucleus, we can interpret covariantly the 4-vector x^i *bound to end on* $\mathscr{E}$ as the 'position operator'. This means three, not four degrees of freedom (for example, x, y, z, as in the non-relativistic problem). This has been possible because, from the start, we have accepted positive and negative frequencies on an equal footing.

Finally, the probability that the particle 'crosses $\mathscr{E}$ at x' is expressed by the flux of the Gordon current $-(i/2k_0)\psi^*[\partial_j]\psi$.

4.1.6. COVARIANT FOURIER ANALYSIS AND THE FIRST-ORDER SPINNING-WAVE EQUATIONS

In the case of light propagation the second-order d'Alembert equation is a consequence of Maxwell's system of first-order equations. Something similar occurs with all spinning-wave equations, general treatments of which have been proposed by quite a few authors.

It has been shown by Gelfand and Naimark and others that a systematic theory of spinning-wave equations can be derived so as to show an intimate connection with that of irreducible representations of the Lorentz group, a one-to-one correspondence binding spinning-wave equations and representations of the Lorentz group. Here again, we are at the very heart of relativistic quantum mechanics. A concise presentation of the question is due to Bargmann and Wigner (1948), who worked it out independently but published jointly.

All spinning-wave equations are of the form

$$(4.1.18)\quad (\gamma_i \underset{\rightarrow}{\partial^i} + k_0)\psi = 0, \qquad \bar{\psi}(-\gamma_i \underset{\leftarrow}{\partial^i} + k_0) = 0,$$

in x space or (remembering formulas (11))

$$(4.1.19)\quad (\gamma_i k^i + ik_0)\theta = 0 \qquad \bar{\theta}(\gamma_i k^i + ik_0) = 0$$

in k space. ψ and θ are column matrices, $\bar{\psi}$ and $\bar{\theta}$ row matrices such that $\bar{\psi} \equiv \psi^+\beta_4$ and $\bar{\theta} \equiv \theta^+\beta_4$; the cross denotes Hermitean conjugation; γ_i and β_i are two sets of self-adjoint, or Hermitean matrices; in Dirac's electron theory $\beta_i = \gamma_i$. These definitions entail relativistic covariance in all calculations; it cannot be denied, however, that there is here something special with the time or the energy coordinate; see in this respect Section 4.4.2 below.

All spinning-wave equations are such that there exists a projector depending on the γ's and the $-i\partial$'s or k's, projecting any solution of the Klein–Gordon equation $\mathscr{G}\psi = \mathscr{G}\theta = 0$ as a solution of the spinning-wave equation $\mathscr{S}\psi = \mathscr{S}\theta = 0$; in other words, $\mathscr{S} = \mathscr{P}\mathscr{G} = \mathscr{G}\mathscr{P}$. In the Dirac theory $\mathscr{P} = (i/2k_0)(\gamma_i k^i - ik_0)$.

Very much as in Section 5, the Parseval equality, defining the Hermitean scalar product $\langle a|b\rangle$ of two state vectors (the single bar reminding us that they are now solutions of a first-order equation) has the covariant expression

$$(4.1.20)\ \langle a|b\rangle = i\iiint_{\mathscr{E}} \bar{\psi}_a\gamma_j\psi_b\,\mathrm{d}u^j = \begin{cases} i\displaystyle\iiint_{\mathscr{H}} \bar{\theta}_a\gamma_j\theta_b\varepsilon(k)\,\mathrm{d}\eta^j \\ \displaystyle\iiint_{\mathscr{H}} \bar{\theta}_a\theta_b\varepsilon(k)\,\mathrm{d}\eta \end{cases}$$

The equality of the two $\mathscr{H}$ integrals follows from (13) and

$$k^j\bar{\theta}_a\gamma_j\theta_b + ik_0\bar{\theta}_a\theta_b = 0,$$

Noticing that the second $\mathscr{H}$ integrals in (12) and (20) are isomorphic modulo the exchange $\psi^* \rightleftharpoons \bar{\psi}$, we can rewrite all formulas of the preceding section with $\bar{\psi}$ in place of ψ^* (implying a summation over components of the ψ), thus establishing the equivalence

$$\langle a||b\rangle = \langle a|b\rangle.$$

This means equality of the integrands of the $\mathscr{H}$ integrals in (12) and (20), but only equivalence of the $\mathscr{E}$ integrals; $\bar{\psi}\gamma_j\psi$ is the well-known conservative Dirac style 4-current.

Therefore all the formulas of Section 5 have corresponding ones, with single bars instead of double bars. Using the projector $\mathscr{P}$ previously mentioned, we redefine the covariant Fourier nucleus and the Jordan–Pauli propagator as

$$\langle k|x\rangle = \mathscr{P}\langle k||x\rangle \qquad \langle x'|x\rangle = \mathscr{P}\langle x'||x\rangle;$$

the solution of the Cauchy problem in terms of $\langle x'|x\rangle$ no longer implies the normal derivative but, instead, requires a multicomponent wave.

Among various consequences of this, I select the following. Using in combination the formalism of the preceding section and this one, we can write

$$\langle k_i \rangle = -\frac{i}{2} \iiint_{\mathscr{E}} \bar{\psi}[\partial_i]\gamma_j \psi \, \mathrm{d}u^j = \iiint_{\mathscr{H}} k_i \bar{\theta}\theta\varepsilon(k) \, \mathrm{d}\eta$$

allowing normalization in terms of momentum energy $p_i = nk_i$. The integrand of the $\mathscr{E}$ integral is known as Tetrode's asymmetric momentum-energy tensor.

4.1.7. PARTICLES IN AN EXTERNAL FIELD

As an example we consider charged particles inside an electromagnetic 4-potential $A_i(x)$; then, in Equations (18), the well-known substitution $i\partial_i \to i\partial_i - QA_i$ is needed, Q denoting the electric charge (in e.m.u.).

Covariant reciprocal Fourier transforms must then be fourfold ones:

$$\psi(x) = (4\pi)^{-2} \overline{\iiiint} \theta(k) \exp(ik_i x^i) \, \mathrm{d}\kappa,$$

$$\theta(k) = (4\pi)^{-2} \iiiint \psi(x) \exp(-ik_i x^i) \, \mathrm{d}\omega.$$

There is in this a difficulty, because the integral $\iiiint \bar{\psi}\psi \, \mathrm{d}\omega$ is unbounded and cannot be used for normalization. However, the associated Parseval integral $\iiiint \bar{\theta}\theta \, \mathrm{d}\kappa$ has an integrand similar to the one inside the last expression (20), so that, in the limit $A_i(x) \equiv 0$, coincidence is obtained. Such forfold Fourier integrals are freely used, as a calculation algorithm, in the Feynman S-matrix scheme, with perfect physical efficiency. Let us continue, after noticing that normalization of the wave function is adequately secured by means of the threefold finite integral $i \iiint \bar{\psi}\gamma_j \psi \, \mathrm{d}u^j$.

If $B^i(k)$ denotes the four-dimensional Fourier transform of the potential $A^i(x)$, the Fourier transform of $iA^j \bar{\psi}\gamma_j \psi$ is a 'convolution product' $i \iiint B^j(k' - k)\bar{\theta}(k)\gamma_j \theta(k) \, \mathrm{d}\kappa$, this being a key element for writing down Feynman transition amplitudes, such as those in Section 4.7.2 below.

An interesting simplification occurs when the external potential $A^i(x)$ is invariant by translation along some specified direction; if it is timelike, the case is termed 'stationary', and occurs quite often. If we take the time axis along this direction, the Fourier integrals factorize into a time or energy integral, and a threefold space or space-frequency integral: thus $\psi(x)$ is expanded as a weighted sum of energy eigenvalues

$$\psi(x, t) = \int \rho(W)\psi(x, W) \exp(iWt) \, \mathrm{d}W$$

with

$$\psi(\mathbf{x}, W) = \iiint \chi(\mathbf{k}, W) \exp(i\mathbf{k} \cdot \mathbf{x}) \, \mathrm{d}\mathbf{k}^3$$

and

$$\theta(k) = \rho(W)\chi(\mathbf{k}, W).$$

expressing solutions of a stationary wave equation in terms of energy eigenvalues W and eigenfunctions; $\psi(\mathbf{x}, W)$ is at the very heart of the celebrated 1926 Schrödinger theory.

By using the threefold Fourier transform $C^i(\mathbf{k})$ of the 4-potential $A^i(\mathbf{x})$, and the corresponding convolution product, the energy problem can be carried out in terms of the eigenvalues $\chi(\mathbf{k}, W)$ (Lévy, 1950).

4.1.8. CONCLUDING THE CHAPTER: QUANTUM AND RELATIVITY THEORIES AS DAUGHTERS OF WAVE PHYSICS

Both the basically reversible calculus of conditional or transition amplitudes, and the concise and efficient symbolic calculus handling solutions of wave equations presented in this chapter, may have convinced the reader that the relativity and the quantum theory are true sisters, and are easily conversant with each other. In this, priority should be given to the formalization, interpretative problems being addressed to later.

The historical truth, however, is that, from de Broglie's 1925 wave mechanics until the brilliant series of articles of Tomonaga, Schwinger, Feynman and Dyson (1946—1949), the relativity and the quantum theories have behaved mostly as quarreling sisters, with only brief agreements from time to time. This is a story to be recalled in the next chapter.

These uneasy days are now well over, and the problems ahead are not radically concerned with relativistic covariance, at least at the level of special relativity. The quantized electrodynamics of Schwinger and Feynman is one of the most successful theories ever produced, with remarkable precision in many of its predictions. From these the conquest goes forward, not without opposition of various kinds of course, but overall in good order, with floating banners proudly displaying the interlaced *h* and *c* devices.

CHAPTER 4.2

1900–1925: THE QUANTUM SPRINGS OUT, AND SPREADS

4.2.1. 1900: MAX PLANCK DISCOVERS THE QUANTUM OF ACTION

Max Planck (1858–1947), in the fateful year 1900, turned away from phenomenological thermodynamics, in which he had great expertise, espousing a statistical approach somewhat in the manner of Boltzmann, whom up to then he had criticized. He thus appears as the last of the great classical and the first of the great modern theoretical physicists. His aim was then to solve the remaining puzzle of the physics of thermal radiation, where Kirchhoff, Stefan, Boltzmann, Wien had made important contributions by using thermodynamical arguments. The overall conclusion, as recalled in Section 3.2.7, was that the radiant energy density at temperature T and frequency ν has the form $u = \nu^3 F(\nu/T)$; deriving the right expression for F was the puzzle.

Experimentation was going on by that time, so something more can be said. Wien showed, in 1896, that for large values of ν/T (high frequencies and low temperatures) the experimental curve is reproduced well by an expression which reads, in post-Planck notation, $8\pi h(\nu^3/c^3)\exp(-h\nu/kT)$, k denoting Boltzmann's constant. On the other hand, for small values of ν/T, a formula derived statistically by Lord Rayleigh in 1900, yielding u as $8\pi k(\nu^3/c^3)(T/\nu)$, fits the experimental curve well.

As is clear from his autobiography, Planck (1948) had been cultivating his own garden, considering himself as the spiritual heir of Clausius, and thus believing that the 'second law' has rigorous validity; he disliked Boltzmann's statistical interpretation of it.

Around 1895 he had become interested in the black-body radiation problem, hoping to derive theoretically the function $F(\nu/T)$. He began convinced that phenomenological thermodynamics would do the job. But a big surprise lay hidden inside the oven.

Not surprisingly, Planck introduced, together with its energy density u, an entropy density s of the radiation, obeying, via thermodynamics, the formula $1/T = \mathrm{d}s(\nu)/\mathrm{d}u(\nu)$.[1]

Then he came to grips with the 'universal' law of heat exchange between the enclosed radiation and the walls of the oven, which he idealized as consisting of electromagnetic harmonic oscillators — a Herculean task needing fifty pages of calculation. These could have been saved, had he hit upon an other approach used by Lord Rayleigh in 1900.

Rayleigh thought of the 'universal' spectral distribution as expanded in 'normal modes' — 'orthogonal' standing waves, in the terms of Fourier analysis — the shapes of which depend upon the shape of the cavity, but without affecting of course the Fourier integral. The easiest choice then is that of a parallelipipedic oven — formally, the three-dimensional extension of a vibrating string — easily yielding $w(\nu) = (8\pi\nu^2/c^3)\bar{u}$, with $\bar{u}$ denoting the mean energy of one normal mode. The calculation produces a 4, but this must be doubled to 8, because there are two orthogonal polarization states per mode). Had Planck reasoned that way he would have been led straight to the 'photon' concept, which Einstein introduced in 1905. At any rate the formula he obtained was a good one.

Planck also used a similar formula for the entropy density, $s(\nu) = (8\pi\nu^2/c^3)\bar{s}$, but, while additivity of the u's is unquestionable, that of the s's is subject to a 'randomness assumption'. Later Planck suggested to Max von Laue, then his student, an elaboration of the general case (Laue, 1906; 1907a).

In 1896 Wien had proposed his expression for the function F, writing $w(\nu) = \alpha\nu^3 \exp(-\beta\nu/T)$ with two adjustable constants. Setting $h = \alpha c^3/8\pi$ and using Wien's formula, Planck wrote down, in 1899, expressions for u and s in which $(\mathrm{d}^2\bar{s}/\mathrm{d}\bar{u}^2)^{-1} = -\beta\nu\bar{u}$. But then something happened.

In 1900, two experimental groups in Berlin, Kurlbaum—Pringsheim and Rubens, measured the energy spectrum at low frequencies, showing that, there, Wien's formula was completely wrong. They found that $\bar{u}$ is proportional to T, to which result Lord Rayleigh was quick to react by pointing that the 'equipartition principle' of statistical mechanics precisely requires that $u = kT$, with $k \equiv R/N$. This may well have triggered Planck's conversion to statistical thinking.

Anyhow what he did first was to write down $(\mathrm{d}^2s/\mathrm{d}\bar{u}^2)^{-1} = -(1/k)\bar{u}^2$ for low values of ν/T, deciding that this expression should be added to the previous one, yielding

$$(4.2.1) \quad -(\mathrm{d}^2\bar{s}/\mathrm{d}\bar{u}^2)^{-1} = \beta\nu\bar{u} + (1/k)\bar{u}^2.$$

From this he derived straightaway the now famous formulas

$$(4.2.2) \quad \bar{u} = h\nu/[\exp(h\nu/kT) - 1], \quad u(\nu) = 8\pi h\nu^3/c^3[\exp(h\nu/kT) - 1]$$

which fit the experimental data wonderfully well. For the entropy per mode he obtained the expression

$$(4.2.3) \quad \bar{s} = k[(1 + \bar{u}/h\nu)\ln(1 + \bar{u}/h\nu) - (\bar{u}/h\nu)\ln(\bar{u}/h\nu)].$$

These results he presented at the Berlin Physical Society, 19 October 1900. This was a very lucky game of guesswork and patchwork, where Planck was clearly stepping aside from the rigorous path of accepted orthodoxy. But, of course, experimentation gathers its own flowers in wilderness, calling theory to follow her there . . .

Having thus established himself as the Kepler of a new field, Planck chose not to stop there but try to become a Newton as well. As Banesh Hoffmann (1959, p. 18) puts it, he had made

> . . . a single suit of clothes by borrowing the trousers from one person and the coat from another. By good fortune and excellent judgment [he] had managed to get [them] to match, the resulting suit being much more valuable than [the separate pieces]. But then he found himself in the position of the schoolboy who, having [stolen] a glance at the answers to the day's homework, [finds] the problems nevertheless difficult.

Planck soon realized that thermodynamics would not help him any more, so that he should have recourse to Boltzmann's sort of statistical methods, as Boltzmann had pressed him to do.

So, imitating Boltzmann's treatment of the momenta and energies of molecules, Planck thought of discrete energy packets, which he labelled $W(\nu)$, and used combinatorial analysis for writing down the 'number of complexions' of 'energy quanta' $W(\nu)$. His formula

$$(4.2.4) \quad W(\nu) = (N + g - 1)!/N!(g - 1)!,$$

where N denotes the number of quanta and g that of accessible cells, put forward with little justification, is certainly alien to Boltzmann's statistics *stricto sensu*, as was made clear later by Natanson (1911), Bose (1924), Einstein and Ehrenfest (1923) and Einstein (1924a).

However, Planck presented his argument at the Berlin Physical Society, 14 December 1900. Adjusting his formula to the most recent experimental data he proposed

$$h = 6.55 \times 10^{-27} \text{ erg s}; \qquad k = 1.346 \times 10^{-16} \text{ erg deg}^{-1},$$

together with derived values for Avogadro's number and the electron charge — the most precise ones produced up to then. So, the newly born 'quantum of action' $h = W/\nu$ had uttered a loud birth cry, nobody suspecting then the very many things it would be lisping, and saying, in the later years.

What happened afterwards is odd.

As for the quantum theory, Planck stopped right there (or almost so), watching skeptically what other physicists would make of his quantum. After 1905 he became an enthusiastic supporter of Einstein's relativity theory, doing and promoting work in relativistic thermodynamics.

As for Einstein, he contributed significantly again and again to the quantum theory, producing, in 1909, an interesting interpretation of Planck's formula (1): by inserting $\bar{u} = \bar{n}h\nu$ and denoting $\Delta\bar{n}^2$ the 'squared statistical fluctuation', he wrote it as

$$(4.2.5) \quad -k(\mathrm{d}^2\bar{s}/\mathrm{d}\bar{n}^2)^{-1} = (\bar{n} + \bar{n}^2)(h\nu)^2 \equiv \overline{\Delta n}^2(h\nu)^2.$$

He showed that the $\bar{n}$ term has a corpuscular, and the $\bar{n}^2$ term a wave, interpretation; this, of course, is connected closely with his 'Lichtquantum' concept.

Notwithstanding his great expertise in statistical mechanics, and his repeated use of it when contributing to the quantum theory, Einstein disliked to the end of his life the 'statistical interpretation of quantum mechanics'.

What had happened before the birth of the h quantum has odd aspects also. Playing with Wien's coefficients α and β, Planck had come to believe that there should be a universal constant having the dimension of an action. At the 1899 meeting of the Prussian Academy of Science he emphasized the existence of three independent universal constants, Newton's G, the velocity of light in vacuum c, and, of course, h. In 1905 the relativity theory established c as a universal constant. Anyhow, Planck declared that this triad of independent universal constants forms a "natural system of units ... necessarily valid for all times and cultures, even non-human and extraterrestrial" — an emphatic, but true statement.

4.2.2. EINSTEIN'S NUMEROUS CONTRIBUTIONS TO THE QUANTUM THEORY: STATISTICS, AND THE PHOTON

The first physicist to recognize that Planck's 1900 proposal was much more than just an ingenious *ad hoc* proposal was Einstein in 1905 – and again, repeatedly, in the following years, until 1925. In all of these papers his reasoning is statistical, often aiming at improving the derivation of Planck's formula, and finally emphasizing Bose's recognition in 1924 that a non-Boltzmannian statistics of indistinguishable particles is implied. Einstein's 'Lichtquantum', termed 'photon' by Lewis in 1926, appears as a byproduct, first in 1905 with its energy $h\nu$, then in 1910 (in a paper written with Hopf) with the corresponding momentum. So, between the momentum energy p^i of a photon propagating *in vacuo* and the 4-frequency $k^i/2\pi$ of the corresponding plane wave, the universal equivalence formula $p^i = \hbar k^i$ holds, with $\hbar \equiv h/2\pi$ in the presently used notation.

The 1905 paper and its 1906 sequel discuss fluctuations in the thermal radiation. Incidentally, Einstein mentions that the photon concept easily explains the 'photoelectric effect' discovered by Hertz in 1887, where the energy of electrons expelled by light from a metal depends only on its frequency, whereas their number per unit time depends on its intensity. The formula $W_e = h\nu - W_0$, where W_0 is a latent energy and W_e that of the ejected electrons, has been verified in careful experiments by Millikan, Maurice de Broglie, and Ellis, in the visible, the X-ray and γ-ray regions of the spectrum, respectively. The reciprocal phenomenon, in the form of an abrupt lower limit in the continuous spectrum of X-ray generators exists also, as shown by Webster and by Maurice de Broglie. These are far more obvious proofs of the existence of the h quantum than were the laws of thermal radiation! In 1923 Compton, in beautiful experiments, observed the momentum-energy conserving 'collisions' between electrons and photons; Einstein comments upon this in . . . the *Berliner Tageblatt*.

Relativistic covariance is, of course, obvious in the $p^i = \hbar k^i$ formula, and Einstein never loses sight of it.

In 1909 two Einstein papers again deal with thermal fluctuations, displaying, in the radiation case, two terms (5) corresponding to those (1) indicated by Planck. The first discusses oscillations of a mirror bombarded by the photons and the second relativistic invariance.

In 1910, in two papers written with Hopf, Einstein again addresses the problem with increased rigor, reducing the number of assumptions. It is there that the photon's momentum appears.

In 1916, in two papers, Einstein discusses stimulated emission and absorption of light, introducing his famous A and B constants, forerunners of a key concept in second quantization. He then, once more, derives Planck's radiation formula and the one for fluctuations. He discusses thermal equilibrium in terms of momentum, relativistic covariance being taken care of.

Finally, in 1924—1925, Einstein proposes in two papers his theory of a 'perfect monoatomic gas'. In the first one Bose's indistinguishability is used, and the so-called 'Bose—Einstein condensation' is derived. In the second paper de Broglie's wave mechanics is used; two terms similar to those found with photons are present. Moreover, what came to be termed, 10 years later, the 'superfluidity' concept, is sketched.

So much for 'Einstein and the photon'.

One significant aspect of Planck's quantal hypothesis is that it represses the energy contribution from low temperatures, as was assumed by the equipartition theorem used by Lord Rayleigh. A similar phenomenon occurs in solids, where the classical Dulong—Petit formula for the specific heat breaks down at low temperatures. This Einstein explained in 1907, in an article dealing with vibrations in crystals which he treats as quantum oscillators.

Einstein's final contributions to the quantum theory were his 1927 intervention at the Fifth Solvay Council and his paper written in 1935 with Podolsky and Rosen, both dealing with the 'paradox' of 'quantal non-separability', to which we shall come back at length in Chapter 4.7.

4.2.3. THE HYDROGEN ATOM OF BOHR (1913) AND SOMMERFELD (1916)

A quotation from Poincaré (1905, Pt 2, Ch. 9) will help us introduce the subject:

> The electron dynamics may be approached from many sides, among which one has been somewhat neglected, although it is one of those promising the most surprises: the motions of the electrons producing emission rays. . . . Why are the spectral lines regularly distributed? These laws have been studied by the experimenters in every detail; they are very precise and rather simple. At first sight one is reminded of the harmonics encountered in acoustics, but the difference is a big one. . . . I believe that there lies one of the most important secrets of Nature. A Japanese physicist, Nagaoka, has recently proposed an explanation: the atoms may be made of a big positive electron surrounded by a ring of numerous small electrons, like the planet Saturn by its ring. This approach should be pursued.

Nagaoka's article appeared in 1904, and Rutherford quotes it when proposing his 1911 atomic model. The Ariadne thread of these luminous labyrinths was found by Bohr in 1913, and the similarity with acoustics was explained by Louis de Broglie in 1924.

In 1885 the Swiss schoolteacher Balmer found a formula expressing the frequencies ν emitted or absorbed by the hydrogen atom, which in 1889 Rydberg recast in the form

$$c\nu = \mathcal{R}(1/n^2 - 1/m^2)$$

where $\mathcal{R}$ is a constant, m and n integers; n assummes the values 1, 2, 3, . . . characterizing 'series' of lines (Lyman, Balmer, Paschen, Brackett, Pfund), and m the values $n + 1$, $n + 2$, . . . , Balmer felt he had hit upon something important, for this is what he writes (1885; Max Jammer's 1966, p. 65, translation):

It appears to me that hydrogen . . . more than any other substance is destined to open new paths in the knowledge of the structure of matter and its properties. In this respect the numerical relations among the wavelengths of the . . . hydrogen spectral lines should attract our attention particularly.

In 1913 the idea occurred to Bohr of quantizing the action along Rutherford's circular orbits, as Planck had done for his linear oscillators. As the calculation is elementary and short I present it.

The Newtonian balance between centrifugal and centripetal forces may be written as $mv^2/r = e^2/r^2$, e denoting the electron charge in e.s.u. Thus from the hypothesis $2\pi rmv = nh$ one gets $r_n = n^2h^2/4\pi^2me^2$. Now, the total energy, kinetic plus potential, of the electron being $mv^2/2 - e^2/r = -e^2/2r$, we find $E_n = -2\pi^2me^4/n^2h^2$, whence

$$E_n - E_m = 2\pi^2h^{-2}e^4m(1/n^2 - 1/m^2).$$

From the hypothesis $E_n - E_m = h\nu$, we recover the Balmer—Rydberg formula with

$$\mathcal{R} = 2\pi^2e^4m/ch^3,$$

in excellent agreement with the experiments.

This was a new triumph for the h quantum, in a domain where it was far less obscure than in the thermal radiation spectrum.

Clearly, not only the action, but also the orbital angular momentum is thus quantized in units $h/2\pi$, or $\hbar$, as it is written now. As a corollary, the magnetic moment of an orbiting electron is quantized in units $e\hbar/2cm$, which Bohr called 'magnetons' (these being not 'universal' quanta, as they depend upon the mass of the orbiting particles).

To account for the recoil of the nucleus of mass M, Bohr proposed the natural formula $mvr/(1 + m/M) = n\hbar$.

The next major triumph in atomic spectroscopy was Sommerfeld's 1916 explanation of the 'fine structure' of the lines, as seen at high spectral resolutions. The dimensionless 'fine structure constant' $\alpha \equiv e^2/\hbar c \simeq 1/137$ made there its first appearance.

Two elements are essential in Sommerfeld's derivation.

The first one uses relativistic dynamics and allows elliptical orbits. It turns out that when viewed in a rotating frame the trajectory is indeed an ellipse, but one whose perihelion advances.

The second element used was the 1916 Wilson—Sommerfeld quantization rules. These state that, in physical systems the coordinates q of which are periodic in time, the action should be quantized by the rule $\oint p\, \mathrm{d}q = nh$, p denoting of course the canonically conjugate momenta; as many such conditions occur as there are periodic coordinates.

Quantizing in this way the angular θ and the radial r motions of the electron, Sommerfeld ends up with his famous formula for the energy levels

$$(4.2.6) \quad W = c^2 m_0 \{1 + \alpha^2/[n_r + (n_\theta^2 - \alpha^2)^{1/2}]\}^{-1/2},$$

which is identical to the one derived later from Dirac's electron theory. In a recent article Biederharn (1983) unravels the underlying mathematical isomorphism explaining this previously not understood coincidence.

This formula for W fits the experimental data extremely well.

4.2.4. THE 'OLD TESTAMENT' OF THE QUANTUM THEORY AND SOMMERFELD'S BIBLE. CORRESPONDENCE PRINCIPLE. TWO NEW IDEAS IN 1925

At work in all this, Bohr, above all, was the spiritual leader, with a leitmotif: his flexible 'correspondence principle' stating that, in the limit of large quantum numbers, the quantal formulas should tend towards the classical ones.

When used in trying to understand other atomic spectra, the Bohr—Sommerfeld theory achieved significant successes, but also met great difficulties. The successes included an explanation of the Stark effect and of the 'anomalous' Zeeman effect resulting from an external electric or magnetic field, respectively. Using in the end four quantum numbers, $n = 1, 2, 3, \ldots; \ell = 0, 1, 2, \ldots; m_\ell = -\ell, -(\ell - 1), \ldots, -1, 0, +1, \ldots, (\ell - 1), \ell; j = \pm\frac{1}{2}$, the theory had become exceedingly complicated, resembling a big construction yard more than a finished monument. A presentation of it was given by Sommerfeld in his 1919 famous treatise *Atombau und Spektrallinien*, often reprinted.

Two very important new ideas were unveiled in 1925. Pauli introduced the fourth quantum number with values $\pm\hbar/2$ and enunciated his 'exclusion principle', according to which no two atomic electrons can have the same set of four atomic numbers; this went beyond Stoner's previous rule fixing the number of electrons per energy level.

In that same year Uhlenbeck and Goudsmit gave meaning to the 'fourth quantum number' by their hypothesis of a 'rotating electron' having both an 'internal angular momentum' or 'spin' and a magnetic moment. This was a queer object: the angular momentum quantum had the values $\pm\hbar/2$, that is, half Bohr's orbital angular momentum quantum; nevertheless, the magnetic moment quantum had the values $\pm e\hbar/2cm_0$, that is, exactly Bohr's 'magneton'. Three years later the Dirac electron theory produced a resolution of this paradox.

4.2.5. BOSE–EINSTEIN AND FERMI–DIRAC STATISTICS

Not long after the promulgation of the Bose–Einstein statistics of indistinguishable particles, Pauli's exclusion principle triggered in 1926 another variety of such statistics, found separately by Fermi and by Dirac, where the 'occupation number' of a state can only be 0 or 1. An approach (Pauli, 1923; Park, 1974, pp. 567–610) fairly appropriate to our forthcoming discussions is as follows.

Let $p_{if} = p_{fi}$ denote the 'intrinsic' transition probability between an 'initial' and a 'final' composite state of particles where they are independent from each other, as in a Boltzmann-like collision. In a predictive calculation using the classical, Boltzmannian, statistics, the p_{if} must be multiplied by the initial occupation numbers n_i of the initial states, whence

$$(4.2.7)\quad P_{if} = \prod(n_i)p_{if}. \qquad \text{(B)}$$

The symmetric statement holds, of course, in a retrodictive calculation.

What is radically new with both the Bose–Einstein and the Fermi–Dirac statistics is that *the occupation numbers of the final state intervene on the same footing as those of the initial state.* So, let us denote by m_f the initial occupation numbers of the final state.

By definition the Bose—Einstein (BE) statistics is such that

$$(4.2.8) \quad P_{if} = \prod(n_i) \prod(1 + m_f) p_{if} \qquad \text{(BE)}$$

and (as a direct consequence of Pauli's exclusion principle) the Fermi—Dirac (FD) statistics is such that

$$(4.2.9) \quad P_{if} = \prod(n_i) \prod(1 - m_f) p_{if} \qquad \text{(FD)}$$

Of course, the reversed statements hold in a retrodictive calculation.

Formulas (7), (8) and (9) clearly show that the two new statistics depart in opposite directions from the classical one.

Now, it is certainly more consonant with the intrinsic symmetry between prediction and retrodiction to consider the final occupation numbers, n_f, of the final state, which are such that $n_f = 1 + m_f$. Then, in both the BE and the FD statistics,

$$(4.2.10) \; P_{if} = \prod(n_i) \prod(n_f) p_{if},$$

displaying *complete time symmetry* (this is read directly in the formula (8) for the BE statistics, and is seen to hold also for the FD statistics, as n_i and n_f can assume only the values 0 and 1).

Viewed in this way, the two new statistics are much closer to the spirit of the Loschmidt symmetry argument than was the classical, Boltzmannian one. In the context of intrinsic past—future symmetry of fundamental physical laws, emphasized throughout the present work, this is much, much more than just an aside. See in this respect Section 3.1.8.

The above rules, (8) and (9), or (10), conform to the classical principle of multiplication of independent probabilities. These, in Jordan's 1926 wavelike probability calculus, must be replaced by a similar rule using amplitudes. To this we shall return.

Thus, in the years 1924—1926 the 'old quantum theory' was singing its swan's song — or rather, like the phoenix, it was dying and rising from its ashes in a fiercely burning fire.

4.2.6 DE BROGLIE'S MATTER WAVES

The ideas contained in Louis de Broglie's 1924 PhD Thesis, published in 1925, are quite obviously among the most important pivots between the 'old' and 'new quantum theory'; in Schrödinger's own words (1926a,

p. 374): "I cannot but mention that I owe the initial impetus from which this work developed mainly to Mr Louis de Broglie's remarkable thesis."

The concept of a 'wave mechanics' as related to the Newtonian, or Hamiltonian mechanics, in a manner similar to the relation of the Young—Fresnel optics to geometrical optics, is quite obviously implied in de Broglie's thesis.[2] Schrödinger expands upon this in the second of his famous series of articles (1926a, b, c), and it is there that the wording 'wave mechanics' occurs for the first time.

Why de Broglie's work is referred to here rather than in the following chapter is because the mathematical machinery constituting the backbone of 'quantum mechanics' is due to other authors, mainly, as is well known, to Heisenberg, Schrödinger and Dirac.

One most remarkable feature of de Broglie's work is that it realizes a perfectly harmonious synthesis of the relativistic requirements and some basic ones of the quantum theory — a feature to be fully recovered only near the end of the first half of this century.

4.2.7. RETROSPECTIVE OUTLOOK

Through a quarter century the 'old quantum theory' had become a huge collection of recipes, either very successful or not quite so. It had become a Ptolemaic sort of theory, adding quantal conditions in place of epicycles.

Time and again it had renewed contact with the relativity theory, each time with flamboyant success.

Also, it had retained its initial faith in statistics, through contributions from Planck, Einstein, Natanson, Bose, Pauli, Fermi and Dirac.

One specific incident should be recalled here, because it traces a link between Planck's 1900 and Born's 1926 contributions: the 1906—1909 Einstein—Ritz controversy, mentioned in Section 4.1.1. Born's and Jordan's 1926 'wavelike probability calculus' — a natural sequel to the Einstein and de Broglie connection $p = \hbar k$ — does indeed require, as explained independently by Fock in 1948 and by Watanabe in 1952, that *the two factlike irreversibility statements of wave retardation and of probability increase are reciprocal to each other.*

Finally, in its first twenty-five years of existence, the quantum theory had covered much ground, in the groping manner of the geometrid caterpillars. Time had come for her to stop, enter a cocoon, and reappear in a brand new, and a winged form.

CHAPTER 4.3

1925–1927: THE DAWN OF QUANTUM MECHANICS WITH A SHADOW: RELATIVISTIC COVARIANCE LOST

4.3.1. LIMINAL ADVICE

These twin chapters, 3 and 4, the subtitles of which are 'Relativistic Covariance Lost' and 'Relativistic Covariance slowly Recovered', do not aim at a balanced presentation of the field of quantum mechanics, of which many significant aspects will not be mentioned. What these chapters aim at is a presentation of those salient aspects of computational methods and of conceptualizing matters which are not only typical of quantum mechanics but are also relevant in our discussion of time, space, motion, probability and information.

4.3.2. 1925: HEISENBERG STARTS THE GAME OF QUANTUM MECHANICS

Like Einstein in 1905, Heisenberg (1925) states that a revolution in thinking is necessary, and that only observables should appear in the mathematics. Such are frequencies and intensities of spectral lines, as he says.

As the atom is a periodic system, recourse to Fourier series is natural. These Heisenberg writes in the exponential complex notation widely used for decades in optics and electromagnetism, where it was considered nothing more than a convenient trick. That it is much more than that, and that it adequately expresses some very deep aspects of Nature, will be one of the many revelations coming from quantum mechanics.

According to Planck's and Bohr's rules, the frequencies of spectral lines are expressed as $\nu = \nu_m - \nu_n$; so, quite naturally, Heisenberg associates the conjugate complex exponentials $\exp[\pm i(\nu_m - \nu_n)]$ to emission and absorption of photons, respectively. He then writes them down in rows and columns labelled by m and n, which, of course, is nothing else than building a Hermitean matrix — but of this Heisenberg was not yet aware.

He thus formalizes the coordinates q and conjugate momenta p of the atomic electrons (unobservable quantities ...) as Hermitean matrices. For these he certainly must define the addition, subtraction (whence differentiation) and multiplication rules. Only the latter raise a problem. What Heisenberg decides is multiplication of the corresponding elements 'rows by columns' — exactly the rule for matrix multiplication. Seeing then that, in general, this sort of multiplication is not commutative, he must decide something for the multiplication of a p by a q matrix. What he does is to write down [1]

$$q_i p_j - p_j q_i = i\hbar\,\delta_{ij} \tag{4.3.1}$$

showing that, as a consequence, the Planck and Bohr formula for energy levels will be recovered; the right-hand side is $i\hbar\,\delta_{ij}$ times a unit matrix. The presence of the imaginary unit i follows necessarily from the assumptions made; also, it is a general rule that the 'commutator' of two Hermitean matrices is an 'anti-Hermitean' matrix.

Expressing the energy of Planck's non-relativistic harmonic oscillator according to these rules, Heisenberg finds, as the energy quantized values,

$$W_n = (n + \tfrac{1}{2})h\nu;$$

that is, Planck's rule, with an additive constant $\frac{1}{2}h\nu$, now termed the 'zero point energy', which is found in all similar quantal calculations.

So, here we have one more example of a major discovery gained by following a very clever, but still simple reasoning. Sophisticated mathematics, of course, will enter the field later. In this particular case, 'later' means in fact 'very soon'.

What lay inside the strongbox Heisenberg had managed to unlock was no less than Hilbert's complex functional space, with its Hermitean operators — the very playground of quantum mechanics.

4.3.3. 1926–1927: BORN AND JORDAN FORMALIZE QUANTUM MECHANICS AS A MATRIX MECHANICS

Born and Jordan were quick to notice that Heisenberg was rediscovering, in his own way, matrix algebra and analysis. This they explained in a paper which is a joint course in matrix calculus and quantum mechanics (Born and Jordan, 1925). As equations of motion they used

Hamilton's relations

$$(4.3.2)\quad \mathrm{d}q/\mathrm{d}t = \partial H/\partial p, \qquad \mathrm{d}p/\mathrm{d}t = -\partial H/\partial q$$

where p, q and H are matrices, the 'Hamiltonian' operator H being expressed as a function of the p's and q's; Heisenberg's commutation formula (1) is of course retained. From this they derive the now famous formulas

$$-ih\dot{q} = Hq - qH, \qquad ih\dot{p} = Hp - pH,$$

deduce Planck's rule $W_m - W_n = h\nu_{mn}$, and recover Heisenberg's formula for the harmonic oscillator.

In a subsequent paper Born, Heisenberg and Jordan (1926) extended the field and showed that their problem was, mathematically speaking, that of finding the eigenvalues of a Hermitean matrix.

Quantization along these lines of the non-relativistic hydrogen atom was performed a little later, by Pauli and Dirac independently, Bohr's initial result being thus recovered.

This matrix route to quantum mechanics was destined to merge with Schrödinger's wave equation route presented in a later section.

As for the question of relativistic covariance, it is obvious that taking as a cornerstone of their theory the Hamilton equations, Born, Heisenberg and Jordan were blocking the way — and blocking it for a long time.

4.3.4. 1925: DIRAC AND THE POISSON BRACKETS

In the first of his many significant contributions to quantum mechanics Dirac showed that Heisenberg's commutators do 'correspond' to the classical Poisson brackets (see Section 2.2.6) and that quantum mechanics can be derived from their use combined with Hamilton's equations — exactly as can be done classically.

This is what Gerald Rosen (1969) terms the 'Dirac passage' to quantum mechanics.

4.3.5. 1926: SCHRÖDINGER FORMALIZES QUANTUM MECHANICS AS A WAVE MECHANICS

De Broglie's PhD Thesis struck a resonance for Schrödinger, who was an expert in the problems of eigenfunctions and eigenvalues of partial differential equations. It seems that he tried first to quantize the hydrogen atom along such lines by using as the electron's wave function what was soon to be called the Klein—Gordon (KG) second-order equation (4.1.9) which expressed directly de Broglie's scheme for a relativistic quantum mechanics; but the energy spectrum he thus found did not fit the observations. So Schrödinger put aside relativistic covariance and wrote down straightaway the now famous energy eigenvalue equation

$$(4.3.3) \quad W\varphi = H\varphi(x, y, z)$$

with

$$(4.3.4) \quad H \equiv (\hbar^2/2m)(\partial_x^2 + \partial_y^2 + \partial_z^2) + V(x, y, z).$$

By setting $V = e^2/r$, this equation exactly solves the non-relativistic hydrogen atom problem. The wave function comes out as the product of a radial function by a spherical Laplace function. All positive values of W are acceptable, yielding a continuous spectrum for an electron scattered by a proton. Negative values of W are acceptable only for integer values of a 'principal quantum number' yielding the Bohr formula; a second quantum number characterizes the Laplace spherical function.

This certainly was a remarkable achievement, and one leading Schrödinger to believe that he had found a 'classical' explanation of the quantization phenomenon.

The operator in Equation (4) is real; its solutions can thus be expressed as real functions of (x, y, z). Schrödinger long remained extremely reluctant to depart from this real character of φ. However, hidden inside the very equation (4), there is a definite reason for moving from the real to the complex field: as this expression transposes *à la* de Broglie the non-relativistic energy-momentum relation

$$W - V(x, y, z) - \mathbf{p}^2/2m = 0,$$

there *must* exist the 'correspondence recipe'

$$(4.3.5) \quad \mathbf{p} = \pm(i/\hbar)\boldsymbol{\partial}$$

soon found, independently, by Schrödinger, and by Born and Wiener; it formalizes the momentum concept in terms of the generator of spatial displacements.

In his next article Schrödinger first discussed, in non-relativisitic terms, the passage from Newtonian to wave mechanics, and then, quantizing Planck's oscillator, he obtained the same result as Heisenberg. He discussed also the rotator and the diatomic molecule.

4.3.6. 1926: MATHEMATICAL 'EQUIVALENCE' BETWEEN HEISENBERG'S AND SCHRÖDINGER'S THEORIES

The following quotation from Schrödinger (1926b) is interesting:

Considering the extreme diversity of starting points and conceptions characterizing ... Heisenberg's quantum mechanics and ... the theory I have called 'wave' mechanics ... [it looks] astonishing that both ... yield the same results, at least in the ... cases presently known, even when these results depart from those given by the old quantum theory. ... This is [all the more] remarkable that everything, starting point, conceptualization, method, mathematical apparatus, looks radically different. Mainly, it seems that these two theories depart from classical mechanics in exactly opposite directions. In Heisenberg's mechanics the classical continuous variables are replaced by ... matrices governed by algebraic equations. This theory [its] authors term 'the true theory of the discrete'. On the other hand wave mechanics improves classical mechanics [by going] in a direction exactly opposite, namely, a theory of the continuous. This theory replaces the mechanical phenomenon, classically described by a finite number of differential equations, by a continuous field ... governed by a unique partial differential equation. ... This ... equation ... replaces both the equations of motion and the quantum conditions. ... In the following pages I will display the intimate connection ... existing between Heisenberg's quantum mechanics and my wave mechanics. Formally, or mathematically, speaking, these two theories are identical. ...

Schrödinger's demonstration proceeds from the remark that the operators q and $\partial/\partial q \equiv \partial_q$ do not commute, as

$$(4.3.6) \quad \partial_q \cdot q - q \cdot \partial_q = 1.$$

A little later Schrödinger observes that this formula must be used in conjunction with (5) to reproduce Heisenberg's quantum condition (1); however, the necessary presence of i for obtaining from ∂_q a Hermitean operator is not explicitly mentioned.

Anyhow, Schrödinger explains that (generally infinite) Hermitean matrices are generated from Hermitean operators, and that the Hamilton equations (2) used in Heisenberg's theory are satisfied by orthogonal solutions of the eigenvalue partial differential equation (4).

This has become the very basic status of (non-relativistic) quantum mechanics: a theory using essentially the machinery of Hilbert's functional space and its Hermitean operators.

In this same year, independently of Schrödinger, Eckart (1926) produced an equivalent demonstration.

Generally speaking, Schrödinger's approach is more convenient than Heisenberg's. It is more 'intrinsic', geometrically speaking, in Hilbert's space; also, it appeals more to the imagination and leaves less to intuitive guesses.

Of course, the rule (5) yielding, in Gerald Rosen's (1969) words, the 'Schrödinger passage' from classical to quantum mechanics is (no more than Dirac's P.B. recipe) an automatic recipe; it leaves open the question of the ordering of non-commuting operators. Of course, if an 'automatic' recipe did exist, leading to the passage, this would mean that the 'new' mechanics is not 'new' at all. . . .

What about the time variation in Schrödinger's theory? The fourth article of this wonderful 1926 series brings it in. Fully accepting by then a complex ψ function, Schrödinger writes his wave equation as

$$(4.3.7) \quad i\hbar\partial_t\psi(x, y, z, t) = H\psi(x, y, z, t)$$

so that, in the energy eigenvalue problem, $\psi = \varphi(x, y, z)\exp(iWt)$. Schrödinger remarks that, as a consequence of the wave equation, the conservation equation

$$\partial_t(\psi^*\psi) = \partial(-\tfrac{1}{2}i\hbar\psi^*[\partial]\psi)$$

holds, allowing him to set the time-independent normalization condition

$$(4.3.8) \quad \iiint \psi^*\psi \, \mathrm{d}x\,\mathrm{d}y\,\mathrm{d}z = 1;$$

this, multiplied by $-e$, expresses conservation of the electric charge.

Various physical problems are tackled by means of this extended formalism.

The 'new quantum mechanics' had thus become a fully viable theory — and an extremely powerful one, as demonstrated by its inventors, and by others who were quick to follow them.

There remains, however, that if, by gaining speed, the flywheel of the Halmiltonian formalism greatly helped in formalizing and solving problems, on the other hand, it rendered more and more difficult the change in steering course needed for regaining de Broglie's initial spirit of a perfectly quantal *and* relativistic theory.

4.3.7. 1926: BORN INTRODUCES, AND JORDAN FORMALIZES, A RADICALLY NEW 'WAVELIKE PROBABILITY CALCULUS'

In treating a particular quantum mechanical problem — collisions — Born (1926) was led to interpret the intensity of a matter wave as the probability that the associated particle occupies it. Jordan (1926) was then quick to see that, quite generally, a *sui generis* wavelike probability calculus should be defined, where partial amplitudes rather than probabilities are added, and independent amplitudes rather than probabilities are multiplied.

This makes one more star in the brilliant 1925—1927 constellation — and the cornerstone of a radically new paradigm in natural philosophy.

4.3.8. NON-COMMUTING POSITION AND MOMENTUM OPERATORS, AND HEISENBERG'S UNCERTAINTY RELATIONS

Heisenberg's (1925) uncertainty relations

(4.3.9) $\Delta x \cdot \Delta p = \hbar$

between the conjugate coordinate x and momentum p of a particle (expressed in Cartesian coordinates) follow directly from the theory of reciprocal Fourier integrals, where the corresponding formula $\Delta x \cdot \Delta k = 1$ holds. Quite similarly the 'fourth uncertainty relation',

(4.3.10) $\Delta t \cdot \Delta W = \hbar$

follows from the reciprocal Fourier transforms between time and frequency. Use of the time—frequency Fourier transforms is common practice in the transmission technique. In quantum mechanics, the expansion of the wave function over the energy eigenfunctions really is a Fourier series or integral over frequencies, as we have seen in Section 4.1.7.

However, by its very nature, the Heisenberg—Schrödinger Hamiltonian formalism views the space—momentum and the time—energy uncertainty relations in essentially different lights.

Various aspects of the 'fourth uncertainty relation' do show up physically, for example in the exponential decay law of higher energy levels or in the Hamiltonian expression of perturbation theory. What can be stated generally is that the 'fourth uncertainty relation' holds between the width $\Delta\nu$ of a spectral line and the time Δt needed for the phenomenon to unfold — or, perhaps, the time alloted by the experimental precedure.

Figure 11 displays a thought experiment where the three uncertainty products $\Delta y \cdot \Delta p_y$, $\Delta z \cdot \Delta p_z$ and $\Delta t \cdot \Delta W$ do appear on exactly the same footing. An inertially moving 'wavicle' falls from the left on a screen placed at the abscissa $x = 0$ with a time-varying aperture, the contours of which obey equations of the form $\mathscr{C}(y, z, t) = 0$. Figure 12 displays the opening and shutting of a typical diaphragm of this sort, the pieces of which move at subluminal velocities. To solve exactly this problem, by expressing the adequate boundary conditions on the plane $x = 0$ and around the aperture, needs some care, especially since one is not allowed to solve a hyperbolic partial derivative equation by choosing the conditions over a timelike surface. The wave existing on the right side, $x > 0$, can be expressed as a Fourier integral over y, z, t, formalizing a 'spacetime diffraction' phenomenon. Both positive and negative frequencies will appear, meaning that the working of the chopping mechanism can in principle create (Figure 11(c)) or destroy (Figure 11(d)) a particle—antiparticle pair.

Coming back to Heisenberg and his 1927 discovery of the un-

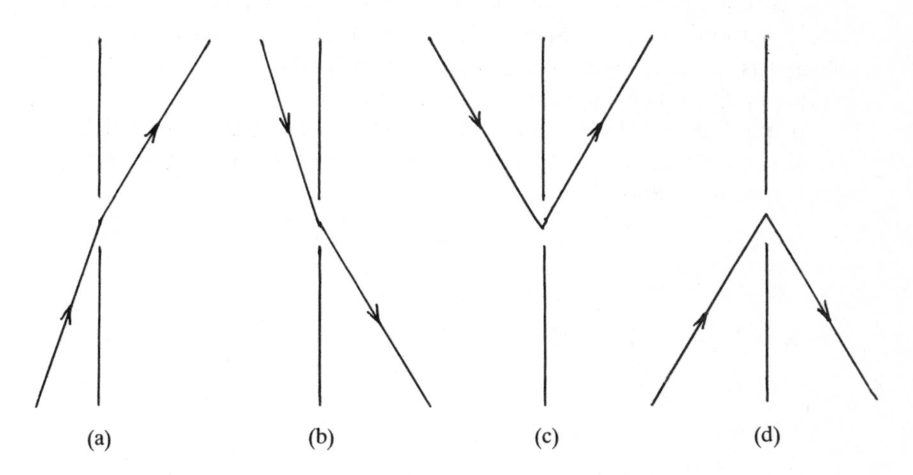

Fig. 11. Spacetime diffraction of a quantized wave by a plane screen $x = 0$ with a variable aperture $\mathscr{C}(y, z, ct) = 0$. Four cases: diffraction of a particle (a) or an antiparticle (b); creation (c) or annihilation (d) of a particle—antiparticle pair.

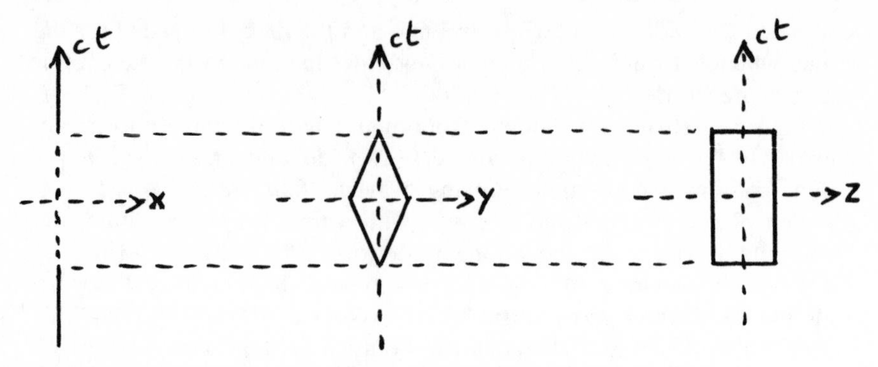

Fig. 12. A possible specification of the varying diaphragm in Figure 11.

certainty relations, let it be recalled that not only did he formalize them in terms of Fourier transforms, but also that he and Bohr discussed various examples of their physical manifestations — a very famous one being the thought experiment of 'Heisenberg's microscope', which happens to be of special interest to us: see below, Section 4.3.11.

4.3.9. NON-COMMUTING ANGULAR MOMENTUM OPERATORS

Non-commutation of elementary rotations inside intersecting biplanes has been discussed in Section 2.5.3. In Euclidean space, these are 'rotations around axes x and y' — which should not be termed 'orthogonal', as their operators do not commute.

In quantum mechanics, denoting u, v, w a circular permutation of the three Cartesian coordinates x, y, z, the 'orbital angular momentum' operators are defined as

$$(4.3.11)\ m_w \equiv i\hbar[x_u\,\partial_v - x_v\,\partial_u] = i\hbar\,\partial\alpha_w$$

thus satisfying the commutation relations

$$(4.3.12)\ [m_u, m_v] = -i\hbar m_w.$$

The operator

$$(4.3.13)\ \mathbf{m}^2 = m_u^2 + m_v^2 + m_w^2$$

commuting with all three m_u's, is ($-\hbar^2$ times) the Laplacian over the unit sphere, the eigenfunctions of which are 'Laplace spherical functions', with eigenvalues $\ell(\ell + 1)$; the $m_\ell = -\ell, -(\ell + 1), \ldots, -1, 0,$

+1, . . . , $(\ell - 1)$, ℓ, quantum number is the eigenvalue of the orbital angular momentum around some chosen axis, say z. A typical difference between the 'new' and the 'old' quantum mechanics is that $\ell(\ell + 1)$ replaces ℓ^2. So, on the whole, at the microlevel, the components of an angular momentum are definitely *not* those of a vector.

Using formulas (12) and (13) in a reciprocal group theoretical reasoning together with the fact that m^2 is positive definite, one is led, after some algebra, to the conclusion that the preceding association of the $\ell(\ell + 1)$ and m_ℓ eigenvalues holds not only for integer, but also for half-integer values of ℓ. These, of course, are not accessible as quantized values of an orbital angular momentum; they do have, however, a physical interpretation, being related to the intrinsic angular momentum, or spin, a 'property' of elementary particles.

For example, the electron has a spin $j = \pm \frac{1}{2}\hbar$, which Pauli formalized in 1927 by using the following set of 2×2 Hermitean matrices:

$$(4.3.14)\quad \sigma_1 = \begin{pmatrix} 0 & 1 \\ 1 & 0 \end{pmatrix}, \quad \sigma_2 = \begin{pmatrix} 0 & -i \\ +i & 0 \end{pmatrix}, \quad \sigma_3 = \begin{pmatrix} 1 & 0 \\ 0 & -1 \end{pmatrix}$$

obeying formulas (12) and (13); $\frac{1}{2}\hbar\boldsymbol{\sigma}$ is the electron's spin.

4.3.10. 1929: ROBERTSON'S FORMALIZATION OF THE UNCERTAINTY RELATIONS

This elegant formalization holds inside the Hamiltonian formalism (Robertson, 1929). R and S denoting two Hermitean operators, r_0 and s_0 two real numbers chosen arbitrarily, $(R - r_0)^2$ and $(S - s_0)^2$ the 'squared deviation' operators with respect to r_0 and s_0, ε and η the 'mean squared deviations', one has

$$\varepsilon \equiv \langle\varphi|\,(R - r)^2\,|\varphi\rangle^{1/2} = ||(R - r_0)\varphi||,$$
$$\eta \equiv ||(S - s_0)\varphi||.$$

Then one obtains, via the 'Cauchy—Schwartz inequality',

$$\varepsilon\eta = |\langle\varphi|\tfrac{1}{2}i[Rs - SR]\,|\varphi\rangle|,$$

$i[RS - SR]$ being, of course, a Hermitean operator.

Placing r_0 at the mean value of R one gets, as usual,

$$\varepsilon^2 = \langle r^2\rangle - \langle r\rangle^2.$$

4.3.11. 1929: HEISENBERG'S MICROSCOPE THOUGHT EXPERIMENT AND STATISTICAL RETRODICTION. 1931: VON WEISZÄCKER'S MODIFIED USE OF IT AND RETROACTION

As a heuristic argument supporting the position—momentum uncertainty relation Heisenberg presented in 1927 his famous microscope thought experiment.

The classical expression $\Delta x \simeq \lambda/2 \sin \alpha$ of a microscope's resolving power, where λ denotes the wavelength used and α the maximum obliquity of entering rays, can be rewritten as $\Delta x \cdot \Delta k_x \simeq 1/2\pi$, $2\pi k$ denoting the spatial frequency; that is, also as $\Delta x \cdot \Delta p_x \simeq h$, p denoting the momentum of a photon. For light scattered from a particle — say an electron observed through the microscope — the combined uncertainty $\Delta x \cdot \Delta p_x$ can be transferred, so to speak, to the electron.

Heisenberg assumed the microscope to be focused onto the object plane, as is appropriate for a position measurement, the accuracy of which is then improved by using short wavelengths and large apertures. However, both of these improvements increase Δp_x, making the knowledge of the electron's momentum more uncertain.

Let it be made clear that *such a position measurement is retrodictive*: from the punctual impact B of a photon in the image plane, the corresponding position A of the electron in the object plane is inferred, in the form of an intensity, or probability distribution: the diffraction pattern ideally produced in the object plane by a point source at B. This is one example showing that '*blind statistical retrodiction*' (in Watanabe's (1955) wording) *does make sense in physics*.

In 1931 C. von Weiszäcker proposed an alternative use of the microscope; by aiming it at infinity, a scattered photon will now measure the momentum of the scattering electron, with a precision that is increased by going to longer wavelengths and smaller apertures, both operations being detrimental with respect to the position. Of course, this also is a retrodictive measurement.

Now, ideally thinking of the microscope as very long, we can conceivably decide which magnitude we choose to measure — position or momentum — while the photon is on its way, *after it has been scattered.* Wheeler (Miller and Wheeler, 1983; Wheeler, 1978) has extensively discussed such 'delayed choice experiments', proposing operational procedures that could be used.

Clearly, deciding 'in the end' which of two *mutually exclusive* magnitudes *has been* measured, is undoubtedly a *sui generis* form of retroaction. . . .

4.3.12. ON THE TIME–ENERGY UNCERTAINTY RELATION IN NON-RELATIVISTIC QUANTUM MECHANICS

There are quite a few instances where a time–energy uncertainty formula, $\Delta t \cdot \Delta W \simeq h$, appears in the Hamiltonian formalization of quantum mechanics. Most usually it expresses a connection between the duration of a measurement, an interaction, or a decay, and the energy uncertainty. Of course if reciprocal Fourier transforms using time and frequency as variables can be significantly introduced, as is common practice in telecommunication problems, such an indeterminacy relation automatically follows. However, this is not a procedure consistent with the instructions for using the Hamiltonian formalism in quantum mechanics, since in this formalism the time t is not represented by an operator, whereas the space coordinates are.

In 1945 Mandelstam and Tamm proposed a formal derivation of the formula $\Delta t \cdot \Delta W \simeq h$ in the Hamiltonian scheme, which many textbooks reproduce in the following abridged form: from the two well-known relations

$$\hbar \, \mathrm{d}\langle R \rangle / \mathrm{d}t = i \langle HR - RH \rangle$$

and

$$\Delta R \cdot \Delta W = \tfrac{1}{2} |\langle HR - RH \rangle|$$

in which H is the Hamiltonian and R some Hermitean operator, one obtains (Jammer, 1966, p. 14) $\Delta t \cdot \Delta W = \hbar/2$, by equating the uncertainty ΔR with the variation of $\langle R \rangle$ in time.

For example, if R is the position operator x, and Δx the length of a wave packet propagating in direction x with the velocity v, then $\Delta t = \Delta x / v$ is the transit time at some place x_0. The energy uncertainty is $\Delta W = \Delta(p^2/2m) = v \, \Delta p$, so that $\Delta t \cdot \Delta W = \Delta x \cdot \Delta p \simeq \hbar/2$, Q.E.D. More precisely, if we consider v as the velocity of the mean value of x, Δx is then equated to the variation of $\langle x \rangle$, as in the general argument above.

Therefore, *operationally speaking*, everything is as consistent as it can be in a non-relativistic formalism.[2]

Of course the 'fourth uncertainty relation' had been discussed more than once before. In 1930, at the 6th Solvay Conference, Einstein told Bohr privately that it would be violated if the energy measurement were replaced by a mass measurement. To this Bohr (1939, p. 226), after a sleepless night, victoriously answered by having recourse to . . . Einstein's formula for the rate change of a clock inside an acceleration or a gravitational field!

Here I present Bohr's argument in a form modified so as to avoid a later criticism by Treder (1971).

Suppose that the pan of some balance moves along the z direction, and that the restoring force, whatever is its physical nature, is vanishingly small, as also is the damping force present in all balances. Then, an error Δm in the mass measurement results in an unknown acceleration g of the pan which, after a time t, imparts to it a velocity gt, assumed to be small with respect to c. According to the Lorentz formulas, $\Delta t/t = c^{-2}vz/t = c^{-2}gz$ then is the error in the assignment of the proper time of an event occurring on the pan, as evaluated in the laboratory. By the change in notation $z \to \Delta z$ we rewrite this as

(4.3.15) $tg\,\Delta z = c^2\,\Delta t.$

Now, at time t, the acceleration g has imparted to the pan an unknown momentum $\Delta p = tg\,\Delta m$ such that the 'third uncertainty relation' $\Delta z \cdot \Delta p_z$ holds, whence

(4.3.16) $\Delta m tg\,\Delta z \geqslant \hbar/2.$

If we assume that Δz is such as to exactly hide all effects of the unknown acceleration g, it must be the same in both relations (15) and (16). Therefore, eliminating $tg\,\Delta z$, we get

$$c^2\,\Delta m\,\Delta t \geqslant \hbar/2, \qquad \text{Q.E.D.}$$

Presented in this way Bohr's argument avoids Treder's criticism and holds 'universally', whatever sort of balance is used.

As for Treder's argument, it rests entirely on the consideration of the damping force, and here it is.

All balances must be damped in order that they can be read. The damping force dissipates an energy ΔW, the statistical fluctuation of which is related to the relaxation time Δt by the 'fourth uncertainty relation' $\Delta t \cdot \Delta W \geqslant \hbar/2$. Therefore, contrary to Einstein's 1930 claim, replacing the energy by a mass measurement is of no use — a consideration that Bohr also had overlooked.

Treder's argument is shorter than Bohr's, but of a less fundamental nature.

Bohr's argument is very closely related to a heuristic Einstein argument of 1911 in which he assumed 'equivalence' between gravity and acceleration for obtaining the so-called 'gravitational Doppler shift'. Therefore it may be of significance for the physical link between the quantal and the gravitational theories.

4.3.13. THE HILGEVOORD–UFFINK CONCEPTION OF THE POSITION–MOMENTUM AND TIME–ENERGY UNCERTAINTIES

This ingenious conceptualization (Hilgevoord and Uffink, 1985) is based on the function known as the 'convolution product' or as the 'characteristic' function' in, respectively, its Fourier associated expressions. Here y denotes a spacetime translation; in fact, a virtual translation of the macroscopic apparatus. We set

$$(4.3.17)\quad F(y) = \iiiint \psi^*(x)\psi(x-y)\,\mathrm{d}\omega = \iiiint \exp(-ikx)|\,\theta(k)|^2\,\mathrm{d}\tau.$$

The bulk of the (real) function $|\,\theta(k)|^2$ of momentum-energy $p = \hbar k$ is taken as contained within a four-dimensional volume; then its (complex) Fourier transform $F(y)$ can also lie mainly within a spacetime volume; explicit discussions are given by the authors. Of course, $\psi^*(x)\psi(x-y)$ is the transition amplitude between two representations of the system separated by a distance y in spacetime. Therefore, with an automatic summation over x,

$$(4.3.18)\ F(y) = \langle a|x\rangle\langle x-y|b\rangle = \langle a|y+x\rangle\langle x|b\rangle \stackrel{\text{def}}{=} \langle a|\,U(y)\,|b\rangle.$$

This conceptualization of the 'complementarity' between spacetime and momentum energy fits quite nicely the Schwinger–Feynman relativistic quantum field theory, which is summarized in Sections 4.4.8 and 4.4.9. If $P = \hbar K$ denotes the (conserved) total momentum energy of interacting fields, formula (18) holds for any of these fields, with [see formula (4.4.18)] $U = \exp(iKx)$ denoting the spacetime translation operator. A similar systematization is of course also possible using the 6-component rotation of the Poincaré–Minkowski tetrapod (rotations and/or uniform translations of the macroscopic apparatus) and the 6-component total angular momentum (angular momentum *stricto sensu* and/or boost). Hilgevoord and Uffink (1985) discuss explicitly the spatial rotation case.

4.3.14. NON-RELATIVISTIC QUANTUM MECHANICS OF MANY PARTICLES

The founders of analytical dynamics — Lagrange, Hamilton, Jacobi — and those of statistical mechanics — Boltzmann, Gibbs, Liapounov — were describing the variations of systems of $3N$ spatial coordinates as functions of 'the' time t; before and after 1900, Darboux, Hertz, Felix Klein, Poincaré thought of these as the $3N$ coordinates of one single point moving in a $3N$-dimensional space — a concept that is, since then, quite commonplace in analytical and in statistical non-relativistic mechanics.

Thus it is that Schrödinger was quite naturally led to write down the wave equation of a system of N interacting particles in the form

$$i\hbar\,\partial_t \psi = \{\sum[(\hbar^2/2m_p)\nabla_p] + V(x_1 \ldots x_{3N})\}\,\psi$$

implying, of course, that Jordan's rules for probability amplitudes replace the classical rules for probabilities. This entails the emergence of highly paradoxical 'interference' or 'beating' phenomena, extremely well substantiated experimentally, which we shall discuss now.

Quantal particles are either 'bosons' or 'fermions', respectively obeying the Bose—Einstein or the Fermi—Dirac statistics. Various authors have pointed out that this can be expressed by using, respectively, a wave function that is symmetric or antisymmetric in an exchange of the particles, a property which is time independent by virtue of the wave equation.

Not only the spatial coordinates but also the spin states of the particles are significant in this exchange.

There are two very typical physical systems, closely resembling each other, where these mathematical and physical facts are displayed. The first one is the helium atom, studied by Heisenberg in 1928; the second one is the hydrogen molecule, studied by Heitler and London in 1927. In both cases there are two distinct spectra, as shown in Figure 13, with only weak transitions between the two.

In both cases the key element consists of a pair of formally 'exchangeable' electrons with energy eigenvalues a and b: those they have

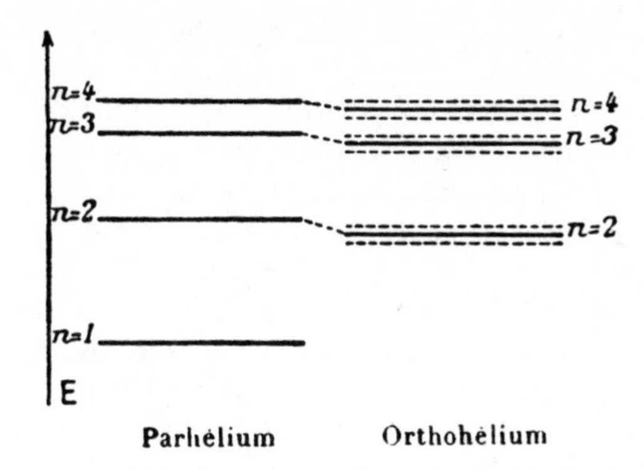

Fig. 13. 'Non-separability' of the two electrons of the helium atom: the spectra of parhelium (a) and orthohelium (b). An analogous but more intricate phenomenon occurs with the hydrogen molecule.

in the Coulomb field of the helium nucleus, or those they have in the fields of their respective protons (this is why Heitler's and London's perturbation method has to be more elaborate than Heisenberg's). For each electron, denoted 1 or 2, there are two possible spin states, denoted as ↑ (up) or ↓ (down). Therefore there is one wave function of the so-called 'para' type (1, 2) [↑↓ − ↓↑], and three wave functions of the 'ortho' type [1, 2] (↑↑), [1, 2] (↓↓) and [1, 2] (↑↓ + ↓↑); a parenthesis means symmetry and a bracket antisymmetry. Explicitly, *a* and *b* denoting energy levels,

$$(4.3.19)\ (1,2) = \varphi_a(1)\varphi_b(2) + \varphi_a(2)\varphi_b(1),$$
$$[1,2] = \varphi_a(1)\varphi_b(2) - \varphi_a(2)\varphi_b(1);$$

Jordan's rules of multiplying independent and adding partial amplitudes are clearly displayed.

If both the *a* and *b* energy levels are the ground states, [1, 2] vanishes but (1, 2) does not; therefore, the ground state seen in the 'para' spectrum has no counterpart in the 'ortho' one.

The single lines of the 'para', and the triple lines of the 'ortho', system do show up in the spectra (Figure 13). The energy levels are slightly lower in the 'ortho' than in the 'para' spectrum owing to the negative sign of the off-diagonal, or beating, contribution to the overall energy *W*, termed 'exchanged energy', and denoted *A*; *C* denotes the usual, Coulomb style energy; $W = C + A$ in the 'para' and $C - A$ in the 'ortho' spectrum.

Finally, very weak transitions occur between the two spectra, as induced by the spin—spin interactions of the electrons.

These 'surprising but adequate explanations' of 'surprising but existing facts' have rightly been hailed as triumphs of the 'new quantum mechanics'.

Why I have presented them in some detail is because they are closely analogous to the even more 'surprising but existing facts' of the so-called EPR correlations, to be discussed at length in Chapter 4.7.

In the helium atom and the hydrogen molecule, the paired electrons behave, so to speak, as Siamese twins that are 'inseparable' and very close to each other. The super-Siamese twins of the EPR correlations are indeed, in the same sense as here, 'inseparable', but they are very distant from each other. Therefore the adequate paradigm is still more surprising than the one that suffices here — and it has to be relativistically covariant.

4.3.15. ENNUPLE QUANTAL CORRELATIONS: GENERAL FORMALISM

Using Dirac's notation, we consider disjoint Hilbert spaces spanned by normalized vectors $|\varphi\rangle, |\psi\rangle, \ldots$, and Hermitean operators $m, p, \ldots$, respectively acting in them.

The 'strict quantal correlation' (of which examples have just been discussed) occurs when state vectors are of the form

$$(4.3.20)\ |\Phi\rangle = \sum c_i |\varphi_i\rangle |\psi_i\rangle \cdots$$

We consider now the more general case where

$$(4.3.21)\ |\Psi\rangle = c^{ij\cdots} |\varphi_i\rangle |\psi_j\rangle \cdots,$$

with an automatic summation on the i's, j's, ...

The mean value of the Hermitean operator

$$(4.3.22)\ M \equiv m \otimes p \otimes \cdots$$

is expressed as

$$(4.3.23)\ \langle\Psi| M |\Psi\rangle = c^{*i'j'\cdots} c^{ij\cdots} \langle\varphi^{i'}| m |\varphi^i\rangle \langle\psi^{j'}| p |\psi^j\rangle.$$

It contains a sum of diagonal terms

$$(4.3.24)\langle\Psi|\, M\,|\Psi\rangle_0 = w^{ij\cdots}\langle m_i\rangle\langle p_j\rangle\cdots,$$

where by definition (no $i, j, \ldots$ summation this time!)

$$w^{ij\cdots} \equiv c^{*ij\cdots}\, c^{ij\cdots},$$

and a sum of off-diagonal terms

$$(4.3.25)\,\Delta\langle\Psi|\, M\,|\Psi\rangle \equiv \langle\Psi|M\,|\Psi\rangle - \langle\Psi|\, M\,|\Psi\rangle_0$$

both of which are of course 'relative' to the representation chosen.

The off-diagonal terms disappear if the representation diagonalizes one [case (20)] or all [case (21)] of the m operators. This is no more than a 'perspective effect', somewhat like one of those three showing a parallelepiped as a rectangle. Even when they are formally absent, the off-diagonal terms are 'potentially present' — as long as no measurement has changed the physical situation.

The essential difference between the classical and the quantal types of probability calculus is seen by comparing formulas (21) and (24); they have similar structures, and respectively display a sum of products of amplitudes and of probabilities. What makes the difference is the existence of the interference terms (23).

The validity of this formalism is not restricted to the Hamiltonian expression of quantum mechanics; it holds also in relativistic quantum mechanics — for example, in its 'S-matrix' algorithm, considered at length in Chapter 5.

4.3.16. THE SCHRÖDINGER, HEISENBERG AND INTERACTION REPRESENTATIONS

In view of a future comparison with the relativistically covariant formalism, a few well-known theoretical facts are here summarized.

If H denotes a time-independent Schrödinger Hamiltonian, the solution of Schrödinger's time-dependent equation (7) may be expressed (using Dirac's notation) as

$$|\psi_2\rangle = |U_{21}\psi_1\rangle \quad \text{or} \quad |\psi_1\rangle = |U_{12}\psi_2\rangle$$

with by definition

$$U_{21} = \exp[iH(t_2 - t_1)] = 1 + iH(t_2 - t_1) + \cdots$$
$$U_{12} = \exp[-iH(t_2 - t_1)] = 1 - iH(t_2 - t_1) + \cdots$$

whence, of course,

$$U_{12} = U_{21}^{+} = U_{21}^{-1}.$$

Therefore, 'as time unfolds', a complete orthogonal set of unit vectors $|\psi_i(x)\rangle$, or $\langle i\,|\,x\rangle$, 'rotates' in the Hilbert space.

This is the 'Schrödinger picture',[3] in which the Hermitean operators R associated with observables are kept constant, so that the variation in time of their mean expected values is expressed as

$$\langle \psi_2 |\, R \,| \psi_2 \rangle = \langle \psi_1 U^{-1} |\, R \,| U\psi_1 \rangle.$$

In the earlier Heisenberg picture the 'state vector' did not show up, meaning implicitly that it was constant; but the Hermitean operators varied in time according to the formula

$$R_2 = U^{-1} R_1 U.$$

Therefore the variation in time of the mean expectation value has the expression

$$\langle \psi |\, R_2 \,| \psi \rangle = \langle \psi |\, U^{-1} R_1 U \,| \psi \rangle$$

so that, of course,

$$\langle \psi_2 |\, R_1 \,| \psi_2 \rangle = \langle \psi_1 |\, R_2 \,| \psi_1 \rangle$$

The 'interaction picture' is useful when one likes to think of 'transitions induced' between the orthogonal eigenfunctions of an 'unperturbed Hamiltonian' H_0 by some 'perturbing Hamiltonian' ΔH. One then writes $H = H_0 + \Delta H$, and treats H_0 *à la* Heisenberg and ΔH *à la* Schrödinger.

4.3.17. NON-RELATIVISTIC PERTURBATION THEORY

Various forms of perturbation theory have been used in quantum mechanics, by its founders and by those following them. A typical one uses the interaction representation as summarized in Section 16, with the perturbing Hamiltonian acting upon the coefficients of the expansion of the state vector over the unperturbed states. This is called the 'method of varying the constants', and has been widely used.

These 'Hamiltonian style' methods, in which the time and the energy concepts are strongly privileged, are thus extremely non-relativistic. For example, in quantum electrodynamics, dealing with emission and absorption of photons by transitions between electronic energy states, rigorous conservation of momentum, but not of energy, was assumed. 'As time unfolds', in this Newtonian sort of paradigm, fluctuating energy defects did show up, later progressively damped. They became negligible after a time interval Δt larger than the photon's reciprocal frequency; this, of course, was an aspect of the 'fourth uncertainty relation'.

Calculations using this scheme were very tedious and inelegant, especially when relativistic covariance was needed in the end result. Obviously, the basic conceptual defect was to treat time and energy differently from length and momentum — and, also, not to require a strict energy conservation.

4.3.18. 'TRANSFORMATION THEORY': DIRAC, 1926; JORDAN, 1927

The so-called 'transformation theory' exploits the full geometric potentialities of the Hilbert space concept, tying them directly to the interpretation of quantum mechanics. Some examples of this have occasionally been met in this chapter. Now, for the sake of completeness, and in view of subsequent discussions, I briefly summarize the matter, not claiming mathematical rigor (far from that!), or completeness.

The Hermitean scalar product $\langle \varphi | \psi \rangle = \langle \psi | \varphi \rangle^*$ of a 'bra' $| \rangle$ and a 'ket' $\langle |$ is interpreted as the *transition amplitude* between a *representation* $|\varphi\rangle$ and a *representation* $|\psi\rangle$; if both are unit vectors ($\langle \varphi | \varphi \rangle = \langle \psi | \psi \rangle = 1$), this is the anolog of a cosine. If both $|\varphi\rangle$ and $|\psi\rangle$ belong to a complete orthogonal set spanning the Hilbert space, the $\langle \varphi_i | \psi_j \rangle$'s can be understood either as the coordinates of the $|\varphi\rangle$'s in the $|\psi\rangle$ representation, or vice versa.

The variables of which the φ's and ψ's are functions can be, say, the space coordinates x in φ and frequency coordinates k in ψ. Therefore, complete orthonormalized sets, such as $\varphi_i(x)$ or $\psi_j(k)$, truly depend upon two variables, say, i and x, or j and k. In other words, they are transition amplitudes between two representations, the i one and the

x one, or the j one and the k one. A familiar example of this is Schrödinger's representation of ψ, in a static problem, as $\psi_i(x)$, with i labeling the energy eigenstates; so these $\psi_i(x)$'s are transition amplitudes between a coordinate x and an energy W representation (clearly, an extremely non-relativistic formalization).

This led Dirac to think of the $|\varphi_i(x)\rangle$'s as Hermitean scalar products $\langle i|x\rangle$ rather than as a collection of kets.[4]

Thus new N-pods of $|i\rangle$'s are brought into the picture, and we can write, with an automatic summation over the repeated index,

$$\langle i|x\rangle = \langle i|j\rangle\langle j|x\rangle, \qquad \langle j|k\rangle = \langle j|i\rangle\langle i|k\rangle.$$

As a physical example we have, in the k representation, the change of a photon's polarization, from one using the two circular, to one using two orthogonal linear polarizations, or vice versa.

Also we can write (again with an automatic summation on the repeated index)

$$\langle x|i\rangle = \langle x|k\rangle\langle k|i\rangle, \qquad \langle k|j\rangle = \langle k|x\rangle\langle x|j\rangle.$$

This means exchanging the x and the k representations — that is, writing reciprocal Fourier transforms.

Geometrically speaking, the composition law of transition amplitudes

$$\langle a|c\rangle = \langle a|b\rangle\langle b|c\rangle \tag{4.3.26}$$

is the composition law of orthogonal transformations (rotations and/or reflections). It clearly has a group structure, which is emphasized by writing, with Landé (1965, p. 156),

$$\langle a|b\rangle\langle b|c\rangle\langle c|a'\rangle = \delta(a, a').$$

The important point for the interpretation of quantum mechanics is that this formula transposes in terms of amplitudes the classical Markovian composition law of probabilities — which is one more expression of the passage from the classical to the Jordan wavelike probability calculus.

A specification of this law that has been met in Chapter 4.1 is

$$\langle x|i\rangle = \langle x|x'\rangle\langle x'|i\rangle$$

with $\langle x|x'\rangle$ denoting a displacement transition amplitude. In the Schrödinger formalism $\langle x|x'\rangle = \delta(x - x')$, with δ denoting 'Dirac's δ'; in the relativistically covariant formalism for free particles, $\langle x|x'\rangle$ is the Jordan—Pauli propagator, behaving as Dirac's δ only if $x - x'$ is spacelike; otherwise, it 'propagates' the field.

The transition amplitude $\langle \psi | \varphi \rangle$ between a prepared state $| \varphi \rangle$ and a measured state $| \psi \rangle$ (or, more explicitly, $\langle \psi | U | \varphi \rangle$ in the Schrödinger picture) is thus a change in representation, just as, say, a Fourier transform $\langle k | x \rangle = \langle x | k \rangle^*$. Being a Hermitean scalar product it is symmetric in the two 'state vectors', in the sense that $\langle \psi | \varphi \rangle = \langle \varphi | \psi \rangle^*$. But, like any scalar product, it can be thought of and computed asymmetrically; that is, either predictively, from $| \varphi \rangle$ to $| \psi \rangle$, by projecting $| \varphi \rangle$ upon $| \psi \rangle$, which is called state vector collapse, or retrodictively, from $| \psi \rangle$ to $| \varphi \rangle$, by projecting $| \psi \rangle$ upon $| \varphi \rangle$, which I propose to call retrocollapse.

These are aspects of the intrinsic reversibility of the probability calculus, still valid in the new Jordan version of it. Of course $|\langle \psi | \varphi \rangle|^2$ is the transition probability between $| \varphi \rangle$ and $| \psi \rangle$.

The previous wording has been purposely presented in an atemporal, *subspecie aeternitatis*, style, so that it fits in advance into the paradigm of the relativistic quantum mechanics.

Finally a few remarks concerning operators are appropriate.

Projectors such as

$$| i \rangle \langle i | = | i \rangle \langle i | i \rangle \langle i | = \cdots$$

have been met repeatedly. Hermitean operators $| R |$ are, by definition, expressible as weighted sums of projectors

$$| R | = | i \rangle \, r_i \langle i |,$$

with their eigenvalues r_i and eigenstates $| i \rangle$ thus displayed. 'Degeneracy' occurs if two or more r_i's are equal; in continuous parts of their 'spectrum', the above sum holds for an integral. One has of course

$$\langle \varphi | \, R \, | \varphi \rangle = \langle \varphi | i \rangle \, r_i \langle i | \varphi \rangle$$

with the $\langle \varphi | i \rangle \langle i | \varphi \rangle$'s, or $c_i^* c_i$'s, denoting the respective weights of the eigenvalues r_i in the $| \varphi \rangle$ state.

Adjoint non-Hermitean operators, A^+ and A, are by definition such that

$$\langle \varphi A^+ | \varphi \rangle = \langle \varphi | A \varphi \rangle.$$

A *Hermitean operator* is self-adjoint, so that $A | = | A$.

Unitary operators are such that, by definition,

$$U^{+} = U^{-1} \qquad \text{or} \qquad U^{+}U = UU^{+} = 1.$$

If R denotes a Hermitean operator, $\exp(\pm iR)$ (defined by means of the power series) are *adjoint unitary operators*.

The very convenient symbolic calculus that has been outlined can be extended to the relativistic quantal formalism, where it greatly helps, as a stenographic notation, in handling the subtle concepts that are at stake.

For an extremely brilliant presentation of the matter, the reader is referred to Dirac's authoritative *Quantum Mechanics*, especially in its third edition presentation.*

4.3.19. 'GRANDEUR AND SERVITUDE' OF THE HALMITONIAN FORMALISM

The Hamiltonian formalism has been to quantum mechanics what the horse collar has been to traditional agriculture.

On the one hand, it has allowed deep plowing, and has thus produced a wonderful harvest of very new, very significant and very useful crops.

On the other hand, it cannot be denied that a horse collar is a sort of yoke, and that it is desirable that unyoked horses are used for running or jumping.

So, just as Newton's 'time instant t' had to expand into the light cone, it has turned out that the Hamiltonian formalism must yield in favor of a new one, essentially using the propagator concept.

* See 'Added in Proof' p. 319ff.

CHAPTER 4.4

1927–1949: FROM QUANTUM MECHANICS TO QUANTUM FIELD THEORY: RELATIVISTIC COVARIANCE SLOWLY RECOVERED

4.4.1. SECOND- AND FIRST-ORDER COVARIANT WAVE EQUATIONS

Maxwell's electromagnetic theory rests on a specific relativistically covariant system of first-order wave equations, entailing as a consequence d'Alembert's equation for the various field components. If charges and currents are present these equations contain source terms.

The extension of d'Alembert's equation for matter waves is known as the 1927 Klein–Gordon (KG) equation (see Klein, 1926; Gordon, 1926), also proposed by other authors. The problem is how to supplement it with physically appropriate source terms — a problem not independent of the one of finding appropriate systems of first-order equations for matter.

As previously said, Schrödinger found that he could not reproduce the hydrogen atom spectrum by using the KG equation with the $i\partial$ operators replaced by $i\partial - Q\mathbf{A}$. Dirac later showed that a group of terms implying a direct coupling between the magneto-electric field and the electron's magneto-electric moment was missing.

Dirac's 1927 discovery of the first-order electron equation, through factorization of the KG equation, has been a major scientific achievement. Due to its relevance to the subject matter of the present book, relativistic invariance of Dirac's equation under rotations and (partial or total) reversal of axes of the Poincaré–Minkowski tetrapod will now be discussed in some detail.

4.4.2. 1927: DIRAC'S FIRST-ORDER EQUATION DESCRIBING JOINTLY AN ELECTRON AND A POSITRON

Dirac (1928), aiming at describing correctly the electron, decided that the wave equation should, like Schrödinger's, be of first order in time, and, by relativistic invariance, also in x, y, z. His reason for this was that he wanted a positive definite expression for the probability density, like Schrödinger's $\psi^*\psi$. Retaining Pauli's (1927) idea that ψ^* and ψ

should respectively be a row and a column matrix, he factored the Klein—Gordon equation by writing[1]

(4.4.1) $(\gamma^i\partial_i - k)(\gamma^i\partial_i + k) \equiv \partial_i^i - k^2$

where $\gamma^i \equiv \gamma_i$ denote four Hermitean, anticommuting matrices:

(4.4.2) $\gamma^i\gamma^j + \gamma^j\gamma^i = 2\delta^{ij}$.

He found this to be possible with rank-4 Hermitean matrices related to Pauli's of rank-2 according to[1]

$$\boldsymbol{\gamma} = \begin{pmatrix} 0 & \boldsymbol{\sigma} \\ -\boldsymbol{\sigma} & 0 \end{pmatrix}, \quad \gamma^4 = \begin{pmatrix} 1 & 0 \\ 0 & -1 \end{pmatrix} \quad \text{or} \quad \gamma^4 = \begin{pmatrix} 0 & 1 \\ 1 & 0 \end{pmatrix}.$$

It turns out that the matrix[2]

(4.4.3) $\gamma^5 \equiv \gamma^1\gamma^2\gamma^3\gamma^4$

is also Hermitean and unitary, and anticommutes with the four γ^i's, so that

(4.4.4) $\gamma^I\gamma^J + \gamma^J\gamma^I = 2\delta^{IJ}, \quad I, J = 1, 2, 3, 4, 5.$

These matrices afford a representation of the 1878 Clifford algebra.

As applied to solutions of the Klein—Gordon (KG) equation $\mathscr{G}\psi = 0$, the Dirac operators $\mathscr{D}_\pm \equiv 2^{-1/2}(\gamma^i\partial_i \pm k)$ are mutually orthogonal projectors. Either of them can be used for writing the Dirac equation $\mathscr{D}_\pm\psi = 0$; the other one, $\mathscr{D}_\mp$, projects any solution of the KG equation as a solution of the Dirac equation.

The conserved 4-current density is defined as

(4.4.5) $j^i \equiv i\bar{\psi}\gamma^i\psi$

where, by definition,[3]

(4.4.6) $\bar{\psi} = \pm\psi^+\gamma^4$

so that, as was required,

(4.4.7) $-ij^4 = \pm\bar{\psi}\gamma^4\psi = \pm\psi^+\psi$.

The freedom of the double sign in (6) is found necessary when discussing reversals of axes. From it the double sign in (7) follows — and of course is required if the 4-vector j^i behaves 'regularly' under reversal of the time axis. As will be seen, this double sign has to do also with the electron—positron exchange, $e \rightleftharpoons -e$.

Presently we discuss the invariance of the Dirac adjoint equations

$$(4.4.8)\quad 0 = \bar{\psi}(-\gamma_i \overleftarrow{\partial}^i + k), \qquad (\gamma_i \overrightarrow{\partial}^i + k)\psi = 0,$$

under rotations of the Poincaré—Minkowski tetrapod and under reversals of its axes. As this question is tied in with that of the charge reversal $-e \rightleftharpoons +e$ denoted C, and also termed particle—antiparticle exchange, the discussion must be extended to the case of an electron subject to an external 4-potential A^i. To this end Dirac made in the (8)'s the usual substitution[4] $i\partial^i \rightarrow i\partial^i + eA^i$ yielding (in units such that $c = 1$ and $\hbar = 1$)

$$(4.4.9)\quad 0 = \bar{\psi}[(-\overleftarrow{\partial}^i - ieA^i)\gamma_i + k], \qquad [\gamma_i(\overrightarrow{\partial}^i - ieA^i) + k]\psi = 0.$$

All sets of (possibly non-Hermitean) matrices γ^i satisfying formulas (2) can be derived from the initially chosen set from either of the two following operations: (1) formally treat the γ^i's as the components of a 4-vector, and apply the orthogonal transformation formula corresponding to rotations of the Cartesian tetrapod, or to reversals of part or all of its axes; as far as the time coordinate is involved, this will not preserve the original Hermiticity of the γ's; (2) perform a possibly non-unitary transformation $\gamma'^i = V^{-i}\gamma^i V$. Equivalence rules exist between these two procedures. Before examining them it must be remarked that the latter procedure has another interpretation: instead of keeping the wave functions ψ and $\bar{\psi}$ fixed and transforming the γ's, one can keep the γ's fixed but transform the ψ according to $\psi' = V\psi$ and $\bar{\psi}' = \bar{\psi}V^{-1}$. This turns out to be the most appropriate way — and the one that is retained.

Coming back to the question of connecting the two transformation procedures of the γ's, we first consider continuous rotations of the tetrapod: either spatial rotations of angle α of, say, the axes 2 and 3, or Lorentz transformations of angle $i\beta$ of the axes 1 and 4 for example. A little algebra then yields the following results: for spatial rotations

$$(4.4.10)\quad V(\alpha) = \cos \tfrac{1}{2}\alpha + i\gamma^2\gamma^3 \sin \tfrac{1}{2}\alpha = \exp(\tfrac{1}{2}\gamma^2\gamma^3\alpha),$$

and for Lorentz transformations

$$(4.4.11)\quad V(\beta) = \cosh \tfrac{1}{2}\beta + i\gamma^1\gamma^4 \sinh \tfrac{1}{2}\beta = \exp(\tfrac{1}{2}i\gamma^1\gamma^4\beta).$$

Clearly, $V(\alpha)$ is a unitary operator, but $V(\beta)$ is not (remember that two γ's anticommute).

Formulas (10) shows that *two complete rotations* ($\alpha = 4\pi$) *are necessary to restore the original* ψ; *one rotation* ($\alpha = 2\pi$) *changes the sign of* ψ. This remarkable fact has been tested in beautiful neutron interferometric experiments performed in Vienna (Rauch, 1975) and in the U.S. Werner *et al.*, 1975. Therefore the 4-component object denoted as ψ is not a vector; it is termed a 'spinor'.

Now we discuss the question of the reversal of axes as formalized by Racah (1937).

Again we consider transformations of the form $\gamma'^i = V^{-1}\gamma^i V$. First we take $V = \gamma^5 = V^{-1}$, yielding $\gamma'^i = -\gamma^i$. This corresponds to reversal of all four spacetime axes. Preservation of Equations (8) would then require a corresponding change $\partial_i \rightleftharpoons -\partial_i$. But, as before, another point of view is possible: changing neither the γ^i's nor the ∂^i's, but setting $\psi' = \gamma^5\psi$. This we denote $\mathscr{PT}$ and call 'Racah spacetime reversal'.

Similarly, setting $V = \gamma^4 = V^{-1}$, we get $\boldsymbol{\gamma}' = -\boldsymbol{\gamma}$ and $\gamma'^4 = \gamma^4$, therefore we denote as $\mathscr{P}$, and call 'Racah space reversal', the operation $\psi' = \gamma^4\psi$.

Finally, setting $V = \gamma^1\gamma^2\gamma^3$, $V^{-1} = \gamma^3\gamma^2\gamma^1$, we get $\boldsymbol{\gamma}' = \boldsymbol{\gamma}$ and $\gamma'^4 = -\gamma^4$. We denote as $\mathscr{T}$ and call 'Racah time reversal' the operation $\psi' = \gamma^1\gamma^2\gamma^3\psi$.

Contrary to the continuous rotations, these 'discrete symmetries' do not automatically preserve the sign of the scalar quantity $\bar{\psi}\psi = \psi^+\gamma^4\psi$; this is the reason why the double sign was needed in the definition (6), in order that $\bar{\psi}\psi$ behave as a true scalar under reflections of the axes. This requirement being satisfied, the five tensors $\bar{\psi}\gamma\psi$ (with by definition $\gamma^{ij\cdots} = \gamma^i\gamma^j \ldots$ if all indexes are different, 0 otherwise; $\gamma^0 = 1$) behave 'regularly'; and the same is true of the five tensors $\frac{1}{2}i\bar{\psi}[\partial_i]\gamma\psi + eA_i\bar{\psi}\gamma\psi$ also appearing in the Dirac theory.[5]

An other interpretation of the $\mathscr{PT}$ operation obviously is 'rest mass reversal', which we denote $\mathscr{M}$, whence identically

$$\text{(4.4.12)} \quad \mathscr{MPT} = 1.$$

Rest mass reversal, which is meaningless with the second-order KG equation, thus takes up a meaning with the first-order wave equation — and an interesting one, as we shall see.

Instead of playing with Equations (8) we now consider Equations (9) where an external potential A^i is present. Denoting as C the 'charge conjugation' operation $e \rightleftharpoons -e$, and Z, or, symbolically, $\psi \rightleftharpoons \bar{\psi}$, the exchange of the two Equations (9),[6] we see by direct inspection that $\mathscr{M}CZ = 1$; whence we set

(4.4.13) $C \equiv \mathscr{M}Z$

and enunciate: charge conjugation is equivalent to the product of rest mass reversal by ψ conjugation.[7] The idea of representing particle–antiparticle exchange by rest mass reversal is due to Tiomno (1955).[8]

Now we define the Wigner (1932) motion reversal T as the product of the Racah time reversal by ψ conjugation. Therefore, setting

(4.4.14) $T = \mathscr{T}Z$ and $P \equiv \mathscr{P}$,

from (12), (13), (14) we get

(4.4.15) $CPT = 1$,

an extremely important theorem, the demonstration of which, from very general assumptions is due to Schwinger (1951), Lüders (1952) and Pauli (1955).

$PT \equiv \mathscr{P}\mathscr{T}Z$ we term *covariant motion reversal*, and conclude from (12) or (15), that *particle–antiparticle exchange and covariant motion reversal are, from the four-dimensional geometrical viewpoint, merely two 'relative' aspects of essentially the same concept.* This is one more instance of a 'geometrization of physics', other examples of which are Einstein's gravitation theory and the Einstein–de Broglie $p = \hbar k$ equivalence formula.

In the presence of an external potential A^i the iterated equation $\mathscr{D}\mathscr{D}\psi = 0$ is a KG equation containing coupling terms $\frac{1}{2}H^{ij}m_{ij}$ between the external field H^{ij} and the electron's magneto-electric moment density m^{ij}. These were the terms missing in Schrödinger's early attack on the hydrogen atom problem.

Successful quantizations of the hydrogen atom were produced in 1928 by Darwin and by Gordon, yielding exactly Sommerfeld's 1916 formula. For a discussion of this extremely surprising fact one is referred to an interesting article by Biederharn (1983).

Many other successful applications of the Dirac theory have been performed. So, once again, alliance of the quantal and the relativistic theories led to a flamboyant success. . . .

4.4.3. 1934–1939: DE BROGLIE, PROCA, PETIAU, DUFFIN, KEMMER: THE COVARIANT SPIN-1 WAVE EQUATION

While the solutions of half-integer spin equations are spinors, those of integer spin equations are tensors. So let us see first how the Klein–Gordon equation can be factored in terms of coupled tensors of ranks '0 and 1', or '1 and 2'.

Both pairs of equations, known as Proca's (1936; 1938) equations,

$$(4.4.16)\ \partial_i \varphi^i = k\varphi, \qquad \partial_i \varphi = k\varphi_i,$$

and

$$(4.4.17)\ \partial_i \varphi^{[ij]} = k\varphi^j, \qquad \partial^i \varphi^j - \partial^j \varphi^i = \kappa\varphi^{[ij]},$$

entail the KG equation for the various φ components; Equations (16) describe a spin-0, Equations (17) a spin-1 particle.

Quantum mechanically speaking, all the above φ components must be taken as complex, so that a second set of equations is needed. These are

$$\partial_i \varphi^{*i} = -k\varphi^*, \qquad \partial_i \varphi^* = k\varphi^*,$$
$$\partial_i \varphi^{*[ij]} = -k\varphi^{*j}, \quad \partial^i \varphi^{*j} - \partial^j \varphi^{*i} = k\varphi^{*[ij]}$$

so that the two conserved 4-current densities $\varphi^*\varphi^i + \varphi\varphi^{*i}$ and $\varphi_j^*\varphi^{[ij]} + \varphi_j\varphi^{*[ij]}$ show up.

Shortly before Proca, de Broglie (1934; 1936) had derived these equations by 'welding' together two spin-$\frac{1}{2}$ Dirac particles. And soon after, Petiau (1936), Duffin (1938) and Kemmer (1939) derived the preceding equations from a pair of Dirac-like equations with matrices β obeying the commutation formulas

$$\beta^i\beta^j\beta^k + \beta^k\beta^j\beta^i = \beta^i\delta^{jk} + \beta^k\delta^{ji}.$$

Two representations exist for the β matrices: one of rank 5 entailing the (16)'s, one of rank 10 entailing the (17)'s.

An auxiliary set of matrices $\eta_i = 2\beta_i^2 - 1$ is introduced, and $\bar{\varphi}$ is defined as $\varphi^+\eta_4$. With spin values higher than $\frac{1}{2}\hbar$ the fourth component of the current $i\bar{\varphi}\beta^i\varphi$ is no longer positive definite, so that it can no longer be interpreted as a probability density. But, of course, it can be interpreted as the density of a mean value of some 'charge'.

The *C*, *P* and *T* reversals are discussed along lines analogous to those of the Dirac equation.[9]

4.4.4. HIGHER ORDER SPIN EQUATIONS. FERMIONS AND BOSONS

Some arbitrariness is left for writing down spinning particle equations with spins 3/2 and higher. Various systematizations have been proposed, among which are de Broglie's (1943) 'welding method'; a comprehensive treatment by Umezawa and Visconti (1956); and two concise schemes, Bhabha's (1945), and Bargmann's and Wigner's (1948).

As previously said, the latter work establishes a one-to-one correspondence between spinning waves equations and irreducible representations of the Lorentz group. Such a demonstration lies at the very heart of 'relativistic quantum rechanics'.

An experimental fact is that half-integer spin particles obey the Fermi—Dirac, and integer ones the Bose—Einstein, statistics; therefore the former are termed 'fermions' and the latter 'bosons'.

4.4.5. 1927—1928: THE JORDAN—KLEIN AND JORDAN—WIGNER 'SECOND QUANTIZED' FORMALISMS

Jordan and Klein (1927) and Jordan and Wigner (1928) produced operator formalisms handling respectively the Bose—Einstein (BE) and the Fermi—Dirac (FD) statistics.

'Bosons', including the photon, can be emitted or absorbed individually; 'fermions', including the electron, only in particle—antiparticle pairs.

In short, the adjoint operators a^+ and a respectively emit a particle *or* absorb an antiparticle, and vice versa. If $|n\rangle$ denotes the state with occupation number n, $|0\rangle$ the vacuum, one has:

$$\begin{aligned}
&\text{FD:}\ a^+|0\rangle = 1, \quad a|1\rangle = 0; \qquad \langle 0|\,a = 1, \quad \langle 1|\,a^+ = 0;\\
&\text{BE:}\ a^+|n\rangle = (\eta+1)^{1/2}|n+1\rangle, \qquad a|n\rangle = n^{1/2}|n-1\rangle;\\
&\qquad \langle n|\,a = \langle n+1|\,(n+1)^{1/2}, \qquad \langle n|\,a^+ = \langle n-1|\,n^{1/2};
\end{aligned}$$

together with the commutation laws:

$$\text{FD:}\ \{aa^+ + a^+a\} = 1; \qquad \text{BE:}\ [aa^+ - a^+a] = 1;$$

and

$$\text{FD \& BE:}\ a^+a = n; \qquad \text{FD:}\ aa^+ = 1 - n; \qquad \text{BE:}\ aa^+ = 1 + n.$$

Matrix representations of these various operators have been given by their authors and are found in the textbooks.

The fundamental coupling between the two exchanges: *emission* $\rightleftharpoons$ *absorption* and *particle* $\rightleftharpoons$ *antiparticle* looks at first like a $CT = 1$ theorem. It must be understood really as a $CPT = 1$ theorem, as should be guessed from the contents of Section 2, and will be more clearly shown in Section 4.7.

Considering the geometry of Hilbert's functional space we can understand emission-in and absorption-from a given state as exterior and contracted tensorial multiplication (Potier, 1957). Also, using the Clifford aggregate concept, we can formalize second quantization in terms of an (infinite order) Clifford algebra (Greider, 1984).

In the second quantized theory the 'state vectors', such as $\bar{\psi}$ and ψ are no longer understood as merely functions of x or k, but rather as emission or absorption operators. Therefore, the above commutation formulas must be generalized so as to include propagators. This has been done by Jordan and Pauli (1928), using the propagator $D(x' - x'') \equiv \langle x' | x'' \rangle$ we have met in Section 4.1.5. For the electron–positron system one has

$$\{\psi(x'), \bar{\psi}(x'')\}_+ = -i\langle x' | x'' \rangle$$

and for the photon (which is indistinguishable from its antiparticle)

$$[A^i(x'), A^j(x'')]_- = ic\hbar\, \delta^{ij}\langle x' || x'' \rangle.$$

Similar formulas hold in the k picture.

Let us recall that the $\langle x' || x'' \rangle = \langle x'' || x' \rangle^*$ propagator is the covariant generalization of Dirac's $\delta(x' - x'')$, but that it behaves like δ only when $x' - x''$ is spacelike. As for $\langle k' || k'' \rangle$, it always behaves like $\delta(k' - k'')$.

For the solutions of wave equations containing source terms, Schwinger (1948) has written down covariant commutation relations less simple than those above.

4.4.6. 1928–1948: THE GROPING YEARS OF THE QUANTIZED FIELDS THEORY

Since its inception at the end of the golden 1924–1927 years, the quantized fields theory has solved numerous problems as best it could, including some very intricate ones pertaining to radiative corrections. However, it never faced squarely the central issue of explicit relativistic covariance. So most of its calculations were lengthy, tedious, and even shaky: exploring galleries with a candle produces many moving shadows! As a mathematician[10] put it, this was very much like "a flood of calculations over a desert of ideas". In the end, however, there came the surprise of a "desert in bloom"!

Most calculations of the quantized fields theory belonged either to the Heisenberg or to the interaction pictures.

In the Heisenberg picture the overall state vector $|\Phi\rangle$, distributing the occupation numbers over a chosen set of orthogonal states of the interacting fields, was by definition fixed, and therefore did not show up in the formulas — except of course when finally expressing the transition amplitude $\langle\Psi|\Phi\rangle$ between a 'prepared' $|\Phi\rangle$ and a 'measured' $|\Psi\rangle$ state.

In quantum electrodynamics the interacting fields were, of course, the Dirac electron—positron field ψ and the photon field A, interpreted as creation or destruction operators. These $\psi(x)$ and $A(x)$ operators respectively obeyed the Dirac equation including the A field, and the Maxwell equations, including as source terms the Dirac 4-current — exactly the operator style transposition of the classical picture. A covariant expression of the commutation relations for the $(\psi, \bar{\psi})$ and the A fields was provided by Schwinger (1948).

The other picture used in the quantized field theory was an interaction picture where the field operators $\psi(x)$, $\bar{\psi}(x)$ and $A(x)$ obeyed the free field equations.[11] There was then 'no problem' with the commutation relations, but there was the problem of finding a covariant generalization of the Schrödinger equation

$$\partial_t |\Phi(n)\rangle = |H\Phi(n)\rangle.$$

This was cleverly done by Tomonaga (1946), and used systematically by Schwinger in 1948. Tomonaga's thinking was based on the hybrid 1932 'multitime formalism' of Dirac, Fock and Podolsky. Tomonaga conceived $|\Phi\rangle$ as a functional depending on an arbitrary surface $\mathscr{E}$, or σ, as we shall rather denote it now, following Tomonaga and Schwinger.

4.4.7. 1948: SCHWINGER'S 'QUANTUM ELECTRODYNAMICS, I: A COVARIANT FORMULATION'

In this epoch-making article Schwinger (1948) conferred full relativistic covariance to the second quantized Heisenberg picture, where the coupling of the electron—positron and the photon fields is expressed, in classical fashion, by source terms in the wave equations. He writes the total conservative momentum-energy density T^{ij} as the sum of Tetrode's tensor for the electron and one implying only the potential A^i for the photon, and then the total momentum energy $P^i = \hbar K^i$ as $\iiint T^{ij}\, d\sigma_j$.

In the next step he replaced the equal time commutation relations, previously used, by

$$\iiint_\sigma [A^i(x'), \partial_i A^j(x)]\, d\sigma^i = -ic\hbar\, \delta^{ij}$$

and

$$\iiint_\sigma \{\psi_\alpha(x'), [\bar{\psi}(x)\gamma_i]_\beta\}\, d\sigma^i = -i\delta_{\alpha\beta},$$

and showed that, for any field magnitude, there follows the Heisenberg style formula

$$i\hbar\, \partial_i F(x) = [P_i, F(x)]$$

displaying P_i as the generator of spacetime displacements:

$$(4.4.18)\ F(x+y) = U(y)F(x) = \exp(iKx)F(x)$$

Schwinger repeatedly used Tomonaga's (1946) definition of the spacetime 'functional derivative' $\delta/\delta\sigma = d/d\omega$, the meaning of which is: functionals $F(\sigma)$ of an arbitrary spacelike surface σ, such as $Q = \iiint_\sigma j^k\, d\sigma_k$ or $P^i = \iiint_\sigma T^{ij}\, d\sigma_j$, undergo a variation δF when (see Figure 14) σ is locally deformed in the vicinity of x, so that a four-dimensional volume $d\omega$ is enclosed; $\delta F/\delta\sigma$ is defined as equal to $dF/d\omega$. Therefore $\delta Q/\delta\sigma = \partial_k j^k$ and $\delta p^i/\delta\sigma = \partial_j T^{ij}$.

The next section of Schwinger's paper defined an 'interaction picture' where the field operators ψ, $\bar{\psi}$ and A obey the free-field equations, and where the commutation relations are (16) and (17). He introduced a unitary operator $U(\sigma)$ that is a functional of σ obeying a Schrödinger-like equation

$$i\hbar\, \delta U/\delta\sigma = \mathscr{H}(x)U(\sigma),$$

$\mathscr{H}$ being some scalar function of the operators ψ, $\bar{\psi}$ and A. He

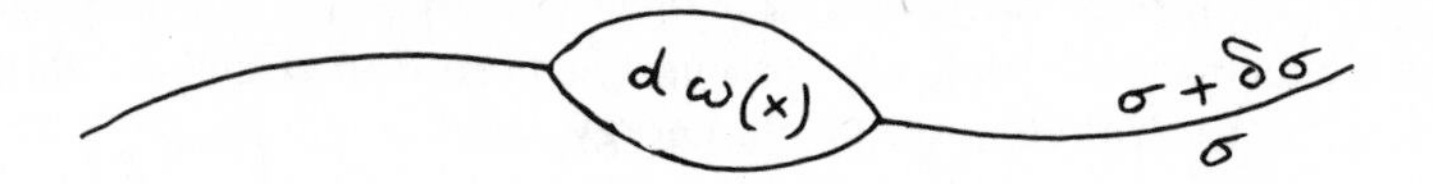

Fig. 14. The Tomonaga—Schwinger elementary deformation of a spacelike surface σ around a point-instant x; by definition, the 'functional derivative' $\delta/\delta\sigma$ has the value $\mathrm{d}/\mathrm{d}\omega(x)$, $\mathrm{d}\omega$ denoting the enclosed 4-volume element.

defined the interaction picture state vector $|\Psi(\sigma)\rangle$ as related to the Heisenberg one $|\Phi\rangle$ via

$$|\Psi(\sigma)\rangle = |U(\sigma)\Phi\rangle.$$

Then any field operator $F(x)$ of the Heisenberg picture will be transformed into an interaction picture operator according to

$$\mathscr{F}(x) = U(\sigma)F(x)U^{-1}(\sigma), \qquad U^{-1}(\sigma)\mathscr{F}(x)U(\sigma) = F(x).$$

This holds for ψ, $\bar{\psi}$, A, and functions of them. Schwinger then showed that, choosing as the Hamiltonian density (an action density, when inserted into a fourfold spacetime integral)

$$\mathscr{H}(x) = -A^k(x)j_k(x)$$

the wave equations and the commutation formulas obeyed by ψ, $\bar{\psi}$ and A are transformed into the free-field ones.

The state vector $|\Psi(\sigma)\rangle$ obviously obeys the Schrödinger-like equation

$$i\hbar\delta\,|\Psi(\sigma)\rangle/\delta\sigma = \mathscr{H}(x)\,|\Psi(\sigma)\rangle.$$

Clearly, it is imperative that

$$|\Psi(\sigma_2)\rangle \equiv |\Pi U(\sigma)\Phi\rangle = |U_{21}\Psi_1\rangle$$

be independent of the order according to which the 'spacetime block' enclosed between the 'pseudo-instant' σ_1 of the 'preparation' $|\Psi_1\rangle$, and the 'pseudo-instant σ_2 of the 'measurement' $|\Psi_2\rangle$, is made of 'bricks' $\mathrm{d}\omega(x)$ (see Figure 14) — provided, of course, that the spacelike character of σ is preserved at each step. The condition for this is that two spatially separated operators $\mathscr{H}(x)$ and $\mathscr{H}(x')$ do commute — which they do. For other significant information I refer to Schwinger's article.

Two other papers follow, discussing higher order effects in the perturbation approach, particularly the 'vacuum polarization' problem and the electron's 'self-energy' problem. These had not been handled properly in the previous non-relativistic calculations. Schwinger produced their theory in excellent agreement with the measurements.

As a final remark, in connection with the 'distant correlations' discussed in Section 4.3.15, let us remark that the transition amplitude $\langle \Psi_2 | \Phi_1 \rangle$ (Heisenberg picture) or $\langle \Psi_2 | U_{21} | \Phi_1 \rangle$ (interaction picture) between a preparation at σ_1 and a measurement at σ_2 implies sums of products of partial state vectors $|\psi\rangle$, $|\bar{\psi}\rangle$ and $|A\rangle$, so that what was said there still holds.

4.4.8. 1949: FEYNMAN'S VERSION OF QUANTUM ELECTRODYNAMICS

The underlying idea, presented by Feynman in his 1948 'Spacetime Approach to Non-relativistic Quantum Mechanics', is expressed by means of an algorithm now called 'Feynman's path integral method'. It is viewed as a mathematical expression of the 'Huygens principle' for wave propagation. Gerald Rosen (1969) terms this the "Feynman passage to quantum mechanics".

In his two famous articles 'The Theory of Positrons' and 'Spacetime Approach to Quantum Electrodynamics', Feynman (1949a, b) presents his covariant perturbation, or iteration method. A key element in it is a systematic use of Stueckelberg's (1941; 1942) proposal that a positron be understood as an electron the trajectory of which is directed backwards in time, or the momentum energy of which points inside the past light cone; the alternative is the one between an x and a k four-dimensional picture, as mutually related by a Fourier transform. 'Feynman graphs' look very much the same in both pictures, and physicists often feel free to speak of the particles 'as if' they had both a trajectory and a momentum energy — without of course being caught by this looseness in speech.

4.4.9. 1949–1950: DYSON'S ARTICLES

Dyson's (1949) brilliant demonstration of the equivalence between Schwinger's and Feynman's formalisms was in fact published between the articles of both authors. A résumé of it will be given at the beginning of Chapter 7, where the S-matrix formalism and the inherent Einstein—Podolsky—Rosen correlations are discussed.

In a second important article Dyson discussed the renormalization problem, which is mentioned here merely for information.

4.4.10. PROVISIONAL EPILOGUE

The covariant calculation methods initiated by Schwinger, Feynman and Dyson were successfully used in difficult problems of 'radiative corrections'. Even in the light of explicit relativistic covariance these problems remain difficult, requiring extreme skill in thinking and in computing. A classical forerunner of this sort of problem was that of the electromagnetic contribution to the electron's rest mass, as discussed by Poincaré among others. The 'classical electron radius' was calculated under the assumption that the electron rest mass was entirely of electromagnetic origin.

In 1947 Lamb and Retherford, using microwaves, found that there exists, in the hydrogen spectrum, an energy difference between the '$2\,^2s_{1/2}$' and the '$2\,^2p_{1/2}$' energy levels, where the Darwin—Gordon (Darwin, 1928; Gordon, 1928) quantization formula finds none. Successively improved measurements ended in 1953 with the following values for $\Delta E = 2\,^2s_{1/2} - 2\,^2p_{1/2}$ as expressed in MC/s (10^6 Hz): $\Delta E_{\text{exp}} = 1057.77 \pm 0.10$ MC/s.

The explanation of this basically consists in an electromagnetic contribution to the electron rest mass, and in the 'polarization of the vacuum' in the presence of the central Coulomb field. Taking account of these, and other effects, in the second and fourth-order graphs, and also of mass corrections, the theoretical estimation of ΔE_H is found as $\Delta E_{\text{theor}} = 1057.19 \pm 2$ MC/s (Jauch and Rohrlich, 1955, p. 359).

'Radiative corrections' also imply a correction to Bohr's magneton, accurately measured in 1952 by Kush and coworkers, and found to have the value

$$\mu_{\text{exp}} = (1.001146 \pm 0.000012),$$

the theoretically predicted value being

$$\mu_{\text{theor}} = (1 + \alpha/2\pi - 2.973\alpha^2/\pi^2) = 1.0011454.$$

This was the beginning of 'normal science' in a new field. . . .

So, once again, use of an explicitly relativistic quantum mechanics meant brilliant success.

In quantum electrodynamics, the covariant iterative methods yield power series in Sommerfeld's fine structure constant $\alpha = e^2/c\hbar \simeq 1/137$. The question is, of course: 'Do such series converge?' Physically this question becomes: 'Are the radiative corrections' to the electron rest mass (Figure 15(a)) and to the (intrinsically zero) photon proper mass (Figure 15(b)) finite?

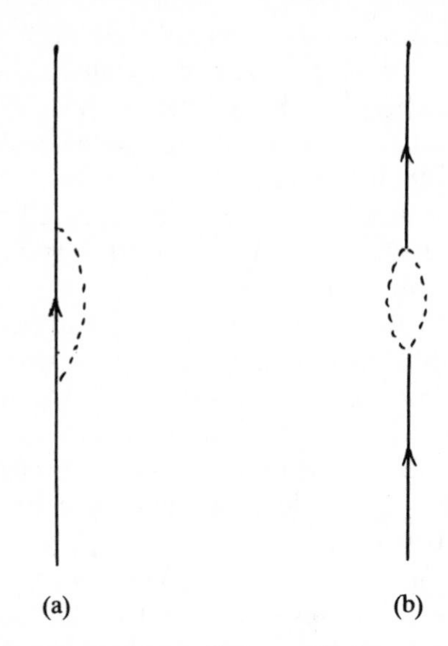

Fig. 15. Graphs of the lowest order radiative corrections to the electron mass (a) and the photon mass (b).

This question has been cleverly circumvented by use of the 'renormalization concept', a subtraction procedure yielding a finite difference between two unknown quantities.

Quantum electrodynamics turns out to be a 'renormalizable' theory. Not all the conceivable quantum field theories share this quality; so 'renormalizability' has become a preliminary requirement for any theory.

Renormalization is a difficult subject lying outside the scope of this book.

CHAPTER 4.5

PARITY VIOLATIONS AND *CPT* INVARIANCE

4.5.1. LIMINAL ADVICE

Lee and Yang's 1956 assumption that Fermi's 1933 'weak interaction' is not invariant under space reflection — 'parity violating', in the jargon — followed by Mrs Wu's (Wu *et al.*, 1957) experimental proof of the fact caused quite a shock; a shock wave, so to speak, the ripples of which are still felt today.

Study of the matter is both exciting and frustrating. Of course, opening of a vast new field is exciting, but failure to uncover a deep and satisfactory reason for symmetry violations is frustrating.

The whole subject is an intricate one, not separable from the study of the respective relations of the four force fields — strong (originally, nuclear), electromagnetic, weak (originally, β-decay), and gravitational. Therefore the ultimate answer may well come with the 'grand unification' actively pursued today — but not touched upon in this book.

So, once again, we shall have to steer a medium course, avoiding many technicalities the expert would ask for, but nevertheless sailing sufficiently off coast that the reefs can be seen. And so, once again, apologies are due to both sides.

4.5.2. CLASSICAL CONNECTION BETWEEN CHARGE CONJUGATION AND SPACETIME REVERSAL

The classical relativistic equation of motion (2.6.15) of a point particle

$$(4.5.1)\quad \mathrm{d}^2x^i/\mathrm{d}\tau^2 = (e/m)H^{ij}\,\mathrm{d}x_j/\mathrm{d}\tau$$

is invariant under the joint symmetries

$$e/m \rightleftharpoons -e/m, \qquad \mathrm{d}\tau \rightleftharpoons -\mathrm{d}\tau.$$

Two interpretations are then possible. In the 'passive' one, the content of spacetime is left unchanged, but the parametrization of the trajectory is reversed. So, if 'initially' the (time-extended) trajectory was explored from $t = -\infty$ to $t = +\infty$ ($t \equiv x^4/ic$), in the transformed presentation the *same* trajectory is explored from $t = +\infty$ to $t = -\infty$.

In the 'active' interpretation the spacetime symmetry $x^i \rightleftharpoons -x^i$ is added, leaving formula (1) unchanged. However, in order to preserve the sign of $\mathrm{d}t/\mathrm{d}\tau$, the two previous symmetries are maintained. So, the transformed trajectory is symmetric in spacetime to the 'initial' one.[2]

Defining the two symmetries:

Charge conjugation C:

$$e/m \rightleftharpoons -e/m;$$

Covariant motion reversal PT:

$$\begin{aligned} &\textit{passive:} \quad \tau \rightleftharpoons -\tau \\ &\textit{active:} \quad \tau \rightleftharpoons -\tau, \quad x^i \rightleftharpoons -x^i, \end{aligned}$$

we end up with the 'classical $CPT = 1$ theorem' concerning *one specific solution* of the equation of motion.

Besides this, formula (1) possesses the three separate symmetries C, P and T in the following sense: changing the sign of e/m can be compensated for by changing the sign of all the charges generating the field $H^{ij}(x)$, which leads to an acceptable solution. Analogously, it can be verified that changing the signs of either $\mathbf{x}$ or x^4 also leads to acceptable solutions.

4.5.3. THE '$\theta - \tau$' PUZZLE RESOLVED: LEE'S AND YANG'S K MESON

In 1956 there existed the so-called 'θ and τ puzzle'.

The θ and τ mesons, both having charge $\pm e$, spin 0, parity -1 (they are pseudoscalars) and the same rest mass ($967 m_e$) could decay via the 'weak interaction' as follows

$$\theta^{\pm} \rightarrow \pi^{\pm} + \pi^0, \qquad \tau^{\pm} \rightarrow \pi^{\pm} + \pi^{\pm} + \pi^{\mp},$$

with the same lifetime (1.3×10^{-8} s). The parity of the π meson was known to be -1 in the strong and the electromagnetic interactions. Therefore, according to the then prevailing dogma of parity conservation, θ had parity $+1$ and τ, -1. So the puzzle was: *Why should so radically distinct particles have exactly the same rest mass and decay lifetime?*

Lee and Yang (1956) boldly proposed to reject the dogma. They assumed that parity is not conserved by the weak interaction, and considered the θ and the τ to be the same particle, today called the K meson.

Reviewing the existing experimental evidence, they found that none tested that crucial point,[1] so that a specific test was needed.

The human industry has a strong preference for right-handed screws, and living beings one for right-handed sugar. Should we not ask if the weak interaction is not similarly biased, so that it more or less ignores one of the two possible signs of a three-dimensional pseudoscalar (say, a 3-volume element)? As a natural test Lee and Yang proposed an experiment using Fermi's β decay, with the nuclear spins all oriented by an external magnetic field $\mathbf{B}$. Since $\mathbf{B}$ is an axial vector and $\mathbf{p}$, the momentum of an ejected electron, a polar vector, $\mathbf{B} \cdot \mathbf{p}$ is a pseudoscalar; the question thus was: Is one of the two signs of $\mathbf{B} \cdot \mathbf{p}$ preferred over the other in β decay?

An unambiguous answer *yes* was given in 1957 by Mrs Wu's experiment using the β decay of cobalt-60: $\mathbf{M}$ denoting the magnetic moment of the neutron, the sign of $\mathbf{M} \cdot \mathbf{p}$ definitely is negative — which means maximal parity violation, or P violation as it is called.

4.5.4. FORGETTING K MESONS: '$V - A$' FORMALIZATION OF THE WEAK INTERACTION

Very soon after Wu's startling result confirmatory evidence came from elementary particle physics. P violations were observed in weak 4-fermions (Fermi) interactions, and also in the weak decay modes of the π meson: $\pi \rightarrow \mu + \nu_\mu$ and $\pi \rightarrow e + \nu_e$, (where $\pi \rightarrow p + \bar{n}$ via the strong, Yukawa, interaction, and then $p + \bar{n}$ enters the 4-fermions interaction); p and n denote of course the proton and the neutron, μ the mu-meson, e the electron, ν_μ and ν_e their respective neutrinos.

Therefore the weak 'universal Fermi interaction' had to be rewritten with a Hamiltonian density comprising both scalar and pseudoscalar contributions — a true heresy, in terms of tensor calculus!

Lee and Yang assumed maximal P violation. Since all four fermions obey Dirac equations, an elegant way for doing this uses the two orthogonal projectors $2^{-1/2}(1 \pm \gamma_5)$ appearing in the Dirac theory (see Section 4.4.2). Denoting the four fermions as a, b, c, d and the 'universal Fermi coupling constant' as g, Lee and Yang proposed a 'pure $V-A$ coupling' in the form

$$H = g[\bar{\psi}_a\gamma_i(1+\gamma_5)\psi_b \cdot \bar{\psi}_c\gamma^i(1+\gamma_5)\psi_d + \\ + \bar{\psi}_d\gamma^i(1-\gamma_5)\psi_c \cdot \bar{\psi}_b\gamma_i(1-\gamma_5)\psi_a]$$

where the two groups of terms are conjugate to each each. Clearly, this Lagrangian density is maximally P and C violating (C denoting charge conjugation) but $PC = T$ conserving (see Section 4.4.2).

Landau (1957) observed that PC invariance offers some consolation for the loss of P invariance: it means that the mirror image of a weak transition physically is an associate weak transition with antiparticles replacing particles, and vice-versa — one more example of a coupling between physics and geometry.

If only weak interactions existed, P and C violations could perhaps be interpreted as 'factlike rather than lawlike': $2^{-1/2}(1 \pm \gamma_5)$ being projectors, one could say that, for some reason, only handed solutions $(1 + \gamma_5)\psi_b$ or $(1 - \gamma_5)\psi_a$ show up — at least in our corner of the Universe. This had been implicitly claimed for the neutrino, as early as 1929, by Weyl. According to an argument presented in Section 4.7.7 below, Weyl's 1929 proposal is in fact not optional, but is a necessary consequence of momentum-energy and of angular momentum conservation.

So it may be that we must come to accept the idea that a Lagrangian density can be the sum of scalars and pseudoscalars, and turn this heresy into a new dogma.

As mentioned in Section 4.4.2, if ψ denotes a solution of a Dirac equation, $\gamma_5\psi$ is a solution of an associated Dirac equation where the sign of the rest mass m is reversed. Of course, if $m = 0$, both ψ and $\gamma_5\psi$ are solutions of the same Dirac equation, which had been Weyl's proposal. By using a representation where γ_5 is diagonal, one thus obtains a 'two-component neutrino theory'. Then the facts are that the neutrino has full left, and the antineutrino full right, helicity.

If m is non-zero, expressions such as $(1 \pm \gamma_5)\psi$ represent an interference, or beating, between two associated solutions, one with $+m$, the other with $-m$.

Of course, $(1 \pm \gamma_5)\psi$ is a solution of the Klein—Gordon equation, but, since m appears there as m^2, rest mass reversal can no longer be used for expressing particle—antiparticle exchange (as was done in Section 4.4.2).

Exploiting the previous scheme Feynman and Gell-Mann (1958) have proposed a two-component fermion theory, also in the presence of an external electromagnetic field, and have shown its equivalence with the original Dirac theory.

Finally we note that Sakurai (1958) has deduced the '$V \pm A$ interaction' by assuming invariance of the interactions under the transformation $\psi \rightleftharpoons a\gamma_5\psi$, $\bar{\psi} \rightleftharpoons -a^*\bar{\psi}\gamma_5$, $m \rightleftharpoons -m$.

4.5.5. ON THE POSSIBILITY OF TIME-REVERSAL VIOLATIONS

Why should T violations be excluded? Consider the very classical case of interacting Amperian current loops; the forces between them are invariant under a combined reversal of the velocities of the flowing charges (interpretable as a T reversal) and of the charges of both the conductors and the current carriers (a C reversal). Thus we end up with a CT-invariance concept.

However, while both the C- and the P-reversal concepts are immediately applicable to the 'state of a system at some time t', the T-reversal concept is not if the d/dt operator is not present. In other words, *the T-reversal concept applies to a process, not to a state.*

This leads us straight back to a main leitmotif in this book: *extended spacetime geometry.* The relativistic S-matrix scheme, to be discussed in Chapter 4.7, essentially deals not with 'evolving states', but with four-dimensional processes connecting a 'preparation' to a 'measurement'. So this scheme offers a unified spacetime framework for discussing consistently the three C-, P- and T-reversal operations.

4.5.6. *CPT* INVARIANCE AND THE SPIN-STATISTICS CONNECTION

The Lüders—Pauli (Lüders, 1952; Pauli, 1955) *CPT* invariance rests on three basic assumptions: spin-statistics connection, as expressed in the creation—destruction operators commutation formulas, locality of interactions, and invariance under the proper Lorentz group of spacetime rotations. Almost simultaneously with Lüders, Schwinger (1951) derived the spin-statistics connection from *CPT* invariance, a reversal of assumptions and consequences.

Locality means that the same point-instant x shows up in interaction

Hamiltonian densities such as, for example, $H(x) = \varphi(x)\bar{\psi}(x)\psi(x)$, concerning a meson φ and a fermion—antifermion system $\bar{\psi}$, ψ.

Invariance under spacetime rotations means that $H(x)$ can be a sum of scalar and pseudoscalar terms — as is entailed by Lee and Yang's discovery, but without requiring CP invariance. The basic remark then is that under the 'strong' spacetime reflection $x \rightleftharpoons -x$, a scalar *and* a pseudoscalar, a vector *and* a pseudovector, a skew-symmetric tensor *and* its dual, behave in exactly the same (three) ways.

We shall now discuss CPT invariance in the same spirit as we did in Section 4.4.2 devoted to Dirac's electron. There we have considered the three operations $\mathscr{M}$, rest mass reversal, $\mathscr{P}$ and $\mathscr{T}$, Racah space and time reversals, which are such that $\mathscr{MPT} = 1$, and defined C, P and T as

$$C = \mathscr{M}Z, \qquad P = \mathscr{P}, \qquad T = \mathscr{T}Z,$$

Z (symbolically) meaning $\psi \rightleftharpoons \bar{\psi}$; T is Wigner's time reversal.

As an example let us choose a meson (φ, φ^i)—fermion $(\bar{\psi}, \psi)$ interaction, the meson obeying Proca's equations (4.4.16) or (4.4.17) and the fermion Dirac's equation. This discussion can be extended to higher tensor or spinor cases.

We define the *strong reflection* as $x \rightleftharpoons -x$ and:

$$(4.5.2)\quad \begin{cases} \varphi(x) \rightleftharpoons i\varphi(-x), & \overset{+}{\varphi}(x) \rightleftharpoons i\overset{+}{\varphi}(-x), \\ \varphi^i(x) \rightleftharpoons -i\varphi^i(-x), & \overset{+i}{\varphi}(x) \rightleftharpoons -i\overset{+i}{\varphi}(-x); \end{cases}$$

$$(4.5.3)\quad \psi(x) \rightleftharpoons i\gamma_5\psi(-x), \qquad \bar{\psi}(x) \rightleftharpoons i\bar{\psi}(-x)\gamma_5.$$

Neglecting the phase factor i, formulas (2) and (3) exactly express $\mathscr{MPT}$ invariance of the Proca and the Dirac equations for free particles.

Now we consider the commutation relations for the free particles

$$(4.5.4)\quad [\varphi(x), \varphi^+(y)]_- = ic\hbar\langle x\|y\rangle,$$

$$(4.5.5)\quad \{\psi_\alpha(x), \bar{\psi}_\beta(y)\}_+ = -i\langle x|y\rangle,$$

where $\langle x\|y\rangle$ denotes the Jordan—Pauli propagator and $\langle x|y\rangle$ its projection onto the spin wave space (see Section 4.1.5). Recalling that $\langle x\|y\rangle$ is purely imaginary and odd in $x - y$, we see that the requirement of invariance of Equation (4) under the strong reflection needs a factor $+i$ (or $-i$) to be inserted in Equations (2). The same is true with formulas (5) and (3), as can be verified by a routine calculation;

incidentally, the strong reflection exchanges the Dirac operator and its associated projector.

So, under the strong reflection, the Hamiltonian interaction density remains invariant: $H(x) \rightleftharpoons H(-x)$. Thus $\mathscr{MPT} = CPT$ invariance has been established in the interaction picture.

Two important remarks are in order before we proceed.

Under the strong reflection the Feynman arrows are reversed, as exemplified in Figure 9, p. 171. Thus, for consistency (that is, for validity of the $\mathscr{MPT} = CPT$ theorem) the order of action of the creation and destruction operators must be 'objectively' maintained in terms of spacetime geometry — meaning that it is reversed in terms of the time coordinate.

The second remark is that, since H enters the transition amplitudes through its square H^+H and since H is the sum $H_s + H_{ps}$ of a scalar and a pseudoscalar contribution, the C-, P-, and/or T-violating contribution consists of the terms that are off-diagonal in H_s and H_{ps}.

Finally, let it be mentioned that the $CPT = 1$ theorem implies that associated particles and antiparticles must have equal rest masses (meaning bare *plus* dressed rest masses) and lifetimes.

4.5.7. BACK TO K MESONS. 1955: GELL-MANN'S AND PAIS'S THEORY OF THE WONDERFUL BEHAVIOR OF K^0 MESONS

One year before Lee and Yang unmasked the $K^\pm$, Gell-Mann and Pais (1955) had unravelled another enigma inside this 'strange' family.

Among the decay modes of the neutral K^0 meson there is

$$|K^0\rangle \to |\pi^+\rangle + |\pi^-\rangle,$$

all three particles having spin 0; however, the corresponding lifetime, some 10^{-10} s, is much longer than the 10^{-20} or so corresponding to the strong interaction. Gell-Mann and Pais assumed the existence of a hindrance, a yet unknown conservation law involving a new quantum number, labelled by them 'strangeness' S; they assigned $S = 1$ to K^0, $S = -1$ to its antiparticle $\bar{K}^0$, $S = 0$ to pions (π mesons). Therefore the K^0 had to decay via the weak interaction.

The enigma then is that also $\bar{K}^0$ decays according to

$$|\bar{K}^0\rangle \to |\pi^+\rangle + |\pi^-\rangle;$$

so *what is the difference between* K^0 *and* $\bar{K}^0$*?*

Wave mechanics is the game played here, as is reminded by the *ket* symbols. The two preceding formulas show that $|K^0\rangle$ and $|\bar{K}^0\rangle$ are coupled via their future common state – precisely the phenomenon we name *inverse EPR correlation* in Section 4.6.12. Therefore, there exists a superposition, or beating process of two oscillating amplitudes, the normal modes of which are $|K^0\rangle + |\bar{K}^0\rangle$ and $|K^0\rangle - |\bar{K}^0\rangle$, as occurs in the classical case of two weakly coupled pendula, which are known to end up swinging either in phase or in opposition.

As Lee and Yang's *CP* invariance was still one year in the future, Gell-Mann and Pais reasoned in terms of *C* and *P* invariance. However, their argument survives the change. So (as everybody does now) we shall summarize it in terms of *CP* invariance.

Up to an arbitrary phase factor we must have

$$CP\,|K^0\rangle = |\bar{K}^0\rangle, \qquad CP\,|\bar{K}^0\rangle = |K^0\rangle,$$

whence, introducing the two normal modes

$$|K_S\rangle \equiv 2^{-1/2}(|K^0\rangle + |\bar{K}^0\rangle) \qquad |K_L\rangle \equiv 2^{-1/2}[|K^0\rangle - |\bar{K}^0\rangle],$$

we get

$$CP\,|K_S\rangle = |K_S\rangle, \qquad CP\,|K_L\rangle = -|K_L\rangle.$$

That is, K_S has $+1$ and K_L has -1 as its 'combined parity' *CP*.

Now, the π mesons, being pseudoscalars, have $P = -1$. As $C\,|\pi^+\rangle = |\pi^-\rangle$ and $C\,|\pi^-\rangle = |\pi^+\rangle$, we see that a $|\pi^+\rangle + |\pi^-\rangle$ pair has $CP = +1$, and that a triplet $|\pi^+\rangle + |\pi^-\rangle + |\pi^0\rangle$.has $CP = -1$. Therefore, if *CP* is conserved, it is the eigenstate $|K_S\rangle$ that decays into $|\pi^+\rangle + |\pi^-\rangle$. Also, it is predicted that the eigenstate $|K_L\rangle$ can decay into the triplet $|\pi^+\rangle + |\pi^-\rangle + |\pi^0\rangle$.

When then of lifetime? Via the 'fourth uncertainty relation', it is coupled to the energy difference involved, which is much larger in the transition to $|\pi^+\rangle + |\pi^-\rangle$ than in the one to $|\pi^+\rangle + |\pi^-\rangle + |\pi^0\rangle$. Therefore K_S is a 'short lived' and K_L a 'long lived' *K* meson – a prediction brilliantly confirmed! The K_L was observed in 1956 by Lederman and coworkers, its signature being its decay into the triplet $|\pi^+\rangle + |\pi^-\rangle + |\pi^0\rangle$.

The conversion formulas between the $|K^0\rangle$, $|\bar{K}^0\rangle$ and the $|K_S\rangle$, $|K_L\rangle$ pairs of orthogonal vectors are exactly similar to those between pairs of circular and of orthogonal linear photon polarizations. And so, just as by inserting a linear polarizer between two crossed linear polarizers, one can 'regenerate' a polarization state previously suppressed before the latter, one can here 'regenerate' the $|K_S\rangle$ state that has faded away inside a K^0 beam, by making the still healthy $|K_L\rangle$ interacts 'strongly' with an appropriately chosen piece of matter.

Such is the fantastic K^0 mesons family, being part of the magic of weak interactions. But the story does not end there.

4.5.8. 1965: CHRISTENSON, CRONIN, FITCH AND TURLAY DISCOVER THE *CP*-VIOLATING DECAY OF K^0 MESONS

By 1965 the physics community had recovered from the shocks of 1955 and 1956. A decade later, pursuing routine 'normal science' experiments on the K^0 decays, Cronin, Fitch and coworkers hit upon one more scandalous behavior of that very 'strange' family: *CP* violation!

They had built an apparatus able to push down to some 1/10 000 the limit of *CP* conservation in the K_L decay. Of course, reducing such limits is typical of 'normal science', in Kuhn's wording. It is also typical that, from time to time, 'paradoxes' are thus unearthed.

What Cronin and Fitch found was that some $(1/300)K_L$'s do decay into a $\pi^+ + \pi^-$ pair — which of course meant *CP* violation! Other laboratories soon confirmed the fact. But, contrary to what had happened in 1956—1957, no elegant, clearcut explanation was found. So we must live with a big question mark unanswered.[2]

Among the other decay modes of K mesons there is the pair

$$|K_L\rangle \to |\pi^\pm\rangle + |e^\mp\rangle + |\nu\rangle;$$

CP violation causes a slight difference in the corresponding transition probabilities, which has been calculated, and experimentally tested.

4.5.9. *T* VIOLATIONS

Via *CPT* invariance, *T* violation is a corollary of *PC* violation. Of course the transition rates of *CPT*-reversed Feynman graphs can be measured. Up to now all such measurements are consistent with *CPT* invariance.

4.5.10. BY WAY OF CONCLUSION, A LITTLE FABLE

Suppose we have a printed movie film. We obtain *parity reversal, P*, by turning the film recto—verso, *motion reversal, T*, by running it backwards, and *covariant motion reversal* as the combined reversal *PT*. Thus, including the 'original' presentation of the sequence, we have four spacetime images of one and the some 'object', the printed film. In the jargon of relativity theory, these are four 'relative' images of one and the same four-dimensional evolution.

If, for instance, the 'original' sequence consists of: *screwing in a right-handed screw*, by *P* we obtain *screwing in a left-handed screw*; by *T*, *screwing out a right-handed screw*; by *PT, screwing out a left-handed screw*.

Suppose now that a physical difference accompanies part or all of these three reversals, meaning that they imply part or all of particle—antiparticle exchange, *C*. Three cases are then of interest: total coupling of *C* with *P*, with *T*, or with *PT*, leading respectively to *CP, CT*, or *CPT* invariance.

In the preceding example, *CP* invariance means that (paired) right- and left-handed screws are particle and antiparticle to each other, regardless of evolution in time. *CT* invariance means that the same screw behaves as particle or antiparticle (or vice-versa) according as it is screwed in or out, meaning that it moves forwards or backwards. The implication then is that there is recognizable difference between fore and aft (as evidenced by the very shape of screws) and that there is motion, or process: moving forwards or backwards. Finally, *CPT* invariance has these combined space-and-time implications.

Let us change our comparison to make it more vivid. If the original sequence displays an American automobile backing out of a garage, we call this process 'emission of an antiparticle'. The *PT*-reversed sequence displays an English automobile, with the wheel on the right side, entering a garage forwards. Then *CPT* invariance calls this 'absorption of a particle'.

Thus *CPT* invariance means that *emission of a particle and absorption of an antiparticle* (and vice-versa of course) *are mathematically equivalent*, and are merely two 'relative' images of intrinsically the 'same' process.

How does this fit with elementary particle physics and the relativistic *S*-matrix scheme? Quite all right, as we show now.

In terms of four-dimensional geometry, an obvious distinction between fore and aft of (free) particles consists of the sign of the scalar product of the 4-velocity and the momentum energy. This exactly is

Stueckelberg's and Feynman's recipe for distinguishing particles and antiparticles.[3]

Handedness of a particle has covariant meaning only in the 'extreme relativistic limit' where the rest mass is negligible; in this sense all particles interacting via the weak interaction do display handedness.

Then *CPT* invariance exactly has the meaning illustrated by means of our previous fable. *CPT* invariance thus says that *particle–antiparticle exchange C* and *covariant motion reversal PT* merely are two different representations of one and the same intrinsic operation. This we have exemplified by means of Feynman graphs in Figure 9, p. 171.

CHAPTER 4.6

PARADOX AND PARADIGM: THE EINSTEIN–PODOLSKY–ROSEN CORRELATIONS

4.6.1. 1927: EINSTEIN AT THE FIFTH SOLVAY CONFERENCE

Electrons and Photons (1928) was the title of the Fifth Solvay Conference held at Brussels, 24—29 October 1927, under the chairmanship of Lorentz. The Founding Fathers of both the 'old' and the 'new' quantum theory were all present,[1] around the very cradle of 'quantum mechanics'. It is there that Einstein, with prophetic sagacity, noticed a sign on the face of the newly born babe, which for more than half a century has been a 'sign of contradiction' — the subject of private puzzlement and of heated discussions among theoreticians, probably heralding a far-reaching change in paradigm. It consists of a seminal 'paradox' lying, as I shall argue, at the very heart of *relativistic quantum mechanics*. It is precisely as such that Einstein cleverly characterized it, as early as 1927.

After Louis de Broglie, who presented his 'double solution' proposal as an interpretation of the new 'wave mechanics', Born and Heisenberg lectured on their matrix mechanics, on the 'transformation theory', and the 'probabilistic interpretation'. That was Born's and Jordan's wavelike probability calculus, which, as Einstein was going to show, implies a startling paradox. They also discussed the indeterminacy relations implied by the wave—particle duality, as 'constituting a universal gauge of indeterminacy', and as unveiling the 'true meaning of Planck's constant'.

Schrödinger, who was the third speaker, lectured on his many particle wave mechanics.

Before opening the general discussion, Lorentz expressed his reluctance to accept that probability is inherent in Nature rather than in our minds as imperfect knowledge. He then called upon Bohr, who presented his 'complementarity' paradigm, as he had a few days before at a Como Conference. Born, intervening after Brillouin and de Donder, addressed Einstein on the subject of 'waves and particles', of α-particles randomly issuing from a radionuclide.

Einstein then rose to speak, beginning thus: "I apologize for not having gone deeply into quantum mechanics. Nevertheless I would like to make some general remarks."

He then described a thought experiment — the first of a long series, the last ones being experiments displaying a specific 'quantal non-separability' of two distant measurements at L and N, correlated through their common past preparation at C.

If a plane wave carrying, for simplicity, just one particle — electron or photon — impinges normally upon, and is diffracted by, a plane screen with a small aperture C in it (Figure 16), and is then received by

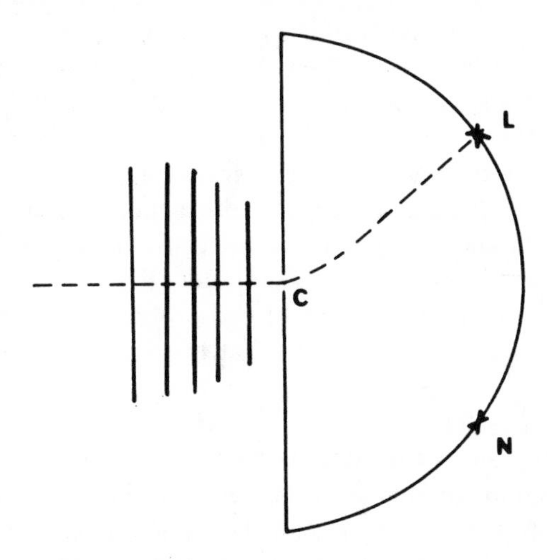

Fig. 16. Einstein's 1927 thought experiment: a wave carrying one particle is diffracted at C, and the particle, if going through the aperture, is detected at L on a semi-spherical film. The question is: How is the information telegraphed between L and any other, non-blackened, grain N of the film? The 'Copenhagen assumption' has it that the chance event occurs at L, not at C.

a hemispherical photographic film centered at C, one pointlike grain, say L, is blackened. How is it, asked Einstein, that any other grain, say N, is prevented from being blackened? It would seem that the Heisenberg–Born conception is 'incomplete', and that it overlooks the existence of "a very specific action-at-a-distance mechanism forbidding that the continuous wave produces effects at two places. ... The [probabilisitic interpretation] of $|\psi|^2$ implies, according to me, a contradiction with the postulate of relativity."

Thus Einstein made his point with wonderful accuracy. '*Quantal non-separability*' (as it is called today) *stems from the wavelike probability calculus* (it does not show up in the classical probability calculus); and *it does contradict the relativistic prohibition against distant spacelike connections* (at least when covariance is defined via the orthochronous Lorentz group, and 'telegraphing into the past' is excluded). To this we shall return.

Let us consider in turn these two points.

Associated with the wave propagation there is a conserved current density (classically, Poynting's current for photons, Schrödinger's current for electrons; relativistically speaking, the appropriate 4-currents have been defined in Chapter 4.4). Is it then not conceivable that, from de Broglie's proposal (soon extended by Madelung (1928)), the particle follows one of these current lines, thus exemplifying Poincaré's concept of chance as 'large, obvious, effects following small, hidden causes'? Even the more sophisticated case of diffraction and interference produced by two parallel slits can be handled in this way, the current lines being obtained with the help of a computer (Bohm and Hiley, 1982). It needs more insight into the core of the problem, and recourse to the 'transformation theory', to find out that this leads to a dead end — as we shall see later.

Concerning Einstein's second objection, that of a 'faster than light interaction' between two distant points L and N, one may wonder why his deep expertise in statistical mechanics did not remind him of Loschmidt's reversibility argument. The phenomenon here discussed mainfestly *is* an elementary one, *at the level where past—future symmetry rules*. It is queer also that he did not recall his bitter 1906—1909 discussion with Ritz, showing an interconnection of the wave and the particle aspects of 'factlike irreversibility' — and, therefore, a binding of the corresponding 'intrinsic symmetries'. If, for the above reasons, causality must be thought of as arrowless at the microlevel, *the zigzag LCN, with a relay at C in the past, is quite acceptable, mathematically and physically speaking, as the link between the two distant chance events at L and N.* A few (non-identical) variants of this idea have been put forward by myself (1977; 1979; 1980), Davidon (1976), Rayski (1979) and Stapp (1975). Incidentally Renninger (1960) has produced the 'negative form' of Einstein's argument, in which the non-detection of the particle at N implies its detection at some other point L. How can a non-measurement have a physical consequence? This could not occur in the classical calculus of probabilities, as the corresponding 'logical inference' took place only in one's mind. In the wavelike probability calculus there is in this a 'paradox' — and one of the 'hard' sort. Heisenberg, in a letter to Renninger, states that *even the absence of a measuring device causes an interference with the measured system, so that 'no record' must be interpreted positively as 'record zero'*.[2]

With this the discussion certainly belongs to what mischievous German students term '*Quantentheologie*'. The point is that there seems to be absolutely no escape from *Quantentheologie*, so that we must worship it, either (as does the majority) by paying lip service and keeping busy with 'hard physics', or (as a fascinated minority does) by trying to relate 'paradox and paradigm'.

All in all, in 1927 Einstein had started a fire for the years to come. A smoldering fire at first, producing more smoke than light, but causing uneasiness in the physicists' minds, private brooding, and low-voiced discussions.

Einstein, Podolsky and Rosen (1935) rekindled the fire. And in 1964 a famous Bell theorem made it burst into flames. Today, in 1984, after hectic days of discussion and experimentation, the watch of the firemen has not yet ended.

4.6.2. 1927—1935: THE BOHR—EINSTEIN CONTROVERSY

In the volume *Albert Einstein Philosopher Scientist*, edited by P. A. Schilpp, N. Bohr (1949, pp. 201—242) recalls very vividly how he answered, one after the other, Einstein's objections to his 'complementarity' interpretation of quantum mechanics, focusing mainly on various aspects of Heisenberg's uncertainty relations. The climax was reached with Einstein's objection against the 'fourth uncertainty relation', where he proposed to replace the energy by a mass measurement, by means of a balance. This we have recalled in Section 4.3.12.

'Defeated but not convinced' Einstein continued brooding, and, in 1935, with the help of Podolsky and Rosen, he produced a formalized version of his 1927 correlation argument, using two particles instead of just one. These, issuing from a common 'preparation' at C, are detected at two distant places L and N where 'measurements' are performed.

4.6.3. 1935: THE EINSTEIN–PODOLSKY–ROSEN ARTICLE 'CAN QUANTUM MECHANICAL DESCRIPTION . . . BE CONSIDERED COMPLETE?"

Believing in 19th-century basic credo that there exists a physical reality independent of observers, the authors (EPR) require that "every element of [it] must have a counterpart in the physical theory". Then they declare that "if, without in any way disturbing a system, we can predict with certainty (i.e. with probability . . . unity) the value of a physical quantity, then there exists a [corresponding] element of physical reality".

Two criticisms must be raised against the latter, long, quotation, one pertaining to matters of chronology which, though important, does not affect the main point, and the other quite fundamental.

As for the first point, the word 'predict' is quite infelicitous, because *the order in time of the two distant measurements at L and N is irrelevant.* Essentially it is a *telediction* that is at stake; apart from this, it can be just as well a prediction or a retrodiction. EPR use the non-relativistic Schrödinger formalism, the 'universal time' dependence of which allows nevertheless non-simultaneous measuremetns to be performed at L and N, and yields for their correlated values *a formula that is essentially symmetric in L and N.*

Significant as it is with respect to 'the physical magnitude time', this critique, however, does not affect the main point aimed at by EPR, to which we come now.

Considering two physical quantities represented by non-commuting operators, EPR argue that either "the quantum mechanical description is not complete or that the two quantities cannot have simultaneous reality". The latter statement, which is inherent in the Copenhagen interpretation, is clearly incompatible with EPR's basic assumption.

Then, using Schrödinger's formalism, EPR construct a system of two particles, a and b, such that by measuring either the position or momentum of a, a precise 'telediction' of the position or momentum, respectively, of b follows — *if such measurement is performed on b.*

The fact is, however, that it is quite possible to measure, say, the position x of a and the momentum p of b; Schrödinger's formalism yields a definite expression $P(x_a, p_b)$ for the correlation probability which is a conditional probability $(x_a | p_b)$ (see Section 6 below).

Therefore *the EPR reasoning is definitely not proof of an incompleteness of quantum mechanics, because the second possibility mentioned by EPR, that "the two quantities cannot have simultaneous reality" remains* (and, of course, is part of the Copenhagen credo).

Of all this EPR were well aware, as they add that, if one insists that "physical quantities can be regarded as simultaneous elements of reality only when they can be simultaneously measured [exactly] . . . the argument breaks down". However, they continue, "according to such a restricted criterion the reality of the position or the momentum of *b* would depend on the . . . measurement performed on *a*" — *which statement definitely is a paralogism disregarding the phenomenology* (we *can* measure x_a *and* p_b); it uses as a premiss the wanted conclusion.

EPR conclude that "such a mutual dependence of distant measurement outcomes", when certainly the one "does not disturb in any way the other", is not acceptable to them, as "no reasonable definition of reality could be expected to do this". Well, as the poet Boileau put it: "It sometimes does happen that truth is unlikely." First, why are EPR so sure that no 'disturbance' can be telegraphed from *L* to *N*, or vice versa? *What* if causality were arrowless at the microlevel, so that the two distant measurements at *L* and *N* could be connected via the zigzag *LCN*, the past preparation at *C* being the relay? And second, how can EPR be so sure that their 'reasonable definition of reality' is also the one accepted by Nature?

In concluding this section we must insist that very far from proving the "existence of an independent reality", the EPR argument sheds crude light on a paradox[2] of the hard sort, and thus greatly helps in characterizing *what 'physical reality' is not.*

The essential merit of the EPR argument is that it extends to large spatial separations the paradoxical sort of non-separability that the theoretical and experimental studies of the helium atom and the hydrogen molecule had already displayed — *a non-separability stemming directly from the 'Born wavelike probability calculus'*.

In this way the EPR argument definitely bears upon the deepest aspects of 'the Nature of Reality' or 'the Reality of Nature'. For one thing, it bears upon the question of quantum mechanical measurements (Schrödinger, 1935), as these rest on an EPR coupling between the 'measured system' and the 'measuring device'. So the study of EPR correlations must be pursued — *and pursued in an explicitly relativistic fashion.*

4.6.4. 1935: ON BOHR'S REPLY TO EPR

Of course Bohr (1935) did not accept the EPR proposal.

First, as EPR had used the commuting magnitudes $x_a - x_b$ and $p_a + p_b$, so did Bohr, remarking that these are appropriate for discussing diffraction by a rigid screen with two parallel slits, the separation of which is $x_a - x_b$. This magnitude is measured *ipso-facto*, while $p_a + p_b$ can be measured if the screen is free to move. This remark in fact strengthens the EPR 'paradox', by emphasizing the 'rigidity' existing in its 'immaterial' connection.

Then Bohr comments that

> there is . . . no question of a mechanical disturbance . . . during the last critical stage of the measuring procedure. But, even at that stage there is . . . the question of an influence of the very conditions which define the possible types of prediction . . . Since these constitute an inherent element of the description of any phenomenon to which the term 'physical reality' can be properly attached . . . the [EPR] argumentation . . . does not justify their conclusion that quantum mechanical description is . . . incomplete.

This is a short expression of the Copenhagen creed, preceded by the *implicit suggestion of delayed choice experiments*. Also, as we shall see, Bohr's rejection of a 'mechanical' aspect in the correlation was probably too strong.

4.6.5. 1935–1936: SCHRÖDINGER'S AND FURRY'S DISCUSSIONS OF THE EPR ARGUMENT

Schrödinger (1935) and Furry (1936a) independently elaborated upon the mathematics of the EPR correlations. Like EPR, Schrödinger found the quantum mechanical consequences unacceptable. In one of his arguments one of the correlated systems is macroscopic, being, for example, a cat which will be either killed or left alive depending on whether or not an ionizing particle issuing from a radionuclide enters a Geiger counter. Is it true, asks Schrödinger, that the cat is in a superposed state $\alpha\,|\text{alive}\rangle + \beta\,|\text{dead}\rangle$ as long as the biophysicist observer has not looked inside the cage? Or, as Putnam asks, does the cat himself, being a qualified observer, collapse his own wave function? This brings in an interesting question, because it so happens that *precisely that sort of experiment has been performed by parapsychologists in the form of a reward-or-punishment rather than a life-or-death experiment*, with the result that the animal can learn how to influence the random outcome.

This brings us right back to the problem of *lawlike symmetry versus factlike asymmetry between the two faces of the information concept*, as discussed in Chapter 3.5.

However the whole matter is not so easily settled, because Schrödinger's macroscopic object need not be a living being. How and when does the transition between the micro-quantal superposition and the macro-separation of states occur? This is a much discussed question — and of course a very important one (Çini, 1983; Gutkowski and Valdes Franco, 1983; Zurek, 1981).

As for Furry, using mathematics quite similar to Schrödinger's, he showed in detail that the quantal formalism *necessarily* entails the existence of the 'paradoxical' correlations, thus implicitly suggesting that appropriate experiments be performed to test them.

4.6.6. MORE THOUGHTS ON THE EPR THOUGHT EXPERIMENT

By using, *à la* Weiszäcker (1931), one of two Heisenberg microscopes pointing straight at each other along a z axis, the original EPR (1935) proposal can be brought much closer to an actual experiment. Suppose for example that, between the two microscopes, a positronium atom P disintegrates into two photons flying in opposite directions, each received in the image plane of one of the microscopes. As explained in Section 4.3.11, one can thus retrodict either the position $\mathbf{r}$ or the momentum $\mathbf{p} = h\mathbf{k}$ of P in the x, y plane. We assume here that one microscope measures $\mathbf{r}$ and the other $\mathbf{k}$.

The conditional amplitude $\langle \mathbf{r} | \mathbf{k} \rangle$ is nothing else than the Fourier nucleus. Therefore, the two prior amplitudes $\psi(\mathbf{r}) = |\mathbf{r}\rangle$ and $\theta(\mathbf{k}) = |\mathbf{k}\rangle$ are Fourier associated, so that the Heisenberg uncertainty relation $\varepsilon \cdot \eta \geqslant \hbar$ is obeyed. Physically this results from the finite apertures of the microscopes. Of course the more refined Hilgevoord—Uffink (1985) analysis we have presented in Section 4.3.13 is significant here.

As for the two photons, a and b, the position of the one, $\mathbf{r}_a$ and the momentum of the other, $\hbar\mathbf{k}_b$, are measured accurately at their final impacts on the photographic plates. But one is *definitely not* allowed to infer (as did EPR) that, in this way, the momentum $\hbar\mathbf{k}_a$ and the position $\mathbf{r}_b$ are indirectly measured! *Values of quantal magnitudes are conditional upon actual measurements.* In the present case the *condi-*

tional amplitude $\langle \mathbf{r} | \mathbf{k} \rangle = (2\pi)^{-1/2} \exp[i(\mathbf{k} \cdot \mathbf{r})]$ is the alpha and the omega of the whole question.

Summarizing: first, *the two distant measurements need not fit each other*; and, second, as a consequence of the wavelike probability calculus, *the results they disclose do not pre-exist in the source.*

4.6.7. 1947: A PERSONAL RECOLLECTION

In 1947, in the theoretical physics group headed by Louis de Broglie at the Institute Henri Poincaré, the 1927 Einstein and 1935 EPR ideas became a subject of interest. One day, after much pondering, discussing the matter with him and Mme Tonnelat, I argued somewhat like this: Einstein of course is right in seing an incompatibility between his special relativity theory and the distant quantal correlations, but *only under the assumption that advanced actions are excluded.*

Certainly there exists no direct connection between the two distant measurements at L and N: both the mathematics and the physics exclude it. But *there is an indirect connection, via the past preparation at C.* This is mathematically and physically quite certain.

Now, as the phenomenon under discussion is a statistical one occurring at the elementary level, Loschmidt's 1876 reversibility argument must hold, *mutatis mutandis*; *Jordan's wavelike transition probabilities, like Loschmidt's classical ones, are reversible.* Therefore, if we accept that causality is arrowless at the micro level, the Einstein and the EPR correlations not only *are understandable*, but *are so without conflict with the relativity theory.*

I dare say that Louis de Broglie was far from willing to accept my proposal. Said Mme Tonnelat, when we walked out of de Broglie's office: "Did you see his glance at you? It was as if he thought you were crazy." In 1953, however, he 'presented' my (1953) 'Note aux Comptes Rendus', explaining this; of course, in the meantime, the Feynman diagrams had come out and were widely used. Today, as the phenomenon of the EPR correlations is very well validated experimentally, and is in itself a 'crazy phenomenon', *any* explanation of it must be 'crazy'. And now mine has been published quite a few times (1977; 1979; 1980).

4.6.8. 1949: WU'S AND SHAKNOV'S EXPERIMENT ON CORRELATED LINEAR POLARIZATIONS OF PHOTON PAIRS ISSUING FROM POSITRONIUM ANNIHILATION

In 1949 Mrs Wu and Shaknov tested at Columbia University Wheeler's 1946 prediction that the spin-zero photon pairs issuing from the annihilation of positronium should exhibit full correlation if received through crossed linear polarizers (see Wu and Shaknov, 1950). Wu and Shaknov extended their measurement to various relative angles, thus verifying the theoretically predicted sine expression of the correlation, to which we shall return in Section 12.

4.6.9. 1951 AND 1957: BOHM'S AND BOHM–AHARONOV'S CORRELATED SPINS VERSION OF THE EPR

In his book *Quantum Theory*, Bohm (1951) established the form taken by the EPR correlation for the measured spins of a spin-zero fermion pair. The calculation is non-relativistic, and the spins are measured as perpendicular to some axis $x'x$, along which the two fermions travel in opposite directions. The spin dependence of the total wave function of the pair comes out as $[\uparrow\downarrow - \downarrow\uparrow]$, the notation being the one explained in Section 4.3.14.

If the spins are measured along two axes orthogonal to $x'x$ with relative angle α, the correlation formulas come out as $\langle +|-\rangle = \langle -|+\rangle = (\frac{1}{4})(1 + \cos\alpha), \langle +|+\rangle = \langle -|-\rangle = (\frac{1}{4})(1 - \cos\alpha)$.

In their 1957 article, Bohm and Aharonov went from this case to that of the correlated linear polarizations of photon pairs, pointing out that such an experiment had been performed by Wu and Shaknov, thereby confirming the theoretical formula and establishing the existence of the EPR correlations.

4.6.10. 1964: BELL'S THEOREM

Bell's (1964) theorem caused an explosion in thinking and experimenting, and blew the smoldering EPR fire into a fiercely burning one.

In the spirit of previous 'impossibility arguments' due to J. von Neumann (1932, Ch. 4), Gleason (1957), Jauch and Piron (1963), Bell asked himself 'can a hidden variables theory duplicate the measurement predictions of quantum mechanics?', and this he did in the very context of the EPR correlations. By definition, hidden variables theories obey the additive and multiplicative rules of the classical probability calculus. It is therefore *a priori* very doubtful that the said duplication can be generally obtained. Therefore the motivation for a search of hidden variables can lie only in a deeply rooted faith in — or a nostalgia for — the classical world view.

Bell's characterization of 'local hidden variables theories', in relation with the EPR correlations, is that the joint probability $P(A \cap B)$ of finding at L the result A and at N the result B has the form

$$P(A \cap B) = \int \rho(\lambda_a, \lambda_b)\mathscr{L}(A, \lambda_a)\mathscr{N}(B, \lambda_b)\, \mathrm{d}\lambda_a\, \mathrm{d}\lambda_b,$$

with λ denoting the hidden variables, and A and B the settings of the measuring devices. Bell has cleverly concocted counterexamples where this expression cannot duplicate the quantum mechanical predictions. Again, this is hardly surprising, due to its 'classical' form. In EPR correlations, the expression of the quantal amplitude is similar, but with amplitudes replacing $\rho, \mathscr{L}, \mathscr{N}$ (and, of course, no λ's).

Comments, extensions and variants of Bell-like criteria have been adduced by many authors, some of them being very clever. This makes for an enormous literature, itself testimony of a deeply rooted nostalgia, and perhaps of a subconscious unease when having to face the consequences of non-locality.

4.6.11. 1967—1982: EXPERIMENTING AND THINKING WITH CORRELATED LINEAR POLARIZATIONS OF PHOTONS

The 1950 Wu—Shaknov experiment, in which correlated polarizations of spin-zero photon pairs issuing from the annihilation of positronium were studied, was later felt as perhaps too heavily 'theory laden', as Compton scattering was the means used for measuring the polarizations. To circumvent this objection C. A. Kocher and E. D. Commins (1967) used as a generator of spin-zero photon pairs an atomic cascade: a 0—1—0 transition in the calcium spectrum, the frequencies of the photons being both in the visible range, so that ordinary linear polarizers could be used. Successively improving upon this scheme, Freedman and Clauser (1972), Clauser (1976), Fry and Thompson (1976), Alain Aspect and coworkers (1981; 1982a, b), have established beyond any doubt the 'reality of the EPR paradox'. Figure 17 displays the remarkable agreement between Aspect's experimental results and the curve (a sinusoid) expressing the quantum mechanical formula; let me emphasize that the curve is the theoretical curve itself, not a least squares fit. Figure 18 is a sketch of Aspect's second experimental setup, where pairs of orthogonal linear polarizations, each corresponding to a 'yes—no' question, are measured on each beam — a very significant improvement over the previous setups.

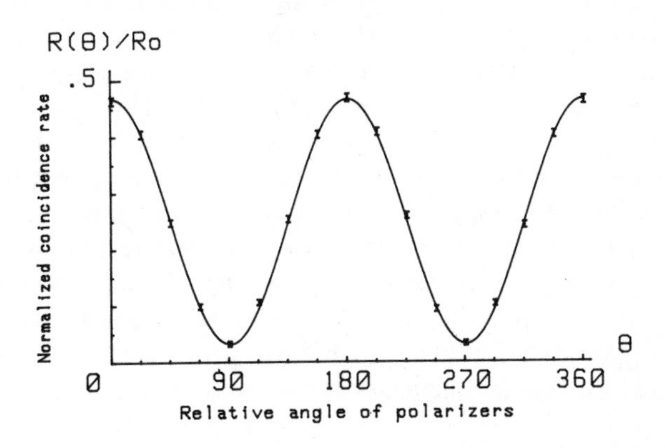

Fig. 17. Aspect's 1981 test of the Einstein—Podolsky—Rosen correlation: the very small error bars fall right over the theoretical sinusoid (not a mean squares adjusted sinusoïd)!

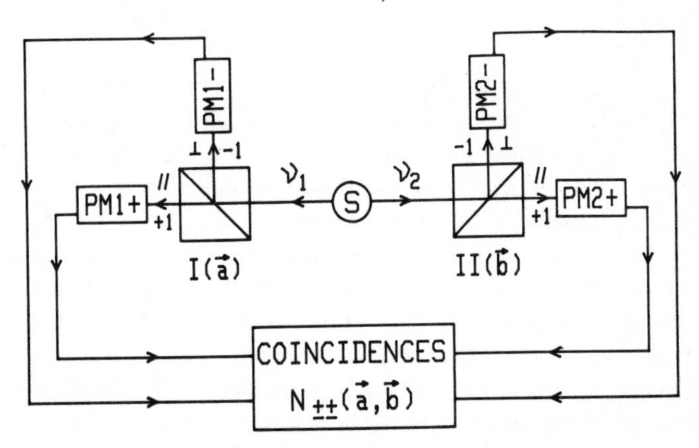

Fig. 18. Aspect's 1982 second experiment using two-channel linear polarizers (a man-made substitute for birefringent cristals).

Thus cascade experiments have settled the EPR question, transforming Einstein's 1927 and EPR's 1935 thought experiments into real, clearcut, experiments. However, the tradition of the e^+e^- annihilation experiments has been pursued, also yielding confirmations and a contribution to the understanding of the problem: Kasday—Ullman—Wu (1975); Wilson—Lowe—Butt in 1976; Bruno—d'Agostino—Maroni in 1977. These experiments are interesting, in that the distances between source and detectors can be made extremely large (up to about 2.5 m), largely above the coherence length of the γ radiation (12 cm). Moreover, these two distances have been varied independently.

All these experiments have been planned, performed and discussed in close interaction with Bell's problematic. In 1969 Clauser and Shimony—Holt independently adapted Bell's criterion to the case of linearly polarized photons, explaining this in a joint article (Clauser *et al.*, 1969). In 1982, Aspect stated that his experimental results differ from the Bell derived value by more than 13 standard deviations.

Before proceeding to the next section, where the correlation formula for linear polarizations of spin-zero photon pairs is established and discussed, I draw attention to a very thought-provoking article by N. D. Mermin (1981), and to the sequel to it I have written (1983b).

4.6.12. DEDUCTION AND DISCUSSION OF THE CORRELATION FORMULA FOR LINEAR POLARIZATIONS OF SPIN-ZERO PHOTON PAIRS

The derivation (Costa de Beauregard, 1977; 1979; 1980) now presented of the transition amplitude for correlated linear polarizations of spin-zero photon pairs essentially belongs to the very core of the Heisenberg—Schrödinger 'new quantum mechanics'. Thus it could have been presented at the 1927 Solvay Conference, yielding an exceedingly strong statement of the 'Einstein paradox'. As will be seen, this derivation is quite simple and easy to grasp; it clearly emphasizes the need of producing a new paradigm and of revising drastically our accepted views concerning 'the physical magnitude time'.

In atomic spectroscopy, 'cascade transitions' are those where an intermediate energy level is very short lived, so that two photons are shot in close succession with no intermediate recording. This is typically a case where Jordan's 1926 'wavelike probability calculus' prescribes that independent amplitudes should be multiplied and partial amplitudes be added.

In the cascades selected for the argument, the 'measured system' consists of a pair of photons with zero total angular momentum. In the laboratory these depart from some small region *C*. For convenience, only those pairs flying in opposite directions along some axis $x'x$ are selected (Figure 19(a)). As these photons do not have equal frequencies, and merely for practical reasons, monochromatic filters L'' and N'' select the frequencies, one on each side.[3]

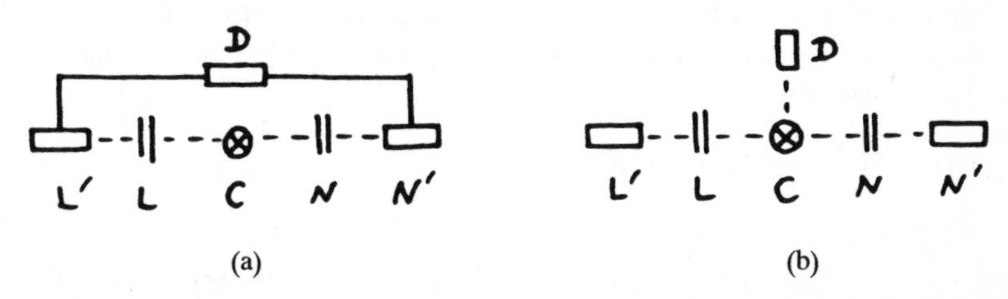

Fig. 19. Scheme of direct (a) and reversed (b) EPR correlation experiments for linear polarizations of photon pairs. *C*, cascading (a) or anticascading (b) atoms; *L*, *N*, linear polarizers of relative adjustable angle α; L', N', photodetectors (a) or lasers (b); *D*, coincidence detector (a) or fluxmeter (b).

Two linear polarizers, with arbitrary orientations *A* and *B* and relative angle $\alpha \equiv A - B$, are placed at *L* and *N*. Finally, two photomultipliers L' and N', working in coincidence,[4] make sure that two photons belonging to the same pair are those analyzed. This sort of technology was not available in 1927, but the argument could have been given as a 'thought experiment'.

When a photon normally falls upon a linear polarizer, either it passes, thus (so to speak) answering 'yes' (denoted 1) 'I take the linear polarization *A*', or it is stopped, answering 'no' (denoted 0), that is 'I take the linear polarization $A \pm \pi/2$'. By using birefringent polarizers or equivalent devices of a more sophisticated nature (Aspect *et al.*, 1982a), both sorts of answers can be displayed in the same experiment.

So, in the cascade experiments we are speaking of, there are 4 possible overall answers, with respective probabilities $(1|1)$, $(0|0)$, $(1|0)$, $(0|1)$, which must be calculated according to Jordan's 1926 rules (see my article, 1977; 1979; 1980).

These, as has been explained in Section 4.1.3, are *conditional probabilities* connecting the possible results of the measurements performed at L and N or, in other words, *transition probabilities connecting two representations of the overall*, spin-zero, *system*.

As expressed in terms of left L or right R circular polarizations, a spin-zero 'pure state' of the (a, b) pair is either L_aL_b or R_aR_b. Then the assumption of parity invariance (or left—right symmetry), holding in electrodynamics, requires that only the symmetrized and the anti-symmetrized 'state vectors' $2^{-1/2}(L_aL_b \pm R_aR_b)$ are retained. These are 'mutually orthogonal' in the sense of quantum mechanics (as were, of course, the L_aL_b and R_aR_b states).

As is well known from classical optics, equivalent expressions of these two states are possible in terms of orthogonal linear polarizations, Y and Z, namely

$$(4.6.1\,\|)\quad 2^{-1/2}(L_aL_b + R_aR_b) = 2^{-1/2}(Y_aY_b + Z_aZ_b),$$

$$(4.6.1\,\perp)\quad 2^{-1/2}[L_aL_b - R_aR_b] = 2^{-1/2}i[Z_aY_b - Y_aZ_b].$$

These are sums of products of probability amplitudes, essentially similar to those encountered in Section 4.3.14, in relation with the hydrogen atom and the helium molecule problems. But, here, the two 'Siamese twins' 'measured at L and N are widely separated. Therefore the sort of chord joining them is very extended; also — as we shall see — it has a very unusual spacetime nature!

The two formulas (1) respectively characterize different types of atomic cascades that have been tested: the '0—1—0', and the 1—1—0' cascades. The numbers denote the spin values of the three energy levels implied: upper, intermediate (or 'virtual'), lower; in the second instance the upper level is a triplet state, but the transition used starts from the 0 value in the triplet.

Now we shall show, using in succession the 'circular polarization' picture and the 'linear polarization' picture, that

$$(4.6.2\,\|)\quad (1|1) = (0|0) = \tfrac{1}{2}\cos^2\alpha, \qquad (1|0) = (0|1) = \tfrac{1}{2}\sin^2\alpha;$$

$$(4.6.2\,\perp)\quad (1|1) = (0|0) = \tfrac{1}{2}\sin^2\alpha, \qquad (1|0) = (0|1) = \tfrac{1}{2}\cos^2\alpha.$$

Calculation in terms of circular polarizations

Turning the polarizer L by ΔA will shift the L_aL_b pair by (say) $+\Delta A$ and (then) the R_aR_b pair by $-\Delta A$. Similarly, turning the polarizer N by ΔB will induce the phase shifts $-\Delta B$ and $+\Delta B$. Thus the two amplitudes at stake are $\exp(i\alpha)$ and $\exp(-i\alpha)$ so that, from Born's rule, we derive formulas (2||) in the form

$$(4.6.3||)\quad (1|1) = (0|0) = \tfrac{1}{8}|\exp(i\alpha) + \exp(-i\alpha)|^2 = \tfrac{1}{4}(1 + \cos 2\alpha),$$
$$(1|0) = (0|1) = \tfrac{1}{8}|\exp(i\alpha) - \exp(-i\alpha)|^2 = \tfrac{1}{4}(1 - \cos 2\alpha).$$

The diagonal contributions

$$(1|1)_0 = (0|0)_0 = (1|0)_0 = (0|1)_0 = \tfrac{1}{4}$$

are none else than those obtained via the classical probability calculus, under the assumption that the photons do 'possess' opposite circular polarizations when leaving the source. The off-diagonal, interference style contributions, $\pm(\frac{1}{4})\cos 2\alpha$ are 'neoquantal corrections', which would have caused quite a shock if unveiled in 1927!

Calculation in terms of orthogonal linear polarizations

From classical optics we know that the probability that a photon successively crosses two linear polarizers of relative angle α is $\cos^2 \alpha$. Hence, by Jordan's rule, the amplitudes at stake are $\cos A \cdot \cos B$, $\sin A \cdot \sin B$, $\cos A \cdot \sin B$ and $\sin A \cdot \cos B$. Therefore again we obtain formulas (2||) in the form

$$(1|1) = (0|0) = \tfrac{1}{2}(\cos A \cdot \cos B + \sin A \cdot \sin B)^2$$
$$= \tfrac{1}{2}(\cos^2 A \cdot \cos^2 B + \sin^2 A \cdot \sin^2 B) + \tfrac{1}{4}\sin 2A \cdot \sin 2B$$

$$(1|0) = (0|1) = \tfrac{1}{2}(\cos A \cdot \sin B - \sin A \cdot \cos B)^2$$
$$= \tfrac{1}{2}(\cos^2 A \cdot \sin^2 B + \sin^2 A \cdot \cos^2 B) - \tfrac{1}{4}\sin 2A \cdot \sin 2B$$

Again, the diagonal contributions $\frac{1}{2}(\ldots)$ express the 'paleoquantal' prediction under the assumption that the photons, when leaving the source, 'possess' either the one or the other of two orthogonal linear polarizations. As these contributions are not rotationally invariant, a randomization would then have been necessary. This is easily done by rewriting the neoquantal correction $\Delta\langle 1, 1\rangle = \Delta\langle 0, 0\rangle = -\Delta\langle 1, 0\rangle = -\Delta\langle 0, 1\rangle$ as $\frac{1}{2}\cos 2\alpha - \frac{1}{2}\cos 2(A + B)$, the mean value of which is $\frac{1}{2}\cos 2\alpha$. Finally one gets

$$\overline{(1|1)_0} = \overline{(0|0)_0} = \tfrac{1}{4} + \tfrac{1}{8}\cos 2\alpha$$
$$\overline{(1|0)_0} = \overline{(0|1)_0} = \tfrac{1}{4} - \tfrac{1}{8}\cos 2\alpha$$

to be compared with the (3)'s.

Summarizing: the 'neoquantal' calculation, as performed either in the 'circular polarization' or in the 'orthogonal linear polarization' pictures, definitely shows that *the two 'correlated photons' leaving an atomic cascade do not 'possess' then definite polarizations* (neither both a left or a right helicity, nor two linear polarizations either parallel or perpendicular to each other, depending on the type of cascade).

The linear polarizations as measured at L and N are borrowed only then (a leitmotif in the new quantum mechanics) — *but they are correlated* (which is the 1935 EPR paradox).

To see this in a cruder light let us set $\alpha = \pi/2$, for example. Then we get, for the type $\|$ cascades, $(1|1) = (0|0) = 0$ and $(1|0) = (0|1) = \frac{1}{2}$ (and the contrary for type $\perp$ cascades). This, of course, is quite all right, *except that the* $\|$ (or $\perp$) *states, as measured, could not exist at the departure from the source* — because, then, we would get $(1|1) = (0|0) = (1|0) = (0|1) = \frac{1}{4}$!

A very striking way for displaying experimentally the 'paradox' is to show that formulas $(2\|)$ or $(2\perp)$ hold even when the orientations of the measuring polarizers are fixed only after the photons have left the source. This has been successfully done by Aspect *et al.* (1982b)[5] (Figure 20).

An *ad hoc* proposal has been made by Selleri (1980) to alleviate the 'paradox', with the result that it is instead reinforced. Selleri assumes that, in each wavefront, there exists a 'hidden direction', and that, in the

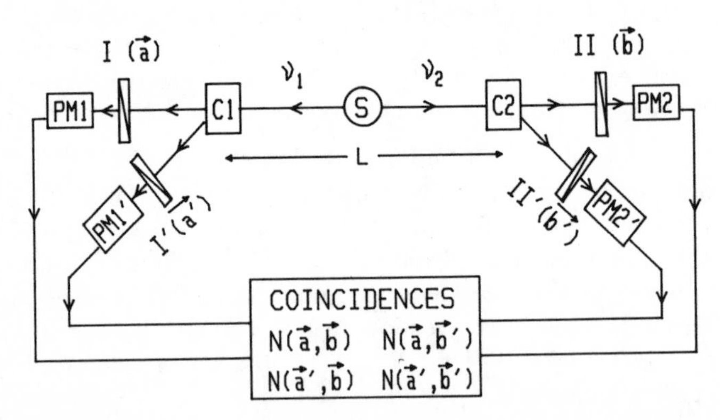

Fig. 20. Aspect's third 1982 experiment: rather than turning the polarizers after the photons have left the source, Aspect directs each of them along one of two channels comprising a fixed polarizer; the two switches are variable gratings, each consisting of an elastic standing wave, the frequency of which is arbitrarily controlled.

photon pairs, these 'hidden directions' are respectively parallel or perpendicular to each other, corresponding with the type, $\parallel$ or $\perp$, of the pair. Selleri assumes that, apart from this correlation, the 'hidden directions' are randomly distributed, and that, when impinging upon a linear polarizer, the photon falls into the nearer of the two orthogonal directions available. Then it is easily seen that the quantal sinusoid $1 \pm \cos 2\alpha$ is replaced by the zigzag shown in Figure 21, cutting the sinusoïd at its maxima, minima and inflexion points.

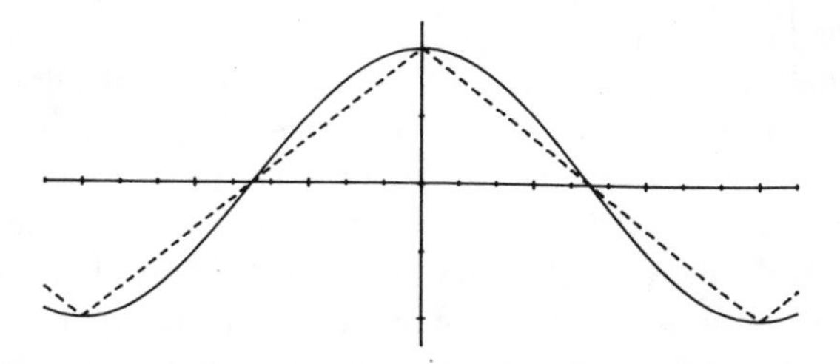

Fig. 21. Selleri's *ad hoc* model of correlated photons yields (broken line) a correlation less strict than the quantal (sinusoïdal) one.

So Bell's theorem is vindicated: Selleri's 'hidden variable theory' does not duplicate the quantal correlation formula. For additional remarks I refer to Mermin's article (1981) and mine (1983b), already quoted.

Putting things in a more literary form: we are playing some sort of an Alice-in-Wonderland dice game, where the very mathematics show that the chance event does not occur when the twin dice are shaken together inside the cup. The twin chance events occur when the dice stop rolling on the table (all right: why not?) — *but they are correlated.* That is the paradox.

4.6.13. DIRECTIONLESS CAUSALITY

If causality has any operational meaning, it implies the idea that *something can be arbitrarily adjusted at some time and place*, inducing effects at other times and places. Then it is the task of physics (experimental and theoretical) *to assess the time and space ordering between cause and effect.*

In the EPR correlation experiments, adjustable parameters exist at L

and N, not at C. Therefore we have at L and N two 'causes' (the adjustable angles A and B of the polarizers) and two 'effects' (the occupation numbers measured by the photodetectors). These numbers, which are functions of A and B, are correlated, the link of the correlation being the spacetime zigzag LCN with a relay at C, in the past of L and N, the vectors CL and CN being lightlike (or timelike, if massive particles were used). Therefore we are *necessarily* led to conclude that, *at this elementary quantal level, causality, as observed in the laboratory, is channelled by timelike (or lightlike) vectors, but is arrowless* — in full agreement with Loschmidt's 1876 argument, which showed it as reversible in time.

This conclusion is reinforced by the two additional remarks that the correlation formula, which is symmetric in L and N, is found (both theoretically and experimentally) to be insensitive to the (space and time) distances CL and LN (be they equal or unequal), and also that it depends only on the orientations of the polarizers while the photons go through them (what they are before or after being irrelevant).

That the spacetime zigzag LCN truly is the link of the correlation is directly proved by the experiments where the distances CL and CN are arbitrarily varied. Let us illustrate this by a little parable.

If some archaeologists discover that similar cultures have flourished along two confluent rivers running through barren country — say, the Tigris and the Euphrates — will they conclude that the connection has been mysteriously tied via some direct influence propagated through the desert? Of course not: they will conclude that the connection has been made by navigating up and down the two rivers. Similarly here: *the only mathematical and physical connection between L and N is the zigzag LCN made of the paths of the two photons. This indeed is the link of the connection.* None other exists, or is conceivable.

Reversed Einstein correlation

As usual, the transition amplitudes and probabilities we have calculated hold just as well for the reversed transitions. At L and N we now have two lasers of appropriate frequencies, shooting straight at each other at C (Figure 19(b)). They are followed by two linear polarizers of arbitrarily adjustable orientations A and B, and produce at C an 'anticascade', or 'échelon absorption', raising atoms from a lower to a higher (via an intermediate) energy level. Today this is a perfectly standard sort of experiment (although this specific one has not yet been performed). Measurement of the transition rate would be made easily by means of a fluxmeter registering the radiation re-emitted along another channel.[6]

Now, the point is that *the very two traits which look so paradoxical in the EPR correlation proper — insensitivity of the correlation formula to the distances CL and CN, and insensitivity of it to turning the polarizers while the photons are on their ways — do look utterly trivial in the reversed correlation.* The reason for this is that, by relying upon macroscopic experience, one feels that each photon does 'retain', until its absorption in C, the polarization which 'has been imparted' to it at L or N. In other words, *retarded causality looks trivial and advanced causality looks paradoxal.*

However, what both the mathematics and the phenomenology do show, is *that causality is directionless at the elementary level.*

In the quantum mechanical jargon the phenomenology we have discussed is termed *non-separability of measurements issuing from a common preparation* (in the EPR correlations proper), and *non-separability of preparations converging into a common measurement* (in the reversed EPR correlations).

As the preceding reasoning does rely upon the fundamental rules of the 'new quantum mechanics', but not, however, on the use of a Hamiltonian, there is no objection against ending with a discussion of relativistic covariance.

All the previous reasoning is invariant with respect to a Lorentz transformation involving the x and the ct variables. Of course, the frequencies of the photons are thus changed — but they do not enter the formulas. The polarization states (either circular or linear), which do enter the formulas, are invariant under such Lorentz transformations. They are even invariant with respect to the arbitrary relative velocities we can impart to the three pieces of the apparatus along the $x'x$ axis (Figure 22(a) and (b)). This confirms that *the time ordering of the two measurements* (EPR proper) *or the two preparations* (reversed EPR) *performed at L and N is totally irrelevant.*

One more aspect of this 'paradoxical' EPR correlation should be considered.

In the EPR correlation proper we are of course not allowed to think that the linear polarizations of relative angle α measured at L and N

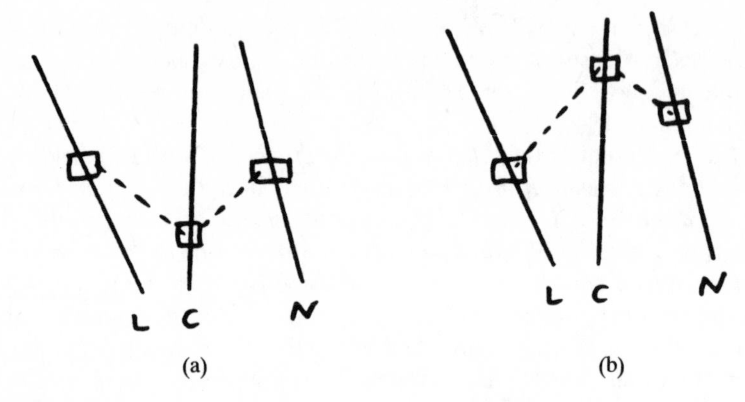

(a) (b)

Fig. 22. Spacetime diagrams for cascades (a) or anticascades (b). *L, C, N*, spacetime trajectories of the pieces of apparatus pictured in Figure 19(a) and (b), the positions and velocities of which are arbitrary inside the (x, ct) plane.

retrodictively fix those that the photons 'had' when leaving the source *C*. This would not be compatible with the assumption that, when leaving *C*, the pair 'has' a total spin zero. In other words, what is prepared at *C* is a pair, and what is measured at *L* and at *N* is a single photon an act of severance.

Therefore, disregarding any question of time ordering and concentrating on the topology, we see that the phenomenon discussed is essentially the same as that of one single photon crossing in succession two linear polarizers. And this is exactly what formulas (2) are saying.

This sort of topological invariance is at the root of the Feynman graphs and algorithms to be discussed in the next chapter.

We shall leave this to be pondered by the perhaps perplexed reader, as Einstein, Schrödinger and other 'towering minds' have been perplexed before him. The whole matter will be reassessed in Chapter 4.7, in the light of explicit Lorentz covariance, including partial or total reversal of the spacetime axes.

The point is that the intermediate summation in formula (4.1.6) is over virtual, not real hidden states, as it was in the classical formula (3.1.8). Thus at *C* we have, in Miller and Wheeler's 1983 wording, a "smoky dragon". In this consists, the 'paradox' of what is called (d'Espagnat, 1971, Part 3; Clauser and Shimony, 1978) *quantal nonseparability*, or, in its geometrical specification, *nonlocality*.

CHAPTER 4.7

S-MATRIX, LORENTZ-AND-CPT INVARIANCE, AND EINSTEIN–PODOLSKY–ROSEN CORRELATIONS

4.7.1. LIMINAL ADVICE

This chapter may well be the most important one of the whole book. All the subjects we have discussed previously – Lorentz invariance; *CPT* invariance as a Lorentz-invariant generalization of Loschmidt's *T*-symmetry; intrinsic physical reversibility versus factlike, macroscopic, irreversibility, and intrinsic particle—antiparticle symmetry versus factlike, macroscopic preponderance of matter over antimatter; Jordan's wavelike probability calculus supplanting the classical one; Einstein—Podolsky—Rosen correlations and the *CPT*-invariant causality concept; intrinsic symmetry between the 'obvious' and the 'hidden' faces of the information concept (gain in knowledge and organizing power) – are all pieces of a motley puzzle that now fall in place, showing the unified picture of a brand new paradigm.

Our aim is not to present here an in-depth and balanced treatise of the *S*-matrix algorithm (which is today highly elaborate) but to sketch it as a framework supporting a synthesis of the essential aspects of 'the physical magnitude time'.

4.7.2. DERIVATION OF FEYNMAN'S S-MATRIX ALGORITHM FOLLOWING DYSON

$U(\sigma_2, \sigma_1)$ denoting Schwinger's unitary evolution operator yielding the transition amplitude $\langle \Psi_2 | U_{21} | \Phi_1 \rangle$, we consider the '*S*-matrix'[1]

$$S \equiv U(+\infty, -\infty) = \prod_{-\infty}^{+\infty} [1 + i\mathscr{H}(x)\,\mathrm{d}\omega]$$

where the product Π is expanded as

$$S = 1 + \sum_{n=1}^{\infty} S_n$$

with

$$S_n = (1/n!) \int_{-\infty}^{+\infty} \mathrm{d}\omega_1 \int_{-\infty}^{+\infty} \mathrm{d}\omega_2 \int_{-\infty}^{+\infty} \mathrm{d}\omega_n .$$

$$\prod_n \{\mathcal{H}(x_1)\mathcal{H}(x_2) \ldots \mathcal{H}(x_n)\}$$

In this product two $\mathcal{H}$'s with timelike separation operate in order (that is, the later one is written first), while the order of two $\mathcal{H}$'s with spacelike separation is arbitrary. The factor $1/n!$ stems from permutations of $\mathrm{d}\omega$ cells. This is the expansion of the symbolic exponential

$$S = \exp i \iiiint \mathcal{H}(x)\, \mathrm{d}\omega.$$

The operator $\mathcal{H}(x)\,\mathrm{d}\omega$ has the dimensions of an action; if expressed in units such that $c = 1$ and $\hbar = 1$, it is thus a pure number. In quantum electrodynamics it is chosen as $A^k j_k$; that is, as $-eA^k\bar{\psi}\gamma_k\psi$. Since $\bar{\psi}\gamma_k\psi$ has dimension L^{-3} and $eA^k/c\hbar$ has dimension L^{-1}, when expressed in units such that $c = \hbar = 1$, S is dimensionless.

The very definition $\mathcal{H}(x) = -eA^k\bar{\psi}\gamma_k\psi$ shows that, in any cell $\mathrm{d}\omega$ taken sufficiently small, three particles interact: a photon A, and two 'electrons-or-positrons' $\bar{\psi}$ and ψ. In ordinary parlance four cases are thus possible, as shown in Figure 23. These are merely four 'relative manifestations' of what is essentially one and the same 'elementary transition'. In other words, 'collision of a photon with an electron' (Figure 23(a)) or 'with a positron' (Figure 23(b)), 'creation' (Figure 23(c)) or 'annihilation' (Figure 23(d)) of an electron—positron pair by absorption or emission respectively of a photon, are aspects of one and the same 'first-order transition'.

Momentum energy cannot be preserved in this first-order transition, each of the three particles involved being endowed with its proper 'rest mass'. Therefore the lowest order in which momentum-energy conservation is possible is the second order.

Two varieties of the second-order transition exist: either the photon, (Figure 24(a)) or the electron (Figure 24(b)) is 'virtual', meaning that its momentum energy is the one exactly balancing the two other ones, which belong to 'free' or 'real' particles. Thus the momentum energy of

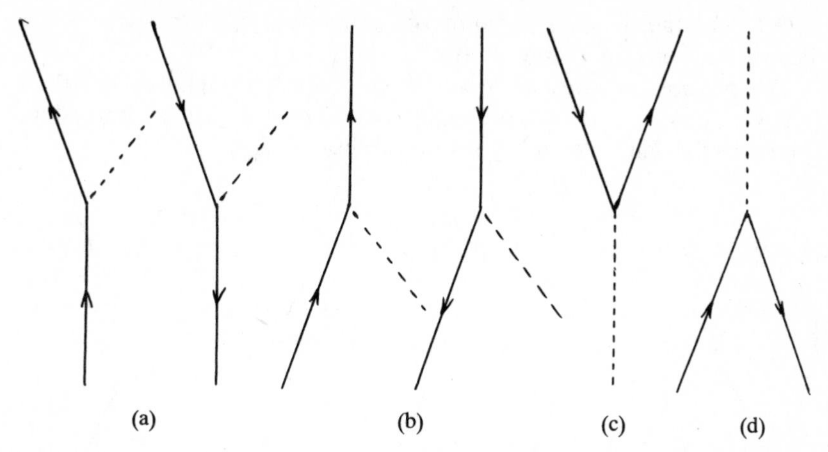

Fig. 23. Various 'relative' aspects of the first-order transition between the electron—positron and the photon systems: the photon is either emitted or absorbed, and the electron—positron system can undergo scattering, creation, or annihilation.

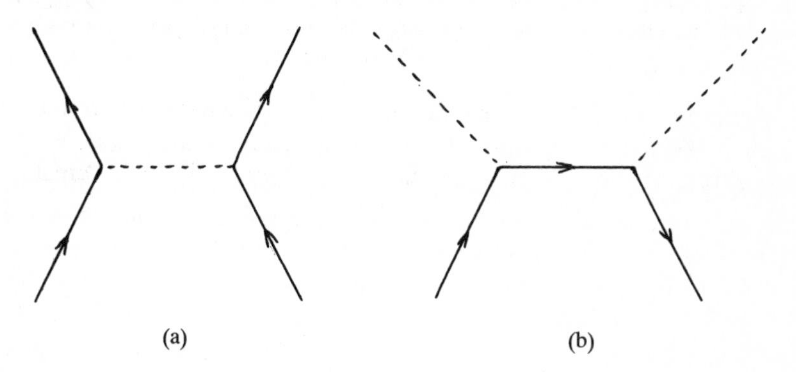

Fig. 24. Second-order transitions between the electron—positron and the photon systems: either the photon (a) or the electron (b) is 'virtual'.

a virtual particle can be any 4-vector (timelike or spacelike), the 'proper mass' being adjusted accordingly (real or imaginary).

This 'Feynman principle' is instrumental in explicit relativistic covariance. Compared to the one previously used in quantum field theory, stating that proper mass, not energy, was conserved, there is the exchange of one constraint and one degree of freedom: instead of the

3 + 1 constraints on momentum and proper mass, there are now the 4 constraints on momentum energy.

The possible manifestations of the case where the photon is virtual (Figure 24(a)) are: 'Coulomb repulsion' (Figure 25(a)) or 'attraction' (Figure 25(b)), and 'electron—positron collision' (Figure 25(c)).

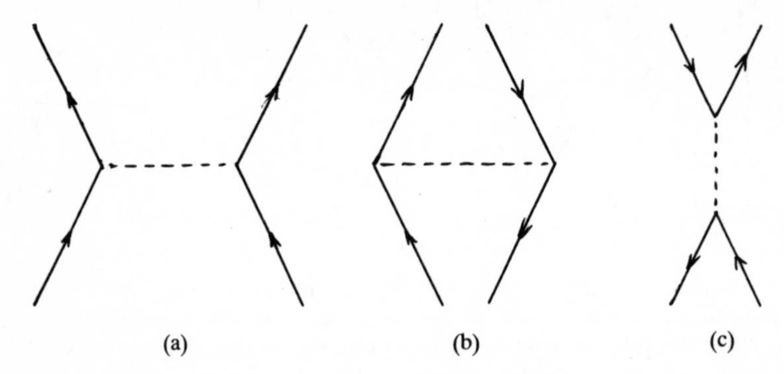

Fig. 25. Relative aspects of the second-order interaction mediated by a photon: attraction (a), repulsion (b), and a variant of the latter existing for a particle—anti-particle system (c).

Similarly, the possible manifestations of the virtual electron case (Figure 26) are: 'Compton effect on an electron' (Figure 26(a)) or 'on a positron' (Figure 26(b)); 'annihilation' (Figure 26(c)) or 'creation

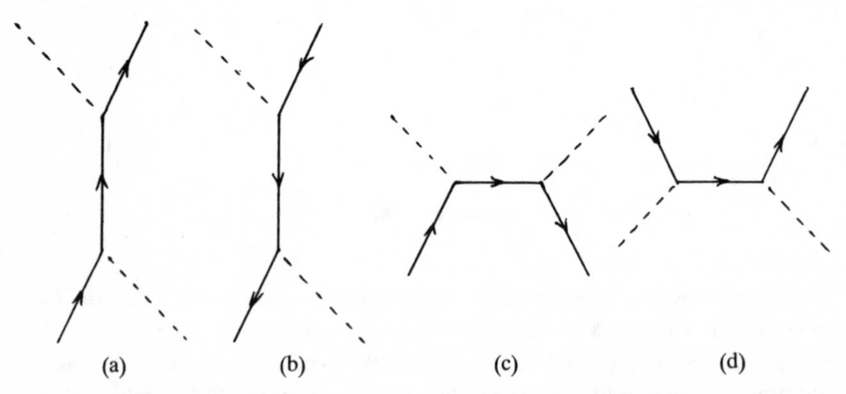

Fig. 26. Relative aspects of the second-order interaction mediated by an electron: Compton scattering of an electron (a) or a positron (b); annihilation (c) or creation (d) of an electron—positron pair.

(Figure 26(d)) of an electron—positron pair' into, or from, a photon pair.

So much for special cases. Generally speaking, an S-matrix transition amplitude is figured as a Feynman graph; that is, a network of links $\langle A|B\rangle = \langle B|A\rangle^*$. Internal links, joining two vertices, are propagators, but not Jordan—Pauli propagators, because the length of the linking momentum energy must remain free. The natural choice then is[2]

$$(4.7.1)\quad \begin{cases} D(k) = (k_i k^i - k_0^2)^{-1}, \\ D(x) = (2\pi)^{-2} \iiiint (k_i k^i - k_0^2)^{-1} \exp(ik_j x^j)\, d\tau \end{cases}$$

(additional comments are given in Section 9). End points — that is, links $\langle E|A\rangle$ with the environment — are wave functions $\langle a|k\rangle$ or $\langle a|x\rangle$, in the x and the k pictures, respectively.

Together with the momentum-energy conservation principle and the 1926 Jordan rules, these 'Feynman rules' produce, at lightning speed, the relevant transition amplitudes. For example, in Figure 25,

$$K_1 = k_c - k_a = k_d - k_b, \qquad K_2 = k_a + k_b = k_c + k_d,$$

and the 'Möller—Bhabha' amplitude is

$$e^2\{|K_1|^{-2}\bar{\theta}_a\gamma_\ell\theta_c \cdot \bar{\theta}_b\gamma^\ell\theta_d + |K_2|^{-2}\bar{\theta}_a\gamma_\ell\theta_b \cdot \bar{\theta}_d\gamma^\ell\theta_c\}.$$

Similarly, in figure 26,

$$k_1 = k_a - K_c = -k_b + K_d, \qquad k_2 = k_a + K_d = k_b + K_c$$

and the 'Klein—Nishina' amplitude is[3]

$$e^2 B_c^i B_d^j\{\bar{\theta}_a\gamma_i(\gamma_\ell k_1^\ell - ik)\gamma_j\theta_b + \bar{\theta}_b\gamma_i(\gamma_\ell k_2^\ell - ik)\gamma_j\theta_a\}.$$

In the general case, as in the two cases just discussed, the Feynman rule is automatic: at each vertex write $-e\gamma_i A^i$, and either the plane wave amplitude or the propagator, according as the corresponding particle is free or virtual. Since ψ and/or $\bar{\psi}$ always appear in pairs, S_n has α^n in factor, α denoting Sommerfeld's 'fine structure constant': $\alpha \simeq 1/137$. Therefore, in quantum electrodynamics, the power series of general term S_n is *a priori* well behaved — that is, well suited for iterative treatments.

4.7.3. CONSISTENCY BETWEEN FEYNMAN'S NEGATIVE ENERGY AND THE COMMONSENSE POSITIVE ENERGY INTERPRETATIONS OF ANTIPARTICLES

Here we no longer discuss the fact that reversal of the sign of e/m can be interpreted either classically, as reversal of e, or *à la* Stueckelberg–Feynman, as reversal of m. This has been examined in Sections 4.1.2 and 4.5.2. We intend to show now that the Stueckelberg–Feynman negative energy interpretation for antiparticles is automatically consistent with the classical positive energy interpretation. To this end, isomorphism between the classical statics of filaments and the relativistic dynamics of point particles will be used once again.

When discussing the equilibrium of a portion of stressed filament subject to a linear force density **f**, we assign an arbitrary sign to the tension **T**; this is because the tensions applied at both sides of a section are opposite. Therefore the *convention* is: *when ideally cutting a stressed filament, conserve the sign of the outgoing, and reverse the sign of the ingoing tension.* This convention is independent of the sign given to the tension.

Similarly with the dynamics of particles treated *à la* Stueckelberg–Feynman. *It is then always found that an energy is 'spent' when emitting, and 'gained' when absorbing, either a particle or an antiparticle.*

In his essay entitled 'Energy and Thermodynamics', Poincaré (1906a, Pt 3, Ch. 8) shows that energy is a constructed concept such that the physical energy, in its variety of aspects, is conserved. In this the sole requirement is 'convenience'. It has long been known that a potential energy can be negative; now, Stueckelberg and Feynman have shown us that a kinetic energy also can be negative.

There is no more intrinsic obligation for an energy to be positive than there is one for time to 'run forward'.

4.7.4. ESSENTIAL *CPT* INVARIANCE OF FEYNMAN'S ALGORITHM

CPT invariance has already been discussed in Section 4.1.2 and, from a classical point of view, in Section 4.5.2.

Lorentz-and-*CPT* invariance of the Feynman algorithm (that is, of the whole scheme of relativistic quantum mechanics as well) expresses an essential trait of Nature, with extremely far-reaching implications clearly displayed in the 'wonderful' phenomenon of the Einstein–Podolsky–Rosen (EPR) correlations, to be discussed now.

4.7.5. A CONCISE DEDUCTION OF THE EPR CORRELATION FORMULA FOR SPIN-ZERO PHOTON PAIRS

Figures 23, p. 171, display, in the (x, ct) or in the $(k_x, \nu/c)$ plane, respectively, the trajectories or the momenta energy of the photon pair and of their source (EPR correlation proper) or sink (reversed EPR correlation). This source or sink is idealized (1983a) as a spin-zero particle, either scalar $|\varphi\rangle$ or pseudoscalar $|\varphi\varepsilon_{ijkl}\rangle$. The two linear polarizations measured (or prepared) at L and N are, relativistically speaking, electromagnetic field strengths $|H^{ij}_a\rangle$ and $|H^{ij}_b\rangle$. In the C, P and T invariant electromagnetic interaction, the transition amplitude must be a relativistic scalar, so that there are two and only two possibilities

$$(4.7.2)\quad \langle\varphi|H^{ij}_a\rangle|H^b_{ij}\rangle \qquad \text{or} \qquad \langle\varphi\varepsilon_{ijkl}|H^{ij}_a\rangle|H^{kl}_b\rangle,$$

or, in prerelativistic notation, using Gaussian units and dropping for simplicity the *bra* and *ket* notation,

$$(4.7.3)\quad \bar{\varphi}(\mathbf{E}_L\cdot\mathbf{E}_N-\mathbf{H}_L\cdot\mathbf{H}_N) \qquad \text{or} \qquad \bar{\varphi}(\mathbf{E}_L\cdot\mathbf{H}_N+\mathbf{H}_L\cdot\mathbf{E}_N),$$

the L, N symmetry of which should be noted.

Taking the axes 1 and 4 inside the plane of the three momenta energies, denoting α the angle between the $\mathbf{E}_a$ and $\mathbf{E}_b$, or the $\mathbf{H}_a$ and $-\mathbf{H}_b$ vectors, and normalizing to unity, we get

$$(4.7.4)\quad 2^{-1/2}\cos\alpha \qquad \text{or} \qquad 2^{-1/2}\sin\alpha,$$

that is, the transition amplitudes exactly fitting the transition probabilities displayed in formulas (4.6.2).

Formulas (2) and (4) are invariant with respect to a Lorentz transformation involving the variables numbered 1 and 4. They are also invariant if arbitrary relative velocities are imparted to the three pieces L, C, N of the apparatus, inside the (1, 4) plane[4]: Figure 22(a) and (b).

What, then, is the link of the 'paradoxical' non-separability of the distant measurements at L and N arising from the common preparation at C (EPR correlations proper), or of the distant preparations at L and N converging into a common measurement at C (reversed EPR correlations)? It cannot be other than the Feynman zigzag LCN. ***This is what the formulas say.***

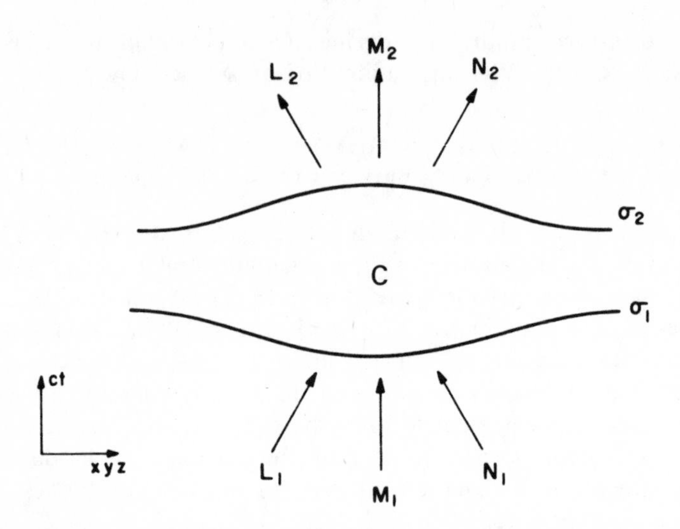

Fig. 27. EPR correlations in the interaction picture: in the remote past and future the particles (or antiparticles) are 'asymptotically free'; the interaction takes place inside a spacetime domain C enclosed between spacelike surfaces σ_1 and σ_2. The 'EPR paradox' consists of non-separability of the 'past preparations' and of non-separability of the 'future measurements', the connection being tied through C (spacelike connections tied by pairs of timelike vectors).

What counts, in the S-matrix scheme, is the definition of the preparing or measuring devices while the photons go through them; what they are before or after is irrelevant. Since adjustable parameters exist at L and N, not at C, if causality has any meaning, it is that 'causes' exist at L and N, not at C. Therefore the whole theoretico-experimental context of our problem evidently shows that causality is directionless at the microlevel. Stated precisely, *it is Lorentz-and-CPT invariant.*

Therefore, *the relativistic S-matrix scheme, as it exists, does have the full theory of the EPR correlations.* Nothing more than this is needed for understanding the EPR correlations — except of course an interpretative wording neatly fitting the formalism.

4.7.6. IRRELEVANCE OF THE EVOLVING STATE VECTOR; RELEVANCE OF THE TRANSITION AMPLITUDE

The transition amplitude $\langle\Psi_2|\, U_{21}\, |\Phi_1\rangle$ between a preparation $|\Phi_1\rangle \equiv \Pi\,|\varphi_1\rangle$ and a measurement $|\Psi\rangle_2 \equiv \Pi\,|\psi_2\rangle$ (Figure 27) holds *iff* each and every one of the incoming particles or antiparticles, and each and every one of the outgoing particles or antiparticles, is respectively *prepared or measured as is written down in the formula.*

Be it as it may, these important (and well-known) statements have been overlooked in quite a few presentation of the EPR correlations. By carelessly reading such expressions as

$$2^{-1/2}(Y_L Y_N + Z_L Z_N) \qquad \text{or} \qquad 2^{-1/2}\varepsilon_{LN}[Y_L Z_N - Z_L Y_N],$$

which are equivalent to (2), it has very often been stated that the first in time of the two distant measurements instantaneously collapses the other subsystem into the strictly associated state. Such a wording is very shocking in three respects. First, it is not symmetric in L and N, and second, it is not relativistically invariant, while the formulas are both. Third, it is self-contradicting in the following sense: if, in some inertial frame, the two measurements are simultaneous (which is allowed, if no energy measurement is performed[5]) and do not fit each other (as, say, two non-parallel or non-orthogonal linear polarizations), then the question is: Which of the two measurements collapses the other substate? What has been forgotten is, of course, that a word *iff* is attached to *both* measurements. A single measurement at L *or* N is definitely not equivalent to two measurements at L *and* N. Only the latter display correlations.[6]

It should be remembered that isomorphism between the formalism and its interpretative wording is the hallmark of a sound theory.[7]

Whence this misconception stems from is use of the evolving state vector concept, which, to begin with, is a non-testable concept. Only two things are tested: the preparation $|\Phi_1\rangle \equiv \Pi\,|\varphi_1\rangle$, and the measurement $|\Psi_2\rangle \equiv \Pi\,|\psi_2\rangle$. *Tampering with them would change the very definition of the operational procedure*; that is (according to a well-known statement by Bohr), *change the phenomenon studied.*

Second, the evolving state vector concept does not have time-reversal invariance: at σ, between σ_1 and σ_2, why should we use the retarded $|U_{\sigma 1}\Phi_1\rangle$ rather than the advanced $|U_{\sigma 2}\Psi_2\rangle$? Both are mathematically equivalent, as $\langle\Psi_2|U_{21}\Phi_1\rangle = \langle\Psi_2 U_{21}|\Phi_1\rangle$.

Finally, the evolving state vector concept is not Lorentz invariant: by changing the σ family we may reverse at will the order of two spacelike separated operators $\mathcal{H}(x)$, which is mathematically irrelevant. Therefore the state vector concept also is irrelevant.

So, *on the whole, being useless* (my article, 1981), *and even misleading* (my article, 1982) *the evolving state vector concept should be discarded* — like the late 'luminiferous aether'. Aharonov and Albert

(1980; 1981) have come to a similar conclusion. In the same vein, as it seems, Wheeler (Miller and Wheeler, 1983) writes that "no elementary quantum phenomenon is a phenomenon until it is registered", adding that it is "'a great smoky dragon'. [Its] mouth is sharp where it bites the counter [and its] tail is sharp, where the photon starts. But of what the dragon does or looks like inbetween we have no right to speak" — not even (as it seems) to write down a state vector!

On the contrary, thanks to its use of propagators, the Lorentz-and-*CPT*-invariant transition amplitude concept discards all spurious problems of time ordering, and strictly sticks to what is operational.

By so doing it emphasizes one of the most ominous traits of the EPR correlations: insensitivity to spatial and to temporal distances; that is, in other words, *direct long-range spacetime 'contacts'*. Fokker (1965), all through his brilliant presentation of the special relativity theory, has emphasized formalizations of this sort; *Feynman's algorithm definitely has a Fokkerian style*; and the EPR phenomenology makes clear that this is not without physical meaning.

Of course, at the macrolevel, the distant EPR correlations are almost completely obliterated by loss of the phase relations; demonstrating their reality with a separation of 12 meters between the detectors is an experimental feat! Incidentally, demonstrating the reversed EPR correlations at great distances would be quite easy. This means that *two CPT-associated Feynman graphs must be thought of as framed pictures, because one cannot CPT-reverse the environment* (including the preparing and measuring devices). Macroscopically speaking, *CPT* invariance is obliterated by the two very large factlike C and T macroasymmetries. These truly *define* the realm of macrophysics, being *necessary conditions* for its existence.

This being said, the transition amplitude $\langle \Psi | \Phi \rangle$ does display the same intrinsic prediction—retrodiction symmetry as has been emphasized classically by Loschmidt. In the Minkowskian '*sub specie aeternitatis*' philosophy, *it merely connects two different representations* of an 'evolution', the $|\Phi\rangle$ one and the $|\Psi\rangle$ one — very much as the Fourier nucleus $\langle k | x \rangle$ connects an $|x\rangle$ and a $|k\rangle$ representation.

A stereotyped statement in many expositions of the quantum theory is that 'there are two ways in which the state vector changes: between preparations and measurements, a continuous one obeying a Schrödinger equation; and a discontinuous one at preparations and measurements'. For the sake of Lorentz-and-*CPT* invariance the state vector concept must yield in favor of the propagator concept, *in which continuous evolution* resides. Also, in the S-matrix scheme, there are sets of partial preparations and measurements — the fixation points, so to speak, of a Feynman cobweb; *in them lies discontinuity*, by severance from the environment. So, on the whole, spacetime evolution is, so to speak, no more than a vivid dramatization of the Jordan algebra of conditional amplitudes and of the quantal stochastic game.

4.7.7. COVARIANT EXPRESSION OF THE EPR CORRELATION FORMULA FOR SPIN-ZERO FERMION PAIRS

Before proceeding in our epistemological discussions I present an S-matrix derivation of the spin correlation of a spin-zero fermion pair. There are two reasons for this. The first one is that this example was the one discussed by Bohm (1951), and that it is still very often discussed in the literature. The second one is that an experiment of this sort, carried out with two protons, has been performed by Lamehi-Rachti and Mittig (1976).

Using an argument quite similar to the one in Section 5 we find that there are two, and only two, amplitudes for a spin-zero fermion pair: $\bar{\varphi}\psi$, and $\bar{\varphi}\gamma_5\psi$; we denote φ an electron state and ψ a positron state.

Choosing units such that $c = 1$, we take the time axis along the overall 4-momentum (in ordinary parlance) so that both particles have the same energy w and opposite momenta, $+p$ for the electron, $-p$ for the positron. In the well-known 'low-velocity representation' of the γ's

$$\bar{\varphi}\psi = \varphi_1^*\psi_1 + \varphi_2^*\psi_2 - \varphi_3^*\psi_3 - \varphi_4^*\psi_4,$$
$$\bar{\varphi}\gamma_5\psi = \varphi_1^*\psi_3 + \varphi_2^*\psi_4 - \varphi_3^*\psi_1 - \varphi_4^*\psi_2.$$

As is well known, the 'large components' φ_1 and φ_2 of the electron, ψ_3 and ψ_4 of the positron, are arbitrary, and the 'small' ones, φ_3, φ_4 and ψ_1, ψ_3 are such that

$$w\varphi_3 = (p_x + ip_y)\varphi_2 + p_z\varphi_1, \qquad w\varphi_4 = (p_x - ip_y)\varphi_1 - p_z\varphi_2$$
$$w\psi_1 = (p_x + ip_y)\psi_4 + p_z\psi_3, \qquad w\psi_2 = (p_x - ip_y)\psi_3 - p_z\psi_4$$

so that, as $w^2 - p^2 = m^2$ and $m^2/w^2 = 1 - \beta^2$,

$$\bar{\varphi}\psi = 0, \qquad \bar{\varphi}\gamma_5\psi = (1 - \beta^2)[\varphi_1^*\psi_3 + \varphi_2^*\psi_4].$$

As φ_1 and ψ_4 are the eigenfunctions of 'spin up', φ_2 and ψ_3 those of 'spin down' along the z axis, and as the bracket is a pseudoscalar, we end up with

$$\bar{\varphi}\psi = 0, \qquad \bar{\varphi}\gamma_5\psi = (1 - \beta^2)[\uparrow\downarrow - \downarrow\uparrow],$$

which is the covariant expression of the well-known formula for correlations of spin-zero fermion pairs.[8]

These expressions (which have been derived solely from relativistic covariance, without appealing to fermion statistics) are invariant with respect to the direction of the opposite momenta $\pm\mathbf{p}$. In particular, if $\mathbf{p}$ is parallel to the z axis, the discussion is in terms of helicities; if it is perpendicular to z, the formula is the one associated with a Stern—Gerlach experiment.

In this latter case it is of course possible to measure the two spins in different directions of relative angle α. Then the joint, or conditional probabilities of paired answers, + or −, are

$$(+|-) = (-|+) = \tfrac{1}{4}(1 + \cos\alpha)$$
$$(+|+) = (-|-) = \tfrac{1}{4}(1 - \cos\alpha).$$

As all four answers do show up, 'angular momentum is not conserved', meaning that there is a reaction of the measuring apparatus upon the measured system. In other words, angular momentum is a property shared between the system and the measuring device.

A final significant remark is as follows. In the extreme relativistic limit, $\beta^2 \to 1$, $\bar{\varphi}\gamma_5\psi \to 0$, p_x, $p_y \to 0$. This means that *fermion and antifermion then are in opposite pure helicity states* — as is well known in the neutrino case: see Section 4.5.4.

So, this quite simple calculation yields, as a bonus, a far reaching consequence: the associated extreme relativistic fermion and antifermion *must* have *pure* and *opposite* helicities.

4.7.8. THE PARADOX OF RELATIVISTIC QUANTUM MECHANICS

Relativistic quantum mechanics, a synthesis of four-dimensional geometry and of a (wavelike) probability calculus (and an extremely operational one) truly is a marriage of water and fire. *How is this possible, and even conceivable?*

Being macroscopic, the preparing and measuring devices are conceived classically as embedded *in* spacetime *and in* the momentum-energy space. But the quantal system, which is in transit between a preparation and a measurement, certainly is neither 'in' spacetime nor 'in' the 4-frequency space, which provide no more than two 'complementary' ways of seeing things. Worse than that, speaking, for example, in terms of spacetime, the 'evolving system' is neither 'in' the retarded state $|U_{\sigma 1}\Phi_1\rangle$ nor 'in' the advanced state $|U_{\sigma 2}\Psi_2\rangle$, *because it is transiting* between $|\Phi_1\rangle$ and $|\Psi_2\rangle$. It is, so to speak, in the aisles of the spacetime theater, where it feels symmetrically the retarded influence of the preparation $|\Phi_1\rangle$ and the advanced influence of the measurement $|\Psi_2\rangle$. *This is what the formula says*, displaying the prediction–retrodiction symmetry.[9]

The $\langle\Psi|\Phi\rangle = \langle\Phi|\Psi\rangle^*$ preparation-measurement symmetry thus requires that *the concept of a 'state vector collapse' be rejected*, as an illegal borrowing from the macroscopic behavior of waves. The transition essentially consists of occupation numbers 'jumping' from retarded states $|\varphi\rangle$ into advanced state $|\psi\rangle$. Figure 28 displays the spatial analog

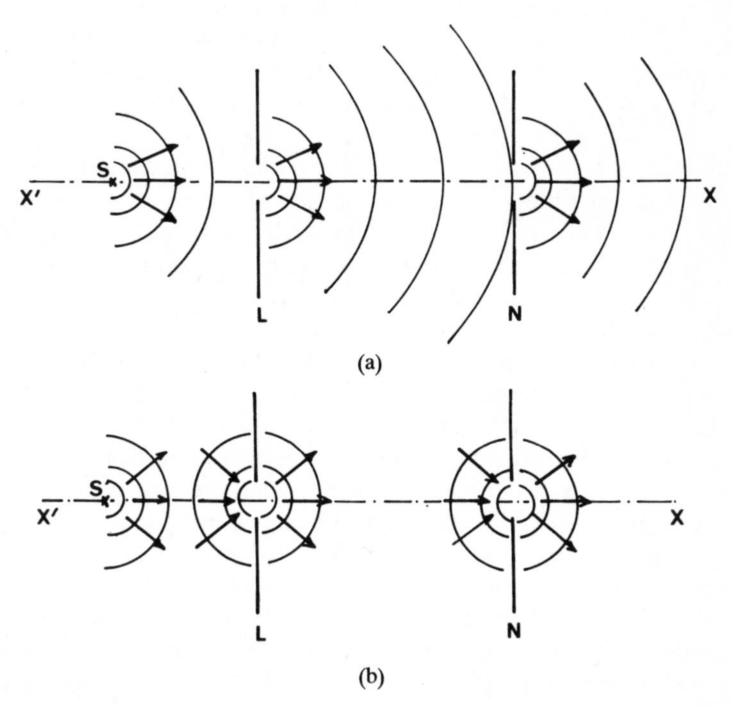

Fig. 28. Spatialized image of the classical wave-collapse concept (a) and of the new collapse-and-retrocollapse concept (b); compare with Figure 10). The wave is emitted from S in permanent regime, and the abscissa x is a substitute for the time variable ct.

of what we are saying: the source S emits, in permanent regime, and the waves are diffracted by successive screens placed at $x = a$, $x = b, \ldots$, so that the space coordinate x is a substitute for the time variable. The macroscopic behavior of the wave is pictured in Figure 28(a), corresponding to the concept of a 'wave collapse' at each 'position measurement' (y, z) at $x = a$, $x = b$, etc. . . . However the elementary phenomenon is x-symmetric, and is pictured as such in Figure 28(b); those particles which go through the apertures in the screens (thus answering 'yes' to the question asked by the measurement) jump, as it were, from the wave diverging from the aperture at $x = a$ to the wave converging into the aperture at $x = b$. A wave converging into, and then diverging from, an aperture, can be thought of as finding there a sink-and-source dipole; it is the spatial analog of the spacetime propagator $\langle x | x' \rangle$.

That relativistic quantum mechanics is a paradoxical marriage of water and fire certainly is true also at the macrolevel (although it is not obvious as such at first sight), because in principle the preparing and measuring devices are quantum mechanically describable. Therefore the 'objectivity' that these devices seem to possess (due to the smallness of Planck's constant h) is illusory — and so is that of 'everything' at the macrolevel. *It thus seems inevitable that a change in the metaphysics is required*, and that the Western metaphysics of a physical world existing in itself, independent of observation procedures, should be replaced by one resembling those of Eastern metaphysics, appropriately accommodating the Hindu 'maya' concept. An immediate question then is, what (if any) operational means do we have for testing the 'maya' character of preparation-and-measurement — of action-and-observation?

Almost all quantum mechanics textbooks state that, due to the finiteness of Planck's constant, 'there is an inevitable reaction of the measuring device upon the measured system'. The implication there goes much farther than is expressed, because *where should one put the cut between the measuring device and the observer*? Should we put it between the dial and the eye? or along the optical nerve? or where? So, what this statement truly implies is that *there is a reaction of the observer upon the observed system*. Wigner (1967, pp. 171—184), also arguing from the symmetries of quantum mechanics and from the general principle that 'to every action there corresponds a reaction', came to an identical conclusion, and then added, tongue in cheek as it seems, that "Every phenomenon is unexpected and most unlikely until it has been discovered — and some of them remain unreasonable for a long time after they have been discovered."

As was explained in Section 3.5.3, the same implication already follows from the finiteness of Boltzmann's constant k. What the finiteness of Planck's constant specifically brings into the picture is substitution of Jordan's wavelike probability calculus for the classical one — thus greatly sharpening the sting of 'paradox'.

Direct reaction of observation upon what is observed has a Janus-faced name: *psychokinesis and/or precognition*, as already discussed in Section 3.5.4.

On the whole, the paradox inherent in relativistic quantum mechanics shapes up into a paradigm where the mutually exclusive x and k four-dimensional theaters of events have, so to speak, side scenes, out of which information flows as knowledge, and into which it flows as organization.

One final remark is in order before we conclude this section.

We have argued that, in the EPR correlations, one single measurement performed at L does definitely not fix the 'corresponding' value at N, if only because it is quite possible (and experimentally demonstrated) to perform at L and N two measurements not fitting each other. In other words, *one single measurement at L is not identical to a couple of measurements at L and N*, and of this there exists a technical demonstration. From this, a very far-reaching conclusion should be drawn concerning the quantum mechanical measurement process.

The quantum mechanical measurement procedure precisely implies that an EPR correlation is established between the 'measured system' S and the 'measuring device' D, as was first noticed by Schrödinger (1935). The implication is that, strictly speaking, one cannot conclude from some reading made upon D that S *has* the strictly associated value. It merely means that *if* that magnitude could be directly measured, then it *would* display that value.

Such a statement lends support to Wheeler's contention (Miller and Wheeler, 1983, p. 151) of an 'information-theoretic' basis to the world we are living in, and is a plea in favor of blending of the Western and the Eastern views concerning cosmology.

4.7.9. A DIGRESSION ON PROPAGATORS. CAUSALITY AND THE FEYNMAN PROPAGATOR

Two propagators have been encountered up to now: the Jordan—Pauli propagator D, defined by formula (4.1.17) as the difference $D_+ - D_-$ between the positive and negative frequency contributions; that is, between the particle and the antiparticle propagators, of which we know that, outside the light cone,

$$(4.7.5)\quad D = 0, \qquad D_+ = D_-;$$

and the $\bar{D}$ propagator defined by formula (1).

This family can be extended by using various integration contours in the complex energy plane for computing the energy part in the integral (1). Two poles, at $\pm m$, exist on the real energy axis. $\bar{D}$, as defined by (1), is Cauchy's 'principal part' of the integral.

The contours yielding D_+ and D_- circle the poles, as shown in Figure 29 — an obviously covariant definition.

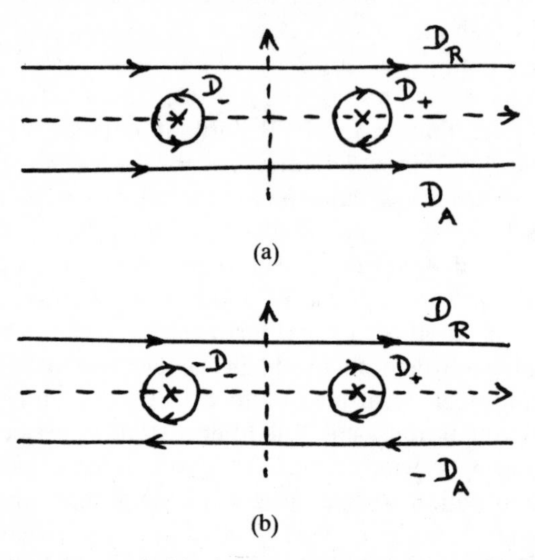

Fig. 29. Integration contours inside the complex energy plane with poles at $\pm c^2 m_0$: retarded D_R and advanced D_A, particle-like D_+ and antiparticle D_- propagators (a); the contour $D_R - D_A = D_+ - D_-$ for the Jordan—Pauli propagator (b).

The retarded D_R and the advanced D_A propagators, non-zero for, respectively, x future and past timelike, are defined via the straight contours in Figure 29 with the covariant prescription: close the D_R contour below for x future-timelike, above otherwise, and the D_A contour above for x past-timelike, below otherwise.

It is clear that

$$(4.7.6)\quad D = D_+ - D_- = D_R - D_A.$$

We consider also

(4.7.7) $D_F = D_R + D_- = D_A + D_+$,

(4.7.8) $D_{AF} = D_R - D_+ = D_A - D_-$,

the corresponding contours being shown in Figure 30(a) and (b), with

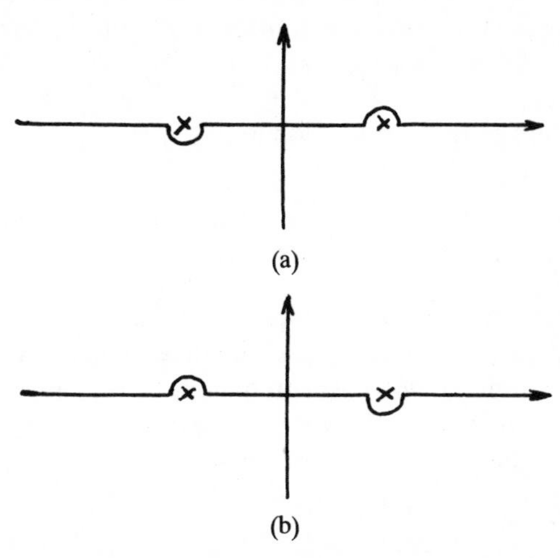

Fig. 30. Integration contours for the Feynman (a) and the anti-Feynman (b) propagators.

the prescription: close below if x is future-timelike, above if x is past-timelike, as you like (see formula (5)) if x is spacelike. Inside the collection D, D_F, D_{AF}, we define the discrete symmetries C and T via

$$C: D_+ \rightleftharpoons -D_-, \qquad T: D_R \rightleftharpoons -D_A;$$

we see that D has the symmetries $C = T = 1$, D_F and D_{AF} the symmetry CT; C or T exchange (up to a sign) D_F and D_{AF}.

D_F has been introduced by Feynman with the remark (see formula (5)) that $D_F = D_+$ if $t > 0$, $= D_-$ if $t < 0$; $D_{AF} = -D_-$ if $t > 0$, $= -D_+$ if $t < 0$. It is D_F rather than $\bar{D}$ that should be used for representing virtual particles in Feynman's algorithm. For example, Jauch and Rohrlich (1955, p. 408) extending a previous, non-covariant, argument by Weisskopf and Wigner, have shown that use of D_F automatically entails an exponential decay of higher energy levels in bound states (while the one of D_{AF} would paradoxically entail a build-up of higher levels).

Quite a few authors more or less have the wrong feeling that there is a causal asymmetry built-in the Feynman propagator; for that reason, they are reluctant to accept the idea that the Feynman zigzag really is the link of the EPR correlations. However, from its very definition (7), it follows that D_F symmetrically yields an exponential decay in predictive calculations and an exponential build-up in retrodictive calculations. So *the use of* D_F *for describing virtual particles is exactly consonant with the whole paradigm of lawlike reversibility and 'factlike irreversibility' as discussed in Part 3 of this book*; use of D_{AF}, on the other hand, would be in contradiction to it.

4.7.10. CONCLUDING THE CHAPTER, AND PART 4 OF THIS BOOK

Relativistic quantum mechanics does entail very drastic consequences concerning our world view. Being 'complementary' to each other, neither the spacetime x not the 4-frequency k pictures can be said to be 'objective'. As they are 'relative' to the experimental procedure, both the 'reality' and the 'objectivity' concepts must yield in favor of 'inter-subjectivity', thus entailing a world view very akin to the Hindu 'maya' concept; that is, to the concept of a sort of common daydream, the illusory character of which is pinpointed by the occurrences of 'para-normal phenomena'.

In itself relativistic quantum mechanics is a 'paradoxical' extremely fruitful marriage of water and fire — *an extended four-dimensional picture blended with a (wavelike) probability calculus.* How is this possible, and even conceivable? Again, 'maya' is the answer. *The quantal transition occurs beyond spacetime, symmetrically feeling (as the formula says) the retarded influence of the 'preparation' and the advanced influence of the 'measurement'.*[9] This is 'process' and 'becoming' at the elementary level, with full Lorentz-and-*CPT* invariance (the latter refining Loschmidt's 1876 *T*-symmetry). Thus *information as cognizance* flows out, and *information as organization* flows in, the material world, with the chain *preparing—evolving—measuring* 'corresponding' to the cybernetist's chain *coding—transmitting—decoding.*

The spacetime and the 4-frequency concepts are essentially symbolic means for computing transition probabilities, and thus dealing with the information extracted from measurements and that injected in preparations. And so *the quantal transition* — the 'elementary phenomenon' par excellence, with its attached reversible, *CPT*-invariant, transition probability — *is the very hinge around which mind and matter interact, holding open the gate through which knowledge flows out, and organization flows in, our 'factlike' and much illusory spacetime.* To think, as one so easily does, that the Aristotelian 'act' and 'potentia' are the one 'past' and the other 'future', separated by a Whiteheadian (1929) advancing spacelike surface σ, is macroscopic 'maya'. It is now well known that the past–future asymmetry is 'factlike, not lawlike', and that the macroscopic spacetime must be some sort of statistical emergence. At the microlevel the σ concept vanishes into utter fuzziness, as is clearly demonstrated by the EPR correlations, with their direct, long-range, arrowless spacetime connections. Therefore the severance between 'act' and 'potentia' consists in that the one is 'inside' and the other 'outside' the 'material world' — meaning, in Jaynes's (1983, p. 86) words, 'which sort of variables you or I choose for experimenting'. So *spacetime is a very porous sort of container* . . .

It is a macroscopic prejudice to think of the elementary phenomenon of 'wave collapse' as implying retarded waves, and thus as being time asymmetric. This again is denied by the very existence of the EPR correlations, with the Lorentz-and-*CPT*-invariant causality they display.[10] *

"Creative acts", in Stapp's (1982) wording, are directly reciprocal to thermodynamical degradation ($\Delta N < 0$) and to its milder form, the cognitive transition ($\Delta N + \Delta I < 0$). As such they are associated with advanced actions and decreasing probabilities, as can be seen, for example, when birds are building nests: repetitive adjustments of appropriate elements, producing a highly improbable contraption, are there the very opposite to 'scattering' or 'dispersion'. At the elementary level past and future are thus on the same footing. So, *in a 'creative act' the future exists no less (but no more) than does the past in a 'cognitive act'*. Neither has full objectivity, but is 'indissolubly objective and subjective'.

Of course one would like to see direct evidence of the $I \rightarrow N$ transition inherent to the core of 'creative act'. This is the 'paranormal phenomenon' termed 'psychokinesis and/or precognition', as examined in Section 3.5.4, where the fine experiments conducted by R. Jahn (see Jahn *et al.*, 1986) at the Engineering Faculty of Princeton University have been quoted. These, following previous experiments by other authors (for example, Schmidt, 1982) do confirm experimentally an open theoretical possibility, thus shedding much light upon the concepts of probability, of information, of 'process' and 'becoming', and upon the problem of mind–matter interaction.

Finally, using the wording of Section 1.1.2, the 'strongly paradoxical'

* See 'Added in Proof' p. 319ff.

phenomenon of the EPR correlations is inherent in the relativistic *S*-matrix algorithm, thus displaying a striking aspect of Lorentz-and-*CPT* invariance. At the macrolevel these are largely blurred by loss of the phase relations, that is, the off-diagonal contribution in the Jordan wavelike probability calculus. *Essentially* speaking, however, they are present, implying direct, long-range spacetime 'contacts', and thus, most likely, the existence of subtle phenomena termed 'psychic' in a broad sense, inside the human, the animal and possibly the vegetal kingdoms.

Quantized wave physics is truly a universal spacetime information-carrying telegraph. 'Information' is encoded as 'occupation numbers' of the waves — arithmetic entities masquerading as 'particles', by conferring weight and life to the covariant veil of the waves. Intrinsic *CPT* invariance means that time reversal of an evolution (of a 'message' connecting a 'preparation' to a 'measurement') in general must imply space—parity reversal and particle—antiparticle exchange (see Section 4.5.10). Again, this is very paradoxical, as shown by the reality of the EPR correlations discussed throughout Chapter 4.6.

The 1926 Jordan 'wavelike probability calculus' is instrumental in this refined sort of cybernetics.

As a final 'mandala' for the meditations of the reader I submit Figure 31, displaying Einstein's prohibition against *direct* faster-than-light telegraphing, Einstein's taboo against *macroscopic* backwards-in-time telegraphing, and (of course) the *factlike* priority of telegraphing into the future.*

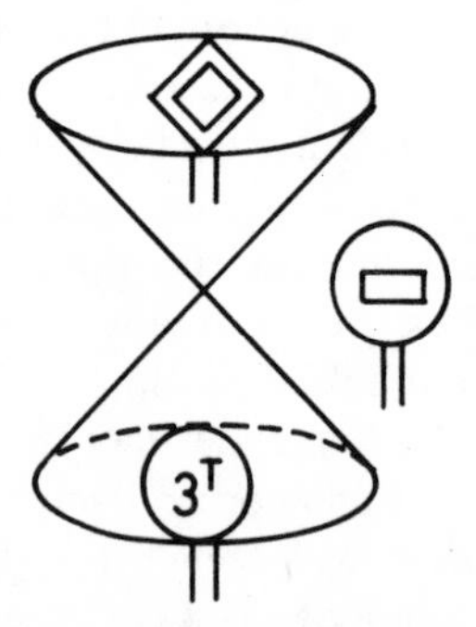

Fig. 31. Lawlike prohibition against telegraphing outside the light cone; factlike, or macroscopic, repression against 'telegraphing backwards in time'; telegraphing forward in time is the priority road. N.B.: telegraphing backwards in time does not mean 'reshaping' history, but shaping it from the future — as do sinks in hydrodynamics.

* See 'Added in Proof' p. 319ff.

PART 5

AN OUTSIDER'S VIEW OF GENERAL RELATIVITY

CHAPTER 5.1

ON GENERAL RELATIVITY

5.1.1. LIMINAL ADVICE

General relativity, Einstein's 1916 theory of gravitation, today is a world in itself, as anyone can see by perusing, for example, the authoritative treatise by Misner *et al.* (1970). So just a short chapter at the end of my small book certainly cannot do justice to the subject. Its very brevity testifies that I am no expert in the field, having almost never contributed to it, so that I can make only few personal remarks. However, it so happens that sometimes an outsider's remarks may make sense.

5.1.2. WHAT IS SO SPECIAL WITH UNIVERSAL GRAVITATION?

What is so special with universal gravitation is that, considering the idealized case of a point particle, it is the same factor that multiplies the acceleration $\boldsymbol{\gamma}$ to produce Galileo's inertial force $m\boldsymbol{\gamma}$, and that multiplies the gravitation field $\boldsymbol{\Gamma}$ to produce the ponderomotive force $\boldsymbol{\Gamma}m$. Therefore, such factors as the e/m which appears when an electromagnetic force acts upon a point particle happen to be 'universally' identical to unity in the gravitation case, m cancelling out of the equation $\boldsymbol{\Gamma} = \boldsymbol{\gamma}$. In other words, *the gravity field is an acceleration field* — a point already quite well understood by Galileo and by Newton.

In Einstein's words this expresses an 'equivalence' between inertial m_i and gravitational m_g mass — a fact extremely well substantiated by the 'universal' success of its use. Due to its importance, however, specific experiments have been devised to test it: Eötvös (1891, 1896 and 1922); Southern (1910), Zeeman (1918), Dicke (1961), with a final value of 1 ± 10^{-11} for the ratio m_g/m_i. Very precise experiments were also performed in 1960—1961 by Hughes *et al.* and by Drever to test the isotropy of gravitation — meaning that m is a scalar (not a rank-2 tensor, as could be implied by Mach's idea that inertia is caused by the gravity field of distant masses).

All experiments until now have substantiated the 'equivalence principle'. What, then, is the deep truth it expresses? Most certainly that this familiar, and even trivial property, inertia, is gravitational in nature. It stems from the fact that the sources of the gravitational potential U of the field $\mathbf{\Gamma} \equiv \mathbf{\partial} U$ do not cancel out in the large (as they do for the other long-range field, electromagnetism). According to Newton, the gravitational field $\mathbf{\Gamma}$ and potential U created by a point source of mass m are, respectively,

$$\mathbf{\Gamma} = Gmr^{-3}\mathbf{r}, \qquad U = Gmr^{-1}$$

with G denoting a constant having the value

$$G = 6.664 \times 10^{-8}\, \text{cm}^3\, \text{g}^{-1}\, \text{s}^{-2}.$$

In these formulas the 'active gravitational mass' m is essentially positive. Therefore, the universal gravitational potential is not zero, having the form

$$U_{\text{total}} = GM/R$$

with M denoting the 'total mass of the Universe' and R a cosmological scaling factor commonly termed the 'radius of the Universe'. U has the physical dimension c^2, so that GM/c^2R is a pure number.

Keeping in mind Mach's well-known idea that inertia is a field effect caused by the world's masses, we find it natural to assume that the universal constant implicit in the right-hand side of the Galileo–Newton formula $\mathbf{F} = m\boldsymbol{\gamma}$ (where it has the dimension zero and the value 1 by joint definition of the force, mass, length and time standards) is nothing else than GM/c^2R. It so happens that this is indeed the case in the original static Einstein model, and also in the once fashionable steady-state model. Regrettably, however, the expanding models cannot posses this attractive feature.

Before proceeding let us refine the statement of the 'equivalence principle'. Considering a system of two self-gravitating point changes of masses m_1 and m_2, the Galileo–Newton formula can be written as

$$\mathbf{r}Gm_1m_2/r^3 = \mathbf{\Gamma}_1 m_2 = \mathbf{\Gamma}_2 m_1,$$

both attractive forces being directly opposite; $\mathbf{\Gamma}_1$ denotes the gravity force as created by m_1 and felt by m_2, $\mathbf{\Gamma}_2$ the gravity force as created by m_2 and felt by m_1. In the two latter expressions m_1 and m_2 denote Galileo's inertial masses. In the first expression m_1 is termed the active gravity mass insofar as it is created at 1 and felt at 2, m_2 the passive gravity mass insofar as it is created at 2 and felt at 1 (and vice versa of course). This may look like hair splitting, but it does make sense conceptually, is needed in some discussions — and is even presently tested in specifically designed experiments.

Up to now there is not the slightest indication that the triple equivalence: *inertial = passive gravity = active gravity mass* is violated. Einstein's gravity theory essentially fulfills this condition.

5.1.3. EINSTEIN'S 1916 FORMALIZATION OF THE 'EQUIVALENCE PRINCIPLE'. GENERAL RELATIVITY THEORY

Consider a frictionless point particle compelled to move inertially inside a hollow rugby football. From the classical, and very beautiful, Gaussian differential geometry of surfaces, and of course classical mechanics, one easily finds that the trajectory is mass independent and is a geodesic followed at constant velocity. A geodesic is the sort of curve with its 'principal normal' normal to the surface. Very obviously, this is the closest conceivable substitute to the concept of 'uniform motion'.

After some deep heuristic thinking expressed in articles in the years 1913—1916, Einstein came to the beautiful idea that the four-dimensional spacetime does have a Gaussian, or Riemannian, curvature, caused by the active gravity masses, and that the trajectory of a test point particle will therefore be a geodesic inside spacetime, regardless of its mass.

Once this brilliant idea is grasped the rest follows naturally. Its formalization proceeds via the Riemann, Ricci and Levi-Civita tensor calculus of 'metric spaces'.

The classical Poisson formula

$$\Delta U = 4\pi G\rho$$

extending to continuous media the formula $U = Gm/r$ must be generalized so that the momentum-energy density T^{ij} shows up on the right-hand side. Since, ∇_i denoting the 'covariant derivative', the condition

$$\nabla_j T^{ij} = 0$$

must be fulfilled (in the absence of non-gravitational forces), U must be replaced by a symmetric second-rank tensor of purely geometric nature. This means that it is expressed in terms of the components of the metric tensor, g^{ij} or g_{ij}, and their derivatives. There is one such tensor: $R^{ij} - \frac{1}{2}Rg^{ij}$, R^{ij} denoting the Ricci curvature tensor (a contraction of the rank-4 Riemann curvature tensor) and R its contraction R^i_i. To this, one can add λg^{ij}, λ being an arbitrary constant termed the 'cosmological constant', chosen as 0 or as non-zero depending on the sort of problem discussed.

Finally, with $\lambda = 0$, the Einstein gravity equation is

$$R^{ij} - \tfrac{1}{2}Rg^{ij} = 4\pi c^{-2}GT^{ij};$$

it is one of the most beautiful jewels of theoretical physics.

Let us particularize T^{ij} in the form $\rho V^i V^j$, as in Sections 2.6.8 and 2.6.9, where the inertial law $V'^i = 0$ for a point particle was deduced from the assumptions $\partial_j(\rho V^i V^j) = 0$ and $\partial_j(\rho V^j) = 0$. Similarly here, with the substitution $\partial_j \rightarrow \nabla_j$, we get $V^j \nabla_j V^i = 0$ — that is, motion along a geodesic, as Einstein had anticipated. As a consequence, no momentum-energy density tensor can exist for the gravity field.

5.1.4. TIME IN GENERAL RELATIVITY

It goes without saying that, since it causes the spacetime manifold to bend, gravity has a direct impact upon the physical behavior of both space and time measurements; not a local one, however, as, at every point-instant, there exists a flat Minkowskian, tangent manifold, where everything goes on as was said in Chapter 2.6, including the isotropy and invariance of c. Minkowskian inertial frames go into 'freely falling' frames, so that questions are raised concerning the use of frames constrained from falling, as is the case with our terrestrial laboratories. This question we shall not touch upon.

Among the non-local effects of general relativity upon space and/or time measurements, the most direct and significant one concerns the frequency change of photons (or other quantal particles) as emitted and received at places where the gravity potential is not the same. Essentially, this is the general-relativistic extension of the Doppler effect. Physically speaking, however, it looks like a *sui generis* effect when both source and receiver are fixed inside a static gravity field — or at least prevented from falling upon each other by a relative transverse motion, as are the Earth and the Sun. As expressed in terms of the

Newtonian potential U, this 'Einstein effect' is

$$d\nu/\nu = c^{-2}\,dU = c^{-2}\mathbf{g}\cdot d\mathbf{r}$$

that is, according to Planck's formula $W = h\nu$ and to Einstein's formula $W = c^2 m$, as

$$dW = m\mathbf{g}\cdot d\mathbf{r}.$$

This exactly is the right expression connecting kinetic and potential energies, and looks like a quasi-miraculous accord between three formulas of quite different origins, something like fraternization of border policemen at a triple point of frontiers. Let us recall that this happens also in Bohr's discussion of the fourth uncertainty relation, as we have seen in Section 4.3.12.

Numerous verifications of this 'Einstein effect' exist, in the form of 'redshifts' of spectral lines emanating from the Sun, from the white dwarf Sirius B, and a few other celestial bodies, and even, thanks to the Mössbauer effect, inside terrestrial laboratories (Pound and Rebka, 1959).

As a corollary to this effect there is, of course, a 'rate change of clocks' ticking at different places, according to the formula

$$dT/T = -c^{-2}\,dU.$$

The so-called 'twins paradox', which was discussed in Section 2.5.7 as a special relativistic effect modulo the assumption $ds = c\,d\tau$, belongs to this category. It expresses the 'non-integrability' of distances as evaluated along curved paths; inside curved spaces it has an extra contribution. Let us reconsider in this respect the Hafele—Keating (1972) effect mentioned in Section 2.5.7, which can be idealized as a clock going around a circle at constant velocity $v = c\beta$. In a flat spacetime, the 'relative gain in aging' thus obtained is $\Delta T/T = \frac{1}{2}\beta^2 = \frac{1}{2}r^2\Omega^2/c^2$, r denoting the radius of the circle and Ω the angular velocity. Clearly, this is nothing else than an 'integrated transverse Doppler shift'. As $r\Omega^2$ is the centrifugal force, $\frac{1}{2}r^2\Omega^2$ is the centrifugal potential U, so that $\Delta T/T = U/c^2$. Now, if the travel is performed around the Earth, there is an other similar contribution coming from the Earth's gravitational field the potential of which is $U = GM/r$. Both effects are taken care of 'indissolubly' by writing down the Riemannian expression of ds — and this is how the real Hafele—Keating effect has been calculated, and quite well validated experimentally.

As an exercise (my article, 1957b) one can use Schwarzschild's spherical metric[1] (written here with units such $c = 1$ and $G = 1$)

$$d\tau^2 = (1 - 2m/r)\, dt^2 - (1 - 2m/r)^{-1}\, dr^2 - r^2(d\varphi^2 + \sin^2 \varphi\, d\theta^2)$$

to calculate the overall $\Delta\nu/\nu$ or $\Delta T/T$ effect for a spectral line emitted from the Sun and received on the circulating Earth, and separate in it the 'static' and 'centrifugal' components. This is an idealization, because the emitter is at the surface of the rotating Sun and the receiver at the surface of the rotating Earth. This also must be accounted for. All this is part of the problem of relating the 'ephemeris time', or 'legal time' to the 'atomic time' measured in laboratories.

Of course the Harress—Sagnac effect, mentioned in Section 2.5.8, is a phenomenon of exactly this type. It exists both for light waves or any material waves. Then, when translated in terms of momentum energies of the associated particles, it explicitly displays the mass of the test particles, thus showing that the 'equivalence principle' breaks down at the quantal level. To this we shall return.

Still another effect of this family is the time delay of radar echoes from planets or man-made satellites, depending on the path they follow within the Solar System. Excellent agreement between theory and measurements is obtained.

5.1.5. BENDING OF LIGHT WAVES. ADVANCE OF PERIASTRONS

That light rays are bent when grazing a strong gravity source follows of course from Newton's theory plus the photon concept. However, as this is an extreme relativistic effect, the correct formula is not obtained in this rudimentary manner. It is obtained by using the Schwarzschild metric, and is verified in Solar eclipses. The first such observation was Eddington's in 1919.

Motion of a planet, as formalized by the Schwarzschild metric, displays an 'advance of the periastron' not present in Newton's theory, but very well confirmed experimentally. Mercury's 'residual' perihelion advance of some 43 seconds of arc per century was known, and unexplained till the advent of general relativity.

5.1.6. QUANTUM MECHANICS IN THE RIEMANNIAN SPACETIME

B. de Witt (1962) was the first to propose a Riemannian style formalization of quantum mechanics. This has to be in the x language, because the Fourier transformation is impossible if spacetime is not flat. With it, quantum mechanics loses the sturdy plough horse that had been so faithful until then.

Writing down covariant tensorial wave equations for integer spin particles poses no problem. One also knows how to handle the spinors of half-integer spin particles: 'fibre bundles' are the recipe, using in fact the tangent Minkowski spacetime.

Propagators in Riemannian spacetime have been used by Hadamard (1923).

Independently, Lichnerowicz (1960; 1961; 1963) and B. de Witt have systematically used Riemannian extensions of the Jordan—Pauli propagator to write down commutation relations for integer and half-integer spinning waves propagating in the Riemannian spacetime. Extension of the Feynman propagator has also been considered by B. and C. de Witt.

With this, it was known in principle how to handle quantal problems within the Riemannian spacetime — for example, the phase shift of falling neutrons, as measured in a famous 1975 experiment by Colella, Overhauser and Werner.

The corresponding formula $\Delta W = h\,\Delta\nu = mg\,\Delta z$ explicitly displays the neutron mass, thus showing that Einstein's 'equivalence principle' is essentially of a macroscopic nature. As Greenberger puts it: "gravity is just another force in quantum theory."

Interaction between quantum particles and the gravity field implies that their momentum-energy density T^{ij} enters the right-hand side of Einstein's gravitation equation.

With all this, only half of the overall problem has been considered. The other half obviously is 'quantization of the gravity field'.

5.1.7. QUANTIZATION OF THE GRAVITY FIELD

This is a tricky problem, for more than one reason. First, the very concept of quantal fluctuations affecting the gravity field is synonymous to fluctuations in the geometry. How can one write or draw on a fluctuating blackboard?

Lichnerowicz's (1961) partial but clever answer to this has been to quantize a perturbation upon a non-quantized gravity background-field.

One other difficulty resulting from the geometrization of gravity is that no momentum-energy density analogous to the Maxwell–Minkowski one of electromagnetism exists in Einstein's theory. Thus it happens that another sturdy plough horse of the quantal theory could not ford the river to general relativity.

5.1.8. GRAVITY WAVES

In the vacuum, that is, if $R_{ij} = 0$, the rank-4 Riemann–Christoffel curvature tensor $R_{([ij],[kl])}$ obeys two equations very reminiscent of Maxwell's

$$\sum_{ijk} \nabla_i R_{jk,\,lm} = 0, \qquad \nabla_i R^i{}_{j,\,kl} = 0.$$

Following Pirani (1957) and Lichnerowicz we find it natural to interpret them as governing the propagation of gravity waves. Those of 'quasi-plane wave' behavior are such that, k_i denoting a lightlike vector tangent to a family of geodesics,

$$\sum_{[ijk]} k_i R_{jk,\,lm} = 0, \qquad k_i R^i{}_{j,\,kl} = 0.$$

Obviously such wave fronts have a quadrupolar structure.

Bel (1958; 1959) has constructed a rank-4 tensor having the same symmetries as R and quadratic in R, quite reminiscent of the Maxwell–Minkowski tensor, namely

$$B_{([ij],[kl])} \equiv \tfrac{1}{2}(R^m{}_i{}^n{}_k R_{mj,\,nl} + R^m{}_i{}^n{}_l R_{mj,\,nk}).$$

He has then shown that

$$T_{ij,\,kl} \equiv B_{ij,\,kl} - \tfrac{1}{2} R g_{ij} g_{kl}$$

obeys the formula

$$\nabla_i T^i{}_{j,\,kl} = 0.$$

Regrettably, however, no physical interpretation of these tensors has yet been found.

Great experimental efforts have been devoted to detecting gravity waves emanating from astronomical objects, without success up to now. These must be exceedingly weak, and detectable only at the quantal level.

CHAPTER 5.2

AN OUTSIDER'S LOOK AT COSMOLOGY, AND OVERALL CONCLUSIONS

5.2.1. GOD SAID: LET THERE BE SELF-GRAVITATING LIGHT! COSMOGENESIS

The so-called 'Big Bang' theory, originally put forward by Friedman and Lemaitre and later named by the latter 'the primeval atom theory', is supported by numerous observational facts, the most obvious ones being darkness of the night sky, the Hubble redshift, and the Penzias–Wilson 2.7 °K thermal radiation. Essentially, this is a cosmological theory based on Einstein's general relativity.

This relativistic 'expanding' model is that of an 'in time expanding space', 'time' and 'space' both being then preferred time and space severances of spacetime; the word 'expansion' has the 'relative' meaning of a time arrow chosen by our prejudices.

From the universal character of gravity it follows that this model can house any physical theory of interacting fields subject to gravitation, because the conserved overall mass–energy comprises the interaction energies, and globally enters the right-hand side of Einstein's equation. And thus it happens that elementary particle physics is at the core of the 'primeval atom theory', explaining how, from a tiny and ultradense fireball of self-gravitating photons, all the elementary (or less elementary) particles must have been generated in extremely short times. For a vivid description of this the reader is referred, for example, to Weinberg's (1977) bestseller *The First Three Minutes*.

Afterwards — that is, for long periods — dynamics and gravitation rule over the Universe at large; photons, neutrinos, and possibly gravitons are the long-range messengers — the first ones quite outspoken, the two others far less. Nuclear and elementary particle physics are restricted mainly to stellar generation and evolution, and to what may happen inside quasars and galactic nuclei.

Statistics — that is, thermodynamics — is a participant with full rights at the cosmology conference table, thus vindicating early claims by Clausius and Boltzmann that the Universe's entropy is a meaningful concept. The entropy of the self-gravitating Universe is going up, so that the primeval fireball 'was given' as a tightly packed tremendous amount of 'information'.

I have no intention to go into the wonderful natural history of the cosmic zoo, neither in its 'early' nor in its 'present' exotic recesses: excellent books and articles exist on the subject. I shall be content with a few remarks.

5.2.2. BLACK HOLES

The Schwarzschild metric obviously is singular at $r_0 \equiv 2Gm/c^2$, but this is only a coordinate singularity, not an intrinsic one. Alternative coordinate systems, 'well behaved' at r_0, have been defined, first by Eddington in 1924 and Finkelstein in 1958. In 1960, independently, Kruskal and Szekeres proposed a new mapping considered today as extremely useful.

The Schwarzschild spherical surface $r = r_0$ acts as a unidirectional membrane. The null geodesics leaving it towards its outside never hit it again, while those hitting it from outside sink inside, never again to emerge. Therefore, very dense spherical gravity sources having no matter outside r_0 act as either perfectly 'white sources' or 'black sinks'. It is easily verified that a radially moving photon, received or emitted at r, has left or will reach the Schwarzschild sphere at $t = \mp\infty$ (respectively), and that it undergoes a total redshift.

Analogous statements hold for radially moving massive test particles. In terms of the proper time of the particle the travel duration is finite, and no signpost is met when crossing the frontier. An extended body meets no signpost either, but it undergoes terrible tidal forces near and inside r_0.

So much for the intrinsic time symmetry of the Schwarzschild metric and of the black hole. As for the factlike time asymmetry, a little fable will help us. Consider an arrow aimed at a target; if it misses the target it will continue flying. Similarly a test body can graze a black hole without falling inside. If, however, the arrow hits the target, it remains stuck there. Its kinetic energy is torn to pieces, and randomly distributed among the target's degrees of freedom.

Reversing the motion *à la* Loschmidt means reversing exactly all internal *and* external motions: 'Can a feathered arrow fly backwards?' And so it is that *in fact* black sinks must largely outweigh white sources.

Inside the sky's botanic garden there may be quite a few of these poisoned, deadly, insect-capturing flowers, ingesting whatever falls within their range. It is suspected that unseen partners in double or multiple star systems may be black holes. No positive identification, however, has yet been made.

There exist also fantastic radiators such as 'quasars' and related objects. One may wonder if, after all, white sources should be excluded altogether.

Let it be mentioned that, using information theoretic arguments, Bekenstein (1973; 1974) has defined the entropy of a black hole as $S = \frac{1}{2} \ln 2\ell_0^{-2}\mathscr{A}$ bits; $\mathscr{A}$ denotes a 'normalized area' of the black hole and ℓ_0 Planck's length (see Section 1.1.3).

5.2.3. SOURIAU'S AND FLICHE'S 'LAYERED UNIVERSE'

Another interesting question concerns particle—antiparticle exchange at the cosmological level. Most cosmological theories argue about the factlike preponderance of matter over antimatter. Here I mention an attractive proposal by Souriau and coworkers (1980) based upon a detailed numerical analysis of astronomical data. They argue that the universal spatial hypersphere consists of two halves separated by a diametral hyperplane, one made of matter, the other of antimatter. They claim they have identified the severing hyperplane. A unique solution of Einstein's equations fits these data — an ever-expanding one, with a non-zero value of the cosmological constant.

5.2.4. BRIEF OVERALL CONCLUSION

General relativity and cosmology are outside my professional field of inquiry, but I could not possibly avoid mentioning them in a book devoted to physical time.

Now we must descend from the cosmological level, not without reminding the reader that generation of ordered systems, *à la* Prigogine is going on throughout the Universe, drawing on the overall negentropy cascade.

Leitmotifs of this book have been relativistic equivalence between space and time, factlike irreversibility as pervasive and largely occulting lawlike reversibility, *CPT* invariance as the appropriate covariant extension of Loschmidt's *T* invariance, Jordan's wavelike probability calculus as supplanting the classical one, intrinsic symmetry of the negentropy $\rightleftharpoons$ information transition implying (as it seems to me) the vindication of 'the claims of the paranormal'.

And now the ball is in the reader's court.

NOTES

CHAPTER 1.1

[1] A 1973 collective book (Colodny, 1973) has exactly this title.
[2] See also my further comments (1983b) on this article.
[3] The explicit definition of k is due to Planck, together with that of his own h.
[4] L. Motz (preprints) proposes a theory where 'unitons' of mass M are the quarks, their (very strong) binding being gravitational. This idea should perhaps not be discarded too lightly.
[5] As by definition $k \equiv R/N$, $k \to 0$ implies $N \to \infty$, that is, going from the discrete to the continuum.

CHAPTER 2.2

[1] Clarification of the 'restricted relativity principle' in the Galileo-Newtonian dynamics is mainly due to Euler, as has been shown by Truesdell.
[2] The Sun's planetary system is in some sense a gigantic clockwork, the theory of which has been established with extreme accuracy by Laplace and other 'celestial mechanicians', using perturbation methods for taking care of the interactions between the planets, and that between Earth and Moon. An unexplained 'residue' in the motion of Mercury has been explained in 1916 by Einstein's general relativity theory.
[3] The first truly satisfactory derivation of Kepler's laws is in fact not Newton's, but Johann Bernouilli's (Weinstock, 1982).

CHAPTER 2.3

[1] Let it be remarked, however, that Roemer was measuring times and Bradley was measuring angles.

CHAPTER 2.4

[1] I refer to J. G. Vargas (1984) for further thinking along Robertson's line, in connection with Brillet's and Hall's experiment.

CHAPTER 2.5

[1] We seek a generalization of the classical additive law (8) that is symmetric and departs minimally from it when A, B and C depart from zero. Thus we write at first order, c denoting a constant, $u + v + w \pm c^{-2}uvw = 0$. By changing the scale we set $c = 1$. Setting $v = -u - \mathrm{d}u$ we get $w = \mathrm{d}u/(1 \mp u^2)$; that is, w being small, $\mathrm{d}A = \mathrm{d}u/(1 \mp u^2)$. The solution $u = \tan A$ is not satisfactory, but $u = \tanh A$ is.

[2] Orthogonal transformations are a subgroup of the 'affine transformations', and of the subgroup of these preserving the equation of the light cone (that is, containing a scalar dilatation factor).
[3] There is a formal homomorphism between this phenomenon and the Bohm—Aharonov effect in quantum electrodynamics: $2\mathbf{\Omega} \rightarrow \mathbf{B}$, the magnetic induction, and $\mathbf{\Omega} \times \mathbf{r} = \mathbf{v} \rightarrow \mathbf{A}$, the vector potential.

CHAPTER 2.6

[1] True, from the concepts of 'temperature' and 'entropy flux' that of a *non conservative* heat flux follows.
[2] Defining covariantly the temperature has been the subject of much discussion. Entropy, as interpretable in terms of probability, must be a scalar, S. Heat, being homogeneous to energy, should show up as a 4-vector, Q^i. Then, according as one privileges the classical Clausius definition in the form $\mathrm{d}Q = T\,\mathrm{d}S$ or $\mathrm{d}S = T^{-1}\,\mathrm{d}Q$, one is led to the 4-vector T^i 'temperature', or θ_i, 'inverse temperature', such that $\mathrm{d}Q^i = T^i\,\mathrm{d}S$ or $\mathrm{d}S = \theta_i\,\mathrm{d}Q^i$; each definition has its advantages, depending on the kind of problem. And a third option also may be useful: 'scalar heat' $Q = (Q_i Q^i)^{1/2}$ and 'scalar temperature' $T = \theta_i T^i$. The whole matter is lucidly examined by P. V. Grosjean (1974), who gives the main references.

CHAPTER 3.1

[1] In this Thomas Aquinas is more in the line of Plato than of Aristotle (H. Barreau, private communication).
[2] In many texts on statistical mechanics, 'chance' has an ambiguous status. At some places it is likened to *lack of detailed knowledge*, at others the statistical laws are considered *objective*. Both ends of the rope should be held fast, entailing that probability and information are *indissolubly objective and subjective* concepts.
[3] See Jaynes (1983), formulas (A6) and (A7), p. 216.
[4] In these examples the probabilities are not normalized to unity; strictly speaking, they are 'numbers of chances'.
[5] Setting $B = A$ in (5) we get $|A)(A| = |A)(A|A)(A|$ teaching three things: $|A)(A|$ is a so-called 'projection operator'; $(A|A) = 1$, that is 'the intrinsic conditional probability of A if A is unity; $|A|A) = |A)$, that is 'the extrinsic conditional probability of A if A equals the prior probability of A'.
[6] Some will argue that biological evolution should be inserted in its context: that of the universally cascading negentropy, upon which-out-of-equilibrium-systems are feeding. There are two rejoinders to this. First, by stating a problem in another way, one changes the set of prior probabilities; as stated in the text, the problem is a perfectly logical one. Second, in Prigogine's thinking, time asymmetry is *assumed*, and therefore the Laplace—Boltzmann problematic is bypassed.

Others will argue that, after all, the eohippus was not so improbable, as one can be seen in the Museum of Natural History in New York. But selecting (among many possibilities) just a significant one *exactly amounts to blind statistical retrodiction.*

CHAPTER 3.3

[1] It so happens that, as pictured in the complex plane, $C(t)$ faithfully represents the physical situation.
[2] In his essay 'Chance', Poincaré (1906a, Pt 1, Ch. 4) gropes around, and almost reaches, this conclusion, finally writing: "All these points would need long comments, that perhaps would help understanding better the world's irreversibility."

CHAPTER 3.4

[1] See in this respect interesting articles by Cocke (1967) and by Schulman (1974; 1976; 1977).
[2] The implicit assumption is that, like a forest of trees, the swarm of stars constantly renews itself. From inside a forest one sees only trees — and only the nearer trees.

CHAPTER 3.5

[1] As I, expressed in bits, is a pure number, the smallness of k displays some sort of 'opacity' of the physical world.

CHAPTER 4.2

[1] Excellent presentations of Planck's work are due to Martin Klein (1970, pp. 218–234) and to Alfred Kastler (1983) (his very last scientific paper).
[2] Quantization of action, in de Broglie's Thesis, is expressed as 'equivalence' between action $\mathscr{A}$ and phase φ according to $\mathscr{A} = \hbar\varphi$. As φ is a pure number, the smallness of $\hbar$ displays some sort of 'opacity' of the physical world.

CHAPTER 4.3

[1] Truly, the δ^{ij} came in later. In Heisenberg's article (1927) it is implicit.
[2] In principle, as pointed out by Aharonov and Bohm (1961), non-relativistic quantum mechanics allows accurate energy measurements in arbitrarily short times. The reasons for this are, first, that measurements are conceived as performed by macroscopic devices, the observables of which commute with those of the system; and, second, that time is a parameter, not an operator, in non-relativistic quantum mechanics. This extremely unsatisfactory conclusion is in itself proof that non-relativistic quantum mechanics is unphysical.
[3] I shall allow myself to freely write 'picture' for 'representation', as it sounds nicer.
[4] This confirms that the concept of state vectors can be dispensed with altogether.

CHAPTER 4.4

[1] This is not exactly Dirac's writing, but one commonly used today.
[2] The expression for γ^5 is the one that has not been retained for γ^4.
[3] The definition (6) privileges in some sense the time coordinate, and this of course is an undesirable feature. Quite recently K. R. Greider (1984) has shown that this defect

can be corrected by an appropriate use of the Clifford algebra, which thus appears as a privileged "spacetime algebrǎ" (Hestenes, 1966).

[4] Beware! $i\partial_i$ is the electron's 'combined momentum energy', in the sense of Section 2.6.13; $i\partial_i + eA_i$ is its proper momentum energy.

[5] These $2 \times 5 = 10$ tensors obey 2 systems of 5 equations first written down by Franz (1935). In the absence of an external A_i they are independent, the one having an 'electric' and the other a 'mechanical' meaning; an external A_i symmetrically couples the two systems (my article, 1943, p. 154). Part, but not all of the 10 Franz equations have been interpreted.

[6] This expresses the 'anti-unitary' transformation of ψ found in the litterature.

[7] So I define both the C and the T operations as anti-unitary, implying the Z, or $\psi \rightleftharpoons \bar{\psi}$ exchange. This is consistent with the 'second quantization' principle that particle—antiparticle exchange and emission—absorption exchange are mathematically equivalent. (In the literature, a transposition of the γ matrices is used to define C as unitary).

[8] In classical electromagnetism, rest—mass reversal and charge conjugation were equivalent. Here, Z must enter the picture.

[9] Let it be mentioned that the projector projecting any solution of the KG equation as a solution of the PDK equation is $(k\beta^i \partial_i + \beta^i \beta^j \partial_i \partial_j)/2k^2$.

[10] V. Lalan, private communication.

[11] Incidentally, the Lorentz condition must be written as $|\partial_i A^i \Phi\rangle = 0$, not as $\partial_i A^i = 0$.

CHAPTER 4.5

[1] The first (overlooked) experimental evidence of parity violation in β-decay is due to Cox and coworkers (1928) and to Chase (1930).

[2] Discussion of the matter, which is going on, is not considered here.

[3] Assuming that the 4-velocity V^i essentially is future-timelike, the formula $p^i =. mV^i$ entails that p^i is future or past-timelike depending on the sign of m.

CHAPTER 4.6

[1] Jordan, however, was absent.

[2] In a recent (1986) article Cramer presents a theory of EPR correlations basically similar to mine. He misunderstands, however (p. 684), my concept of a transiting system jointly feeling the retarded influence from the preparation and the advanced influence from the measurement. By unduly charging me with a 'propagation in time' concept (which I *never* use, deeming it unphysical, as implying the idea of some 'supertime') he accuses me of violating energy conservation. *Not at all.* As I say (p. 159) my view of an evolution in spacetime is a *static* one, analogous to the classical view of a steady hydrodynamic flow, the velocity field of which is determined jointly by the pressure from the sources and the suction from the sinks. I no more have 'positive energy shot backwards in time' than the classicists had 'positive momentum shot upstream'.

[3] Jammer (1966, p. 186) points out that, as it seems, Schrödinger was the first to use the word 'paradox' in the EPR problem.
[4] These can be placed either before or after the polarizers.
[5] In principle no wiring is necessary (Mermin, 1981; my article, 1983b). Practically the wiring is needed for characterizing pairs of photons.
[6] In fact this is only partly true: Aspect's variable gratings were acoustical standing waves independently excited.
[7] In fact the two lasers are beating at the frequency $\nu_a - \nu_b$ and, as there is a phase coherence condition, the absorption occurs in pulses; this does not affect the experimental result.

CHAPTER 4.7

[1] The S-matrix was introduced in non-relativistic quantum mechanics by Wheeler in 1937 and discussed by Heisenberg in 1943—1946.
[2] As remarked by Landé (1965) the presence of the Fourier nucleus in propagators follows necessarily from the Born—Jordan algebra *and* translational invariance in spacetime.
[3] Let it be recalled that the wiring pictured in Figure 17(a) and (b) is not essentially, but only practically needed.
[4] $B^i(k)$ stands for the Fourier transform of $A^i(x)$.
[5] As monochromators are inserted on each beam the photon's energy is known to a good accuracy; but it is not required that the timing of the detectors preserves this accuracy, and so the detection is not an energy measurement.
[6] Showing this explicitly is a matter of simple routine. Representing two strictly correlated dichotomic magnitudes by matrices $z = \begin{pmatrix} 1 & 0 \\ 0 & -1 \end{pmatrix}$ indexed a and b, and using the unit matrix $I = \begin{pmatrix} 1 & 0 \\ 0 & 1 \end{pmatrix}$, the matrices $z_a \otimes I_b$ and $I_a \otimes z_b$ differ from each other *and from their half sum*.
[7] A typical example of this is Einstein's special relativity theory as opposed to Lorentz's and Poincaré's.
[8] The latter formulas are consistent with the fact that the positronium atom can decay into the state $\uparrow\downarrow - \downarrow\uparrow$ but not into the state $\uparrow\downarrow + \downarrow\uparrow$.
[9] This past—future symmetry of preparation and measurement has some similarity with the classical one displayed in the stationary integrals of Fermat and of Hamilton.
[10] See in this respect the end of Section 4.1.5 and Figure 10.

CHAPTER 5.1

[1] One may remark that the Schwarzschild radius $r = 2c^{-2}Gm$ is related to the gravity source m just as the 'classical electron radius' is related to the electron charge.

BIBLIOGRAPHY

Abelé, J.: 1952, 'La vitesse, grandeur qualitative, et la mécanique relativiste', *C.R. l'Acad. Sci.* **235**, 1007–1009.

Abraham, R. and Marsden, J. E.: 1978, 'Analytical Dynamics', in *Foundations of Mechanics* (2nd edn), Benjamin Cummins, Reading Mass., Chapter 3.

Adams, E. N.: 1960, 'Irreversible Processes in Isolated Systems', *Phys. Rev.* **120**, 675–681.

Aharonov, Y. and Bohm, D.: 1961, 'Time in the Quantum Theory and the Uncertainty Relation for Time and Energy', *Phys. Rev.* **122**, 1649–1658.

Aharonov, Y., Bergmann, P. G. and Lebowitz, L.: 1964, 'Time Symmetry in the Quantum Process of Measurement', *Phys. Rev.* **134B**, 1410–1416.

Aharonov, Y. and Albert, D. Z.: 1980, 'States and Observables in Relativistic Quantum Field Theory', *Phys. Rev.* **D21**, 3316–3324.

Aharonov, Y. and Albert, D. Z.: 1981, 'Can we make Sense out of the Measurement Process in Relativistic Quantum Mechanics?', *Phys. Rev.* **D24**, 359–370.

Albert, D. Z., Aharonov, Y. and d'Amato, S.: 1985, 'Curious New Statistical Prediction of Quantum Mechanics', *Phys. Rev. Letters* **54**, 5–7.

Arzelies, H.: 1965, 'Transformation relativiste de la température', *Il Nuovo Cimento* **35**, 792–804.

Arzelies, H.: 1966, *Relativistic Kinematics*, Pergamon Press, Oxford.

Arzelies, H.: 1972, *Relativistic Point Dynamics*, Pergamon Press, Oxford.

Aspect, A. *et al.*: 1981, 'Experimental Tests of Relativistic Local Theories via Bell's Theorem', *Phys. Rev. Letters* **47**, 460–463.

Aspect, A. *et al.*: 1982a, 'Experimental Realization of Einstein–Podolsky–Rosen–Bohm Gedankenexperiment: A New Violation of Bell's Inequalities', *Phys. Rev. Letters* **49**, 91–94.

Aspect, A. *et al.*: 1982b, 'Experimental Test of Bell's Inequalities using Time Varying Analysers', *Phys. Rev. Letters* **49**, 1804–1807.

Baird, K. M.: 1983, 'Frequency Measurements of Optical Radiation', *Physics Today* (January), 52–57.

Balmer, J. J.: 1885, 'Notiz über die Specktrallinien des Wasserstoffs', *Verhandlungen des Naturforschenden Gesellschaft in Basel* **7**, 548–560.

Bargmann, V. and Wigner, E. P.: 1948, 'Group Theoretical Discussion of Relativistic Wave Equations', *Proc. Nat. Acad. Sci.* **34**, 211–223.

Bay, Z., Luther, G. G. and White, J. A.: 1972, 'Measurement of an Optical Frequency and the Speed of Light', *Phys. Rev. Letters* **29**, 189–192.

Bekenstein, J. D.: 1973, 'Black Holes and Entropy', *Phys. Rev.* **D7**, 2333–2346.

Bekenstein, J. D.: 1974, 'Generalized Second Law of Thermodynamics in Black Hole Physics', *Phys. Rev.* **D9**, 2392–3300.

Bel, L.: 1958, 'Sur la radiation gravitationnelle', *C.R. l'Acad. Sci.* **247**, 1094–1096.

Bel, L.: 1959, 'Introduction d'un tenseur du quatrième ordre', *C.R. l'Acad. Sci.* **248**, 1297—1300.

Belinfante, F. J.: 1975, *Measurements and Time Reversal in Objective Quantum Theory*, Pergamon Press, Oxford.

Bell, J. S.: 1964, 'On the Einstein—Podolsky—Rosen Paradox', *Physics* **1**, 195—200.

Bergmann, H.: 1929, *Der Kampf um das Kausalgesetz in der Jüngsten Physik*, Herder, Braunschweig.

Bergson, H.: 1907, *L'Evolution Creatrice*, Alcan, Paris.

Bertrand, J.: 1888, *Calcul des Probabilités*, Gauthier Villars, Paris, pp. 4—5.

Bhabha, H. J.: 1945, 'Relativistic Wave Equations for the Elementary Particles', *Rev. Modern Physics* **13**, 203—232.

Biederharn, L. C.: 1983, 'The "Sommerfeld Puzzle" Revisited and Resolved', *Found. Physics* **13**, 13—34.

Blaney, T. G. *et al.*: 1974, 'Measurement of the Speed of Light', *Nature* **251**, 46.

Bohm, D.: 1951, *Quantum Theory*, Prentice Hall, Englewood Cliffs, N.J., pp. 614—622.

Bohm, D. and Aharonov, Y.: 1957, 'Discussion of Experimental Proof for the Paradox of Einstein, Rosen and Podolsky', *Phys. Rev.* **108**, 1070—1075.

Bohm, D. and Hiley, B.: 1982, 'The de Broglie Pilot Wave Theory and the Further Development of New Insights Arising out of It', *Found. Physics* **12**, 1001—1016.

Bohr, N.: 1935, 'Can Quantum Mechanical Description of Reality be Considered Complete?', *Phys. Rev.* **48**, 696—702.

Bohr, N.: 1949, 'Discussion with Einstein on Epistemological Problems' in P. A. Schilpp (ed.), *Albert Einstein Philosopher Scientist* (The Library of Living Philosophers), Evanston, Illinois, pp. 199—242.

Boltzmann, L.: 1964, *Lectures on Gas Theory* (trans. by S. Brush), Univ. of California Press, pp. 446—448.

Borel, E.: 1914, *Le Hasard*, Alcan, Paris (new edn) 1932, §115, p. 300.

Borel, E.: 1925, *Mécanique Statistique Classique*, Gauthier Villars, Paris, pp. 59—60.

Born, M. and Jordan, P.: 1925, 'Zur Quantenmechanik', *Z. Physik* **34**, 858—888.

Born, M., Heisenberg, W. and Jordan, P.: 1926, 'Zur Quantenmechanik II', *Z. Physik* **35**, 557—615.

Born, M.: 1926, 'Quantenmechanik der Stossvorgänge', *Z. Physik* **38**, 803—827.

Bose, S. N.: 1924, 'Wärmegleichgewicht im Strahlungsfeld bei Anwesenheit von Materie', *Z. Physik* **27**, 392—393.

Brillet, A. and Hall, J. L.: 1979, 'Improved Laser Test of the Isotropy of Space', *Phys. Rev. Letters* **42**, 549—552.

Brillouin, L.: 1967, *Science and Information Theory* (2nd edn; 3rd printing), Academic Press, New York.

Brillouin, M.: 1902, 'Sur la loi de distribution des vitesses et sur la quantité *H* de Boltzmann', in Boltzmann, L.: 1902, *Leçons sur la Théorie des Gaz* (French translation), Gauthier Villars, Paris, pp. 194—197.

Broglie, L. de: 1925, 'Recherches sur la théorie des quanta', *Ann. Physique* **3**, 22—128.

Broglie, L. de: 1934, *Une Nouvelle Conception de la Lumière*, Hermann, Paris.

Broglie, L. de: 1936, *Nouvelles Recherches sur la Lumière*, Hermann, Paris.

Broglie, L. de: 1943, *Théorie Générale des Particules à Spin*, Gauthier Villars, Paris.

Bruno, M., d'Agostino, M. and Maroni, C.: 1977, 'Measurement of Linear Polarisation of Positronium Annihilation Photons', *Il Nuovo Cimento* **40B**, 143—152.

Büchel, W.: 1960, 'Das *H*-Theorem und seine Umkehrung', *Philosophia Naturalis* **6**, 168–180. Reprinted in: 1965, *Philosophische Probleme der Physik*, Herder, Freiburg, pp. 79–113.

Callen, H.: 1974, 'Thermodynamics as a Science of Symmetry', *Found. Physics* **4**, 423–443.

Carnot, S.: 1927, *Biographie et Manuscrit publiés sous le Haut Patronage de l'Académie des Sciences*, Gauthier Villars, Paris.

Chase, C. T.: 1930, 'The Scattering of Electrons by Metals', *Phys. Rev.* **36**, 984–987 and 1060–1065.

Christenson, J. H., Cronin, J. W., Fitch, V. L. and Turlay, R.: 1964, 'Evidence for the 2π Decay of the K_2^0 Meson', *Phys. Rev. Letters* **13**, 138–141.

Cini, M.: 1983, 'Quantum Theory of Measurement without Wave Packet Collapse', *Il Nuovo Cimento* **73B**, 27–56.

Clauser, J. F., Horne, M. A., Shimony, A. and Holt, R. A.: 1969, 'Proposed Experiment to Test Local Hidden Variable Theories', *Phys. Rev. Letters* **23**, 880–884.

Clauser, J. F. and Shimony, A.: 1978, 'Bell's theorem: Experimental Tests and Implications', *Reports on Progress in Physics* **41**, 1881–1927.

Clauser, J. F.: 1976, 'Experimental Investigation of a Polarization Correlation Anomaly', *Phys. Rev. Letters* **36**, 1223–1226.

Cocke, W. J.: 1967, 'Statistical Time Symmetry and Two-Time Boundary Conditions in Physics and Cosmology', *Phys. Rev.* **160**, 1165–1170.

Colodny, R. G. (ed.): 1973, *Paradox and Paradigm*, University of Pittsburgh Press, Pittsburgh.

Corben, H. C.: 1968, *Classical and Quantum Theories of Spinning Particles*, Holden Day, San Francisco.

Costa de Beauregard, O.: 1943, 'Contribution à l'étude de la théorie de Dirac', *J. Math. Pures et Appliquées* **22**, 85–176.

Costa de Beauregard, O.: 1953, 'Une réponse à l'argument dirigé par Einstein, Podolsky et Rosen contre l'interprétation bohrienne des phénomènes quantiques', *C.R. l'Acad. Sci.* **236**, 1632–1634.

Costa de Beauregard, O.: 1955, 'Covariance relativiste à la base de la mécanique quantique', *J. Physique* **16**, 770–780.

Costa de Beauregard, O.: 1957a, 'Symétrie microscopique et dissymétrie macroscopique entre avenir et passé', *Revue de Synthèse*, no. 5–6, 7–34.

Costa de Beauregard, O.: 1957b, 'Relation entre le temps propre d'une horloge terrestre et le temps astronomique de Schwarzschild', *J. Physique* **18**, 17–21.

Costa de Beauregard, O.: 1961, 'La constante de Planck comme carré de la charge gravitationnelle', *Comptes Rendus* **252**, 849–851.

Costa de Beauregard, O.: 1964, 'Irreversibility problems' in *Proc. Int. Congress for Logic, Methodology and Philosophy of Science*, North Holland, Amsterdam, pp. 313–342.

Costa de Beauregard, O.: 1966, *Précis of Special Relativity*, Academic Press, New York.

Costa de Beauregard, O.: 1975, 'Tenseurs d'impulsion-énergie asymétriques et moment angulaire à 6 composantes dans le problème des deux impulsions du photon et de l'effet magnétodynamique', *Canadian J. Physics* **53**, 2355–2368.

Costa de Beauregard, O.: 1977, 'Time Symmetry and the Einstein–Podolsky–Rosen Paradox', *Il Nuovo Cimento* **42B**, 41–64.

Costa de Beauregard, O: 1979, 'Time Symmetry and the Einstein–Podolsky–Rosen Paradox, II', *Il Nuovo Cimento* **51B**, 267–279.

Costa de Beauregard, O.: 1980, 'The 1927 Einstein and 1935 E.P.R. Paradox', *Physis* **22**, 211—242.

Costa de Beauregard, O.: 1981, 'Is the Deterministically Evolving State Vector an Unnecessary Concept?', *Lettere al Nuovo Cimento* **31**, 43—44.

Costa de Beauregard, O.: 1982, 'Is the Deterministically Evolving State Vector a Misleading Concept?', *Lettere al Nuovo Cimento* **36**, 39—40.

Costa de Beauregard, O.: 1983a, 'Lorentz and CPT Invariances and the Einstein—Podolsky—Rosen Correlations', *Phys. Rev. Letters* **50**, 867—869.

Costa de Beauregard, O.: 1983b, 'Running Backwards the Mermin Device', *Amer. J. Physics* **51**, 513—516.

Costa de Beauregard, O.: 1984a, 'CPT-Revisited: a Manifestly Covariant Presentation' in *The Wave Particle Dualism* (S. Diner *et al.* eds), D. Reidel, Dordrecht, pp. 485—497.

Costa de Beauregard, O.: 1984b, 'Simple Remarks Concerning Relativistic Thermodynamics', *Lettere al Nuovo Cimento* **40**, 190—192.

Costa de Beauregard, O.: 1985, 'On Some Frequent but Controversial Statements Concerning the Einstein—Podolsky—Rosen Correlations', *Found. of Physics* **15**, 871—887.

Costa de Beauregard, O.: 1986, 'Bohr's Discussion of the Fourth Uncertainty Relation Revisited', *Found. of Physics* **16**, 937—939.

Costa de Beauregard, O.: 1986, 'On Carmeli's Exotic Use of the Lorentz Transformation and on the Velocity Composition Approach to Special Relativity', *Found. of Physics* **16**, 1153—1157.

Cox, R. T. *et al.*: 1928, 'Apparent Evidence of Polarization in a Beam of β-Rays', *Proc. Nat. Acad. Sci.* (U.S.A.) **14**, 544—549.

Cox, R. T.: 1946, 'Probability, Frequency and Reasonable Expectation', *Amer. J. Physics* **14**, 1—13.

Cox, R. T.: 1961, *The Algebra of Probable Inference*, The Johns Hopkins Press, Baltimore.

Cramer, J. G.: 1986, 'The Transactional Interpretation of Quantum Mechanics', *Rev. Mod. Physics* **58**, 847—887.

Darwin, G. C.: 1928, 'The Wave Equations of the Electron', *Proc. Roy. Soc. London* **A118**, 654—680.

Davidon, W. C.: 1976, 'Quantum Physics of Single Systems', *Il Nuovo Cimento* **36B**, 34—40.

Davies, P. C. W.: 1974, *The Physics of Time Asymmetry*, Surrey Univ. Press, U.K.

Davies, P.: 1981, *Other Worlds*, Dent, London.

Demers, P.: 1944, 'Les démons de Maxwell et le second principe de la thermodynamique', *Canadian J. Research* **22A**, 27—51.

Demers, P.: 1945, 'Le second principe et la théorie des quanta', *Canadian J. Research* **23A**, 47—55.

Denbigh, K.: 1981, 'How Subjective is Entropy', *Chemistry in Britain* **17**, 168—185.

Descartes, R.: 1971, *Correspondance*, Vol. 3, in *Oeuvres Complètes* (C. Adam and P. Tannery, eds), re-edition, Vrin, Paris, Letter 302, p. 663.

Descartes, R.: 1974, *Correspondance*, Vol. 5, in *Oeuvres Complètes* (C. Adam and P. Tannery, eds), re-edition, Vrin, Paris, Letter 525, p. 219.

Dettman, J. W. and Schild, A.: 1954, 'Conservation Theorems in Modified Electrodynamics', *Phys. Rev.* **95**, 1057—1060.

de Witt, B. S.: 1962, 'Invariant Commutators for the Quantized Gravitational Field', in *Recent Developments in General Relativity*, Pergamon Press, Oxford, pp. 175—190.

Dirac, P. A. M.: 1927, 'The Quantum Theory of the Emission and Absorption of Radiation', *Proc. Roy. Soc. London* **A114**, 243—265.

Dirac, P. A. M.: 1928, 'The Quantum Theory of the Electron', *Proc. Roy. Soc. London* **A117**, 610—624 and **A118**, 351—361.

Dirac, P. A. M.: 1947, *The Principles of Quantum Mechanics* (3rd edn), Clarendon Press, Oxford.

Documents Concerning the New Definition of the Meter: 1984, *Metrologia* **19**, 163—178.

Duhem, P.: 1954, *The Aim and Structure of Physical Theory* (trans. by P. P. Wiener), Princeton Univ. Press, Princeton, N.J.

Duffin, R. J.: 1938, 'On the Characteristic Matrices of Covariant Systems', *Physical Review* **54**, 1114.

Dyson, F. J.: 1949, 'The radiation theories of Tomonaga, Schwinger and Feynman', *Phys. Rev.* **75**, 486—502.

Eckart, C.: 1926, 'Operator Calculus and the Solutions of the Equations of Quantum Dynamics', *Phys. Rev.* **28**, 711—726.

Ehrenfest, P. and T.: 1911, 'Begrifflich Grundlagen des statistischen Auffassung in der Mechanik', *Encyklopädie des Mathematischen Wissenschaften*, Barth, Leipzig IV, **2**, II, pp. 41—51.

Einstein, A.: 1905, 'Uber einen die Erzeugung und Verwandlung des Lichtes betreffenden heurististischen Gesichtpunkt', *Ann. Physik* **17**, 132—148.

Einstein, A.: 1911, 'Einflüss des Schwerkraft auf die Ausbreitung des Lichtes', *Ann. Physik* **35**, 898—908.

Einstein, A.: 1912, 'Lichtgeschwindigkeit und Statik des Gravitationsfelds', *Ann. Physik* **38**, 355—369.

Einstein, A.: 1924a, 'Quantentheorie des Einatomigen idealen Gases', Preussische Akademie des Wissenschaften, Phys.-Math. Klasse, Sitzungsberichte, pp. 261—267.

Einstein, A.: 1924b, Note appended to the paper by Bose, (1924), *Z. Physik* **27**, 392—393.

Einstein, A.: 1925, 'Quantentheorie des Einatomigen idealen Gases 2', Preussische Akademie des Wissenschaften, Phys.-Math. Klasse, Sitzungsbrichte, pp. 3—14.

Einstein, A. and Hopf, L.: 1910, 'Uber einen Satz der Wahrscheinlichkeitsrechnung und seine Anwendung in der Strahlungstheorie', *Ann. Physik* **33**, 1096—1104.

Einstein, A. and Ehrenfest, P.: 1923, 'Quantentheorie des Strahlungsgleichgewicht', *Z. Physik* **19**, 301—306.

Einstein, A., Podolsky, B. and Rosen, N.: 1935, 'Can Quantum Mechanical Description of Physical Reality be Considered Complete?', *Phys. Rev.* **47**, 777—780.

Electrons et Photons, Rapports et Discussions du Cinquieme Conseil Solvay: 1982, Gauthier Villars, Paris.

Elsasser, W. M.: 1937, 'On Quantum Measurements and the Role of Uncertainty Relations in Statistical Mechanics', *Phys. Rev.* **52**, 987—999.

Espagnat, B. d'.: 1976, *Conceptual Foundations of Quantum Mechanics*, 2nd ed. Benjamin, London.

Evanson, K. M. *et al.*: 1972, 'Speed of Light from Direct Frequency and Wavelengths Measurements of the Methane-Stabilized Laser', *Phys. Rev. Letters* **19**, 1346—1349.

Feynman, R. P.: 1948, 'Spacetime Approach to Non-relativistic Quantum Mechanics', *Rev. Mod. Physics* **20**, 367—387.

Feynman, R. P.: 1949a, 'The Theory of Positrons', *Phys. Rev.* **76**, 749—759.

Feynman, R. P.: 1949b, 'Spacetime Approach to Quantum Electrodynamics', *Phys. Rev.* **76**, 769—789.

Feynman, R. P. and Gell-Mann, M.: 1958, 'Theory of the Fermi Interaction', *Phys. Rev.* **109**, 193—198.

Fock, V.: 1948, 'On the Interpretation of the Wave Function Directed Towards the Past', *Doklady Akad Nauk SSSR* **60**, 1157—1159.

Fokker, A. D.: 1929, 'Ein invarianter Variationsatz für die Bewegung mehrerer elektrischen Massenteilschen', *Z. Physik* **58**, 386—393.

Fokker, A. D.: 1965, *Time and Space, Weight and Inertia: A Chronogeometrical Introduction to Einstein's Theory*, Pergamon Press, Oxford.

Föppl, A.: 1894, 'Die Elektrodynamik bewegter Leiter'. Einfuhrung in *Die Maxwellsche Theorie der Electrizität*, Teubner, Leipzig, Chapter 5.

Franz, W.: 1935, 'Zur Methodik der Dirac Gleichung', *Sitzber. Math. Abteilung der Bayerische Akademie* **3**, 379—435.

Freedman, J. F. and Clauser, J. F.: 1972, 'Experimental Test of Local Hidden-Variable Theories', *Phys. Rev. Letters* **28**, 938—941.

Fry, E. and Thompson, R. C.: 1976, 'Experimental Test of Local Hidden Variables Theories', *Phys. Rev. Letters* **37**, 465—467.

Furry, W. H.: 1936a, 'Note on the Quantum-Mechanical Theory of Measurement', *Phys. Rev.* **49**, 393—399.

Furry, W. H.: 1936b, 'Remarks on Measurement in Quantum Theory', *Phys. Rev.* **49**, 476.

Gal-Or, B.: 1981, *Cosmology, Physics and Philosophy*, Springer, New York.

Gariel, J.: 1986, 'Tensorial Local-Equilibrium Axiom and Operators', *Il Nuovo Cimento* **94**, 119—139.

Gell-Mann, M. and Pais, A.: 1955, 'Behavior of Neutral Particles under Charge Conjugation', *Phys. Rev.* **97**, 1387—1389.

Giacomo, P.: 1983, 'The New Definition of the Metre', *European J. Physics* **4**, 190—197.

Gibbs, J. W.: 1902, 'Elementary Principles in Statistical Mechanics', reprinted in *Collected Works and Commentary*, Yale Univ. Press, New Haven, U.S.A., 1936.

Gleason, A. M.: 1957, 'Measures on the Closed Subspaces of a Hilbert Space', *J. Math. Mech.* **6**, 885—893.

Gold, T.: 1958, 'The Arrow of Time' in *La Structure et l'Evolution de l'Univers*, Stoops, Brussels.

Gold, T.: 1966, 'Cosmic Process and the Nature of Time' in *Mind and Cosmos* (R. G. Colodny, ed.), Univ. of Pittsburgh Press, Pittsburgh, pp. 311—329.

Gonseth, F.: 1964, 'Autofondation' in *Le Problème du Temps*, Edition du Griffon, Neuchâtel (Switzerland), pp. 327—347.

Gordon, W.: 1926, 'Der Comptoneffekt nach der Schrödingerschen Theorie', *Z. Physik* **40**, 117—133.

Gordon, W.: 1928, 'Die Energieniveaus des Wasserstoftatoms nach der Diracschen Quantentheorie des Elektrons', *Z. Physik* **48**, 11—14.

Grad, H.: 1961, 'The Many Faces of Entropy', *Comm. Pure and Applied Math.* **14**, 323—354.

Grad, H.: 1967, 'Levels of Description in Statistical Mechanics' in *Delaware Seminar in The Foundation of Physics* (M. Bunge, ed.), Springer, Berlin.

Greider, K. R.: 1984, 'A Unifying Clifford Algebra Formalism for Relativistic Fields', *Found. Physics* **14**, 467—507.

Grosjean, P. V.: 1974, 'A propos des températures relativistes', *Bull. Soc. Royale des Sciences de Liège* **43**, 260—265.

Grünbaum, A.: 1962, 'Temporally Asymmetric Principles, Parity between Explanation and Prediction, and Mechanism versus Teleology', *Phil. Science* **29**, 146—170.

Grünbaum, A.: 1963, *Philosophical Problems of Space and Time*, Knopf, New York.

Gutkowski, D. and Valdes Franco, M. V.: 1983, 'On the Quantum Mechanical Superposition of Macroscopically Distinguishable States', *Found. Physics* **10**, 963—986.

Hadamard, J.: 1923, 'Lectures on Cauchy's Problem' in *Linear Partial Differential Equations*, Yale Univ. Press, New Haven, U.S.A.

Hadamard, J.: 1930, *Cours d'Analyse Professé à l'Ecole Polytechnique*, Vol. 2, pp. 335—339.

Hafele, J. C. and Keating, R. E.: 1972, 'Around the World Atomic Clocks', *Science* **177**, 166—170.

Hahn, E. L.: 1950, 'Spin Echoes', *Phys. Rev.* **80**, 580—594.

Halley, E.: 1720, 'Of the Infinity of the Sphere of Fix'd Stars', *Phil. Trans.* **31**, 22—24.

Hamilton, W. R.: 1931, 'Theory of Systems of Rays' in *The Mathematical Papers of Sir William Rowan Hamilton* (A. W. Conway and J. L. Synge, eds), Vol. I, *Geometrical Optics*, pp. 1—294.

Hargreaves, R.: 1908, 'Integral Forms and their Connexion with Physical Equations', *Trans. Cambridge Philosophical Society*, pp. 107—122.

Heisenberg, W.: 1925, 'Uber quantentheoretische Umdeutung kinematischer und mechanische Beziehungen', *Z. Physik* **33**, 879—893.

Heisenberg, W.: 1927, 'Uber der Anschaulischen Inhalt der Quantentheoretischen Kinematik und Mechanik', *Z. Physik* **43**, 621—646.

Hestenes, D.: 1966, *Space-Time Algebra*, Gordon and Breach, New York.

Hilgevoord, J. and Uffink, J. B. M.: 1985, 'The Mathematical Expression of the Uncertainty Principle' in G. Tarozzi (ed.), *Proc. 1985 Urbino Conference, Microphysical Reality and Quantum Formalism*, D. Reidel, Dordrecht.

Hobson, A.: 1971, *Concepts in Statistical Mechanics*, Gordon and Breach, New York.

Hoffmann, B.: 1959, *The Strange Story of the Quantum*, Dover Publications, New York.

Holton, G.: 1973, 'Poincaré on Relativity' in *Thematic Origins of Scientific Thought*, Harvard Univ. Press, Cambridge, Mass., pp. 185—195.

Hoyle, F. and Wickramasinghe, C.: 1983, *Evolution From Space*, Paladin Books London.

Israel, W. and Stewart, J. M.: 1980, 'Progress in Relativistic Thermodynamics and Electrodynamics of Continuous Media' in *General Relativity and Gravitation*, Vol. 2 (A. Held, ed.), Plenum Publishing Corp., New York.

Jahn, R.: 1982, 'The Persistent Persistent Paradox of Psychic Phenomena: an Engineering Perspective', *Proc. IEEE* **70**, 136—170.

Jaki, S. L.: 1969, *The Paradox of Olbers' Paradox*, Herder and Herder, New York.

Jammer, M.: 1966, *The Conceptual Development of Quantum Mechanics*, McGraw-Hill, New York.

Jauch, J. M. and Rohrlich, F.: 1955, *The Theory of Photons and Electrons*, Addison-Wesley, Cambridge, Mass.

Jauch, J. M. and Piron, C.: 1963, 'Can Hidden Variables be Excluded in Quantum Mechanics?', *Helv. Phys. Acta* **36**, 827—837.

Jaynes, E. T.: 1983, *Papers on Probability, Statistics and Statistical Physics*, D. Reidel, Dordrecht.

Jeffreys, H.: 1961, *Theory of Probability* (3rd edn), Oxford Univ. Press, Oxford.

Jordan, P.: 1926, 'Uber eine neue Begründung der Quantenmechanik', *Z. Physik* **40**, 809—838.

Jordan, P. and Klein, O.: 1927, 'Zum Mehrkörperproblem der Quantentheorie', *Z. Physik* **45**, 751—765.

Jordan, P. and Pauli, W.: 1928, 'Zur Quantenelecktrodynamic ladungsfreier Felder', *Z. Physik* **47**, 151—173.

Jordan, P. and Wigner, E.: 1928, 'Uber das Paulische Äquivalenzverbot', *Z. Physik* **47**, 631—651.

Kabir, P. K.: 1968, *The CP Puzzle*, Academic Press, New York.

Kasday, L. R., Ullmann, J. D. and Wu, C. S.: 1975, 'Angular Correlation of Compton-Scattered Annihilated Photons', *Il Nuovo Cimento* **25B**, 633—661.

Kastler, A.: 1983, 'Max Planck et le concept de quantum d'énergie lumineuse', *Ann. Fond. Louis de Broglie* **8**, 827—303.

Katz, A.: 1967, *Principles of Statistical Mechanics, The Information Theory Approach*, Freeman, San Francisco.

Katz, A.: 1969, 'Alternative Dynamics for Classical Relativistic Particles', *J. Math. Physics* **10**, 1929—1931.

Kemmer, N.: 1939, 'The Particle Aspect of Meson Theory', *Proc. Roy. Soc. London* **173**, 91—116.

Keynes, J. M.: 1921, *A Treatise on Probability*, Macmillan, London.

Klein, F.: 1890, 'Ueber neure englishe Arbeiten zur Mechanik', *Jahresbericht des Deutschen Mathematiker Vereinigung* **1**, 35—36.

Klein, O.: 1926, 'Quantentheorie und fünfdimensionale Relativitätstheorie', *Z. Physik* **37**, 895—906.

Klein, M. J.: 1970, 'The Essential Nature of the Quantum Hypothesis' in Paul Ehrenfest, *The Making of a Theoretical Phycisist*, North Holland, Amsterdam, Vol. 1, pp. 218—234.

Kocher, C. A. and Commins, E. D.: 1967, 'Polarization Correlation of Photons, Emitted in an Atomic Cascade', *Phys. Rev. Letters* **18**, 575—577.

Komar, A.: 1978, 'Constraint Formalism of Classical Mechanics', *Phys. Rev.* **D18**, 1881—1886.

Kruskal, M. D.: 1960, 'Maximal Extension of Schwarzschild Metric', *Phys. Rev.* **119**, 1743—1745.

Kuhn, T.: 1970, *The Structure of Scientific Revolutions* (2nd edn), Univ. of Chicago Press, Chicago.

Lamehi-Rachti, M. and Mittig, W.: 1976, 'Quantum Mechanics and Hidden Variables: A Test of Bell's Inequality by the Measurement of the Spin Correlation in Low Energy Proton—Proton Scattering', *Phys. Rev.* **14**, 2543—2555.

Landau, L.: 1957, 'On the Conservation Law for Weak Interactions', *Nuclear Physics* **3**, 127—131.

Landé, A.: 1965, *New Foundations of Quantum Mechanics*, Cambridge Univ. Press, Cambridge, U.K.

Laplace, P. S.: 1891, 'Mémoire sur la probabilité des causes par les évènements' in *Oeuvres Complètes*, Gauthier Villars, Paris, Vol. 8, pp. 27—65.

Larmor, J.: 1900, *Aether and Matter*, Cambridge Univ. Press, Cambridge, U.K.

Laue, M. von: 1906, 'Zur Thermodynamik der Interferenzerscheinungen', *Ann. Physik* **20**, 365—378.

Laue, M. von: 1907a, 'Die Entropie von partiell kohärenten Strahlenbündeln', *Ann. Physik* **23**, 1—43 and 795—797.

Laue, M. von: 1907b, 'Die Mitführung der Lichtes durch bewebter Körper nach der Relativitätsprinzip', *Ann. Physik* **23**, 989—990.

Laue, M. von: 1911, *Das Relativitätsprinzip*, Vieweg, Braunschweig.

Lee, T. D. and Yang, C. N.: 1956, 'Question of Parity Conservation in weak Interactions', *Phys. Rev.* **104**, 254—258.

Levy, M.: 1950, 'Wave Equations in Momentum Space', *Proc. Roy. Soc. London* **A204**, 146—169.

Levy-Leblond, J. M.: 1976, 'One More Derivation of the Lorentz Transformation', *Amer. J. Physics* **44**, 271—277.

Levy-Leblond, J. M. and Provost, J. P.: 1976, 'Additivity, Rapidity, Relativity', *Amer. J. Physics* **47**, 1045—1049.

Lewis, G. N.: 1930, 'The Symmetry of Time in Physics', *Science* **71**, 570—577.

Lichnerowicz, A.: 1960, 'Ondes et radiations électromagnétiques et gravitationnelles en relativité générale', *Ann. Matematica* **50**, 2—98. Reprinted in *Choix d'Oeuvres Mathématiques*, 1982, Hermann pp. 33—127.

Lichnerowicz, A.: 1961, 'Propagateurs et commutateurs en relativité générale', *Cahiers de Physique* **7**, 11—21. Reprinted in *Choix d'Oeuvres Mathématiques*, 1982, Hermann, pp. 33—127.

Lichnerowicz, A.: 1963, 'Champs spinoriels et propagateurs en relativité générale', *Bulletin de la Société Mathématique de France* **92**, 11—100. Reprinted in *Choix d'Oeuvres Mathématiques*, 1982, Hermann, pp. 139—228.

Loschmidt, J.: 1876, 'Uber der Zustand das Warmegleichgewichtes eines Systems von Körpern mit Rücksicht auf die Schwerkraft', *Sitzbe. Akad. Wiss. in Wien* **73**, 139—145.

Loys de Cheseaux, J. P.: 1744, 'Sur la force de la lumière et sa propagation dans l'éther, et sur la distance des étoiles fixes' in *Traité de la Comète qui a paru en 1743 et 1744*, Marc-Michel Bousquet, Lausanne et Genève.

Lüders, G.: 1952, 'Zur Bewegungsumkehr in quantisierten Feldtheorien', *Z. Physik* **133**, 325—339.

Ludwig, G.: 1954, *Grundlagen der Quantenmechanik*, Springer, Berlin, pp. 178—182.

Madelung, E.: 1928, 'Quantentheorie in Hydrodynamischer Form', *z. Physik* **40**, 322—326.

Malvaux, P.: 1952, 'Recherche d'une loi intrinsèque de composition des vitesses', *C.R. l'Acad. Sci.* **235**, 1009—1011.

Marchildon, L., Antippa, A. A. and Everett, A. E.: 1983, 'Superluminal Coordinate Transformations; Four-Dimensional Case', *Phys. Rev.* **D27**, 1740—1751.

Mascart, E.: 1872, 'Sur les modifications qu'éprouve la lumière par suite du mouvement de la source lumineuse et du mouvement de l'observateur', lère Partie, *Annales Scientifiques de l'Ecole Normale Supérieure* **1**, 157—214.

Mascart, E.: 1874 [same title], 2ème Partie, *Annales Scientifiques de l'Ecole Normale Supérieure* **3**, 363—420.

Maxwell, J. C.: 1859, '*Illustrations of the Dynamical Theory of Gases*'. Reprinted in *Collected Works*, 1890 (W. D. Niven, ed.), London, pp. 713—741.

McConnell, A. J.: 1929, 'Strain and Torsion in Riemannian Space', *Ann. Mat. Pura e Applicata* **6**, 207—231.

McLennan, J. A.: 1959, 'Poincaré recurrence times', *Physics of Fluids* **2**, 92—93.

McLennan, J.: 1960, 'Statistical Mechanics of Transport in Fluids', *Physics of Fluids* **3**, 493—502.

Mehlberg, H.: 1961, 'Physical Laws and the Time Arrow' in *Current Issues in the Philosophy of Science* (H. Feigl and G. Maxwell, eds), Holt, Rinehart and Winston, New York, pp. 105—138.

Mermin, N. D.: 1981, 'Bringing Home the Atomic World: Quantum Mysteries for Anybody', *Amer. J. Physics* **49**, 940—943.

Miller, A. I.: 1981, *Albert Einstein's Special Theory of Relativity*, Addison Wesley, Reading. Mass.

Miller, W. A. and Wheeler, J. A.: 1983, 'Delayed Choice Experiments and Bohr's Elementary Quantum Phenomenon', in *Proc. Int. Symp. Foundations of Quantum Mechanics in the Light of New Technology*, Tokyo Physical Society of Japan, pp. 140—152.

Minkowski, H.: 1907, 'Das relativitätsprinzip', *Ann. Physik* **47**, 927—938.

Minkowski, H.: 1908a, 'Die Grundgleichungen für die electromagnetischen Vorgänge in bewegten Körpern', *Göttingen Nachrichtung* **53**, 111—145.

Minkowski, H.: 1908b, 'Raum und Zeit', *Physik. Z.* **20**, 104—111.

Misner, W., Thorne, K. S. and Wheeler, J. A.: 1970, *Gravitation*, Freeman, San Francisco.

Natanson, L.: 1911, 'Uber die statistische Theorie der Strahlung', *Physik. Z.* **12**, 659—666.

Neumann, J. von: 1932, *Mathematische Grundlagen der Quantentheorie*, Springer, Berlin.

Newton, I.: 1687, *Philosophiae Naturalis Principia Mathematica*, London.

Nicolis, G. and Prigogine, I.: 1977, 'Self-Organization' in *Non-Equilibrium Systems; From Dissipative Structures to Order through Fluctuations*, Wiley, New York.

Olbers, W.: 1823, 'Uber die Durchsichtigkeit des Weltraums', *Astronomisches Jahrbuch für das Jahr 1826* **51**, 110—121.

Papapetrou, A.: 1951, 'Spinning Test-Particles in General Relativity', *Proc. Roy. Soc. London* **A209**, 248—258.

Park, D.: 1974, *Introduction to the Quantum Theory*, McGraw-Hill, New York.

Pauli, W.: 1923, 'Uber das thermische Gleichgewicht zwischen Strahlung und freuen Elektronen', *Z. Physik* **18**, 272—286.

Pauli, W.: 1927, 'Zur Quantenmechanik des Magnetischen Elektrons', *Z. Physik* **37**, 263—281.

Pauli, W.: 1955, 'Exclusion Principle, Lorentz Group and Reflexion of Space-Time and Charge' in W. Pauli (ed.), *Niels Bohr and the Development of Physics*, Pergamon, London, pp. 30—51.

Pauli, W.: 1958, *Theory of Relativity*, Pergamon Press, London.

Pauling, L.: 1987, 'Schrödinger's Contributions to Chemistry and Biology', in *Schrödinger, Centenary Celebration of a Polymath*, C. W. Kilmister ed., Cambridge University Press, pp. 225—233.

Penrose, R.: 1959, 'The Apparent Shape of a Relativistically Moving Sphere', *Proc. Cambridge Philosophical Society* **55**, 137—139.

Penrose, O. and Percival, I. C.: 1962, 'The Direction of Time', *Proc. Physical Society* **79**, 605—616.

Perutz, M. F.: 1987, 'Erwin's Schrödinger's "What is Life" and Molecular Biology', in *Schrödinger, Centenary Celebration of a Polymath*, C. W. Kilmister ed., Cambridge University Press, pp. 234—251.

Petiau, G.: 1936, 'Contribution à la théorie des équations d'ondes corpusculaires', *Académie Royale de Belgique, Classe des Sciences* **16**, 3—116.

Pflegor, R. L. and Mandel, L.: 1967, 'Interference of Independent Photon Beams', *Phys. Rev.* **159**, 1084—1088.

Pflegor, R. L. and Mandel, L.: 1968, 'Further Experiments of Interference of Independent Photon Beams at Low Light Levels', *Phys. Rev.* **58**, 946—950.

Picard, E.: 1905, *La Science Moderne*, Flammarion, Paris.

Pirani, F. A. E.: 1957, 'Invariant Formulation of Gravitational Radiation Theory', *Phys. Rev.* **105**, 1089—1098.

Planck, M.: 1948, *Wissenschaftliche Selbstbiographie*, Barth, Leipzig.

Poincaré, H.: 1905, *La Valeur de la Science*, Flammarion, Paris. English translation: 1958, *The Value of Science*, Dover, New York.

Poincaré, H.: 1906a, *La Science et l'Hypothèse*, Flammarion, Paris. English translation: 1952, *Science and Hypothesis*, Dover, New York.

Poincaré, H.: 1906b, 'Sur la dynamique de l'électron', *Rend. Circolo Mat. Palermo* **21**, 129—175. Reprinted in: 1934—1953, *Oeuvres*, Gauthier Villars, Paris, pp. 551—586.

Poincaré, H.: 1908, *Science et Méthode*, Flammarion, Paris. English translation: 1908, *Science and Method*, Dover, New York.

Popper, C.: 1956—1957, 'The Arrow of Time', *Nature* **177**, 538—540; **178**, 382—384; **179**, 1296—1298.

Potier, A.: 1874, 'Conséquences de la formule de Fresnel relative à l'entrainement de l'éther par les milieux transparents', *J. Physique* **3**, 201—204.

Potier, R.: 1957, 'Sur la double covariance (quantique et relativiste) de la seconde quantification', *J. Physique* **18**, 422—433.

Pound, R. V. and Rebka, G. A.: 1959, 'Gravitational Redshift in Nuclear Resonance', *Phys. Rev. Letters* **3**, 439—441.

Proca, A.: 1936, 'Sur la théorie ondulatoire des électrons positifs et négatifs', *J. Physique* **7**, 347—353.

Proca, A.: 1938, 'Théorie non relativiste des particules à spin entier', *J. Physique* **9**, 61—66.

Racah, G.: 1937, 'Sulla simmetria tra particelle e antiparticelle', *Il Nuovo Cimento* **14**, 322—326.

Ramakrishnan, A.: 1973, 'Einstein, a Natural Completion to Newton', *J. Math. Analysis* **42**, 377—380.

Rauch, H. *et al.*: 1975, 'Verification of Coherent Spinor Rotation of Fermions', *Phys. Letters* **A54**, 425—427.

Rauch, H.: 1983, 'Tests of Quantum Mechanics by Matter Wave Interferometry', *Foundations of Quantum Mechanics in the Light of New Technology*, Physical Society of Japan, pp. 277—288.

Rayski, J.: 1979, 'Controversial Problems of Measurements within Quantum Mechanics', *Found. Physics* **9**, 217—136.

Recami, E. and Rodrigues, W. A.: 1982, 'Antiparticles from Special Relativity with Orthochronous and Antichronous Lorentz Transformations', *Found. Physics* **7**, 709—718.

Reichenbach, H.: 1956, *The Direction of Time*, Univ. of California Press, Berkeley.

Renninger, M.: 1960, 'Messungen ohne Störung des Messobjekts', *Z. Physik* **158**, 417—421.

Rhine, G. B.: 1964, *Extrasensory Perception*, 2nd ed. Bruce-Humphries, Boston.

Riesz, M.: 1946, 'Sur certaines notions fondamentales en théorie quantique relativiste', *Comptes Rendus du Dixième Congrès des Mathématiciens scandinaves*, Copenhagen, pp. 123—148.

Rietdijk, C. W.: 1981, 'Another Proof that the Future can Influence the Present', *Found. Physics* **11**, 783—790.

Rindler, W.: 1969, *Essential Relativity*, Van Nostrand Reinhold, New York.

Ritz, W. and Einstein, A.: 1909, 'Zum Gegenwärtigen Stande des Strahlungsproblems', *Physik. Z.* **10**, 323—324.

Robertson, H. P.: 1929, 'The Uncertainty Principle', *Phys. Rev.* **34**, 163—164.

Robertson, H. P.: 1949, 'Postulate versus Observation in the Special Theory of Relativity', *Rev. Mod. Physics* **21**, 378—382.

Rosen, G.: 1969, *Formulations of Classical and Quantum Dynamical Theories*, Academic Press, New York.

Rothstein, J.: 1958, *Communication, Organization and Science*, The Falcon's Wing Press, U.S.A.

Sakurai, J. J.: 1958, 'Mass Reversal and Weak Interactions, *Il Nuovo Cimento* **7**, 629—660.

Schawlow, A. L.: 1982, 'Lasers and Physics: a Pretty Good Hint', *Physics Today* (December), pp. 46—51.

Schlesinger, G.: 1964, 'Is it False that Overnight Everything has Doubled in Size?', *Philosophical Studies* **15**, 65—80.

Schlick, M.: 1932, 'Causality in Everyday Life and in Recent Science', *University of California Publications in Philosophy* **15**, 99—126.

Schmidt, H.: 1982, 'Collapse of the State Vector and Psychokinetic Effect', *Found. Physics* **12**, 565—581.

Schopenhauer, A.: 1883, *The World as Will and as Idea*, London.

Schrödinger, E.: 1926a, 'Quantisierung als Eigenwertproblem', *Ann. Physik* **79**, 361—376 and 489—527; **80**, 437—490; **81**, 109—139.

Schrödinger, E.: 1926b, 'Über das Verhältnis der Heisenberg—Born—Jordanschen Quantenmechanik zu der meinen', *Ann. Physik* **79**, 734—756.

Schrödinger, E.: 1926c, 'Der stetige Übergang von der Micro- zur Macro-mechanik', *Die Naturwissenschaften* **14**, 664—666.

Schrödinger, E.: 1935, 'Discussion of Probability Relations between Separated Systems', *Proc. Cambridge Philosophical Society* **31**, 555—562.

Schrödinger, E.: 1944, *What is Life? The Physical Aspects of the Living Cell*, Cambridge University Press.

Schrödinger, E.: 1950, 'Irreversibility', *Proc. Royal Irish Academy* **53A**, 189–195.

Schulman, L. S.: 1974, 'On Deriving Irreversible Thermodynamics' in B. Gal-Or (ed.), *Modern Development in Thermodynamics*, Keter Publishing House, Jerusalem, pp. 81–88.

Schulman, L. S.: 1976, 'Normal and Reversed Causality in a Model System', *Phys. Letters* **57A**, 305–306.

Schulman, L. S.: 1977, 'Illustration of Reversed Causality with Remarks on Experiment', *J. Stat. Physics* **16**, 217–231.

Schwinger, J.: 1948, 'Quantum Electrodynamics, I: A Covariant Formulation', *Phys. Rev.* **74**, 1439–1461.

Schwinger, J.: 1951, 'Theory of Quantized Fields, I', *Phys. Rev.* **82**, 914–927.

Sciama, D. W.: 1958, 'On a Non-symmetric Theory of the Pure Gravitational Field', *Proc. Cambridge Philosophical Society* **54**, 72–80.

Selleri, F.: 1980, 'Photon Coincidences with Crossed Polarizers', *Epistemological Letters* **25**, 39–46 (Association F. Gonseth, P.O. Box 1081, CH-2501 Biel).

Shadowitz, A.: 1968, *Special Relativity*, Saunders, Philadelphia.

Shannon, C.: 1948, *Bell System Technical Journal* **27**, 379–424 and 623–669. Reprinted in C. E. Shannon and W. Weaver: 1949, *The Mathematical Theory of Communication*, Univ. of Illinois Press, Urbana, pp. 3–93.

Smoluchowski, M. von: 1912, 'Esperimentall nach weisbare, der üblicher Thermodynamik widersprechende Molekularphänomene', *Physik. Z.* **13**, 1069–1080.

Sommerfeld, A.: 1919, *Atombau und Spektrallinien*, Vieweg, Braunschweig.

Souriau, J. M.: 1980, 'Stratification de l'univers', in *Einstein, Colloque du Centenaire*, Collège de France, Editions du C.N.R.S., Paris, pp. 197–239.

Souriau, J. M.: 1982, 'Physique et géométrie', in *La Pensée Physique Contemporaine* (S. Diner, D. Fargue and G. Lochak, eds), Editions Augustin Fresnel, Hiersac, pp. 343–364.

Stapp, H. P.: 1975, 'Bell's Theorem and World Process', *Il Nuovo Cimento* **29B**, 270–276.

Stapp, H. P.: 1982, 'Mind, Matter and Quantum Mechanics', *Found. Physics* **12**, 363–400.

Stueckelberg, E. C. G.: 1941, 'Remarques à propos de la création de paires de particules en théorie de la relativité', *Helv. Phys. Acta* **14**, 588–594.

Stueckelberg, E. C. G.: 1942, 'La mécanique du point matériel en théorie de relativité et en théorie des quanta', *Helv. Phys. Acta* **15**, 23–37.

Synge, J. L.: 1934, 'Energy Tensor of a Continuous Medium', *Trans. Roy. Soc. Canada* **28**, 127–171.

Szekeres, G.: 1960, 'On the Singularities of a Riemannian Manifold', *Publications in Math. (Debreczen)* **7**, 285–301.

Szilard, L.: 1929, 'Entropieverminderung in einem thermodynamischen System bei Eingriffen intelligenter Wesen', *Z. Physik* **53**, 840–856.

Terletsky, J. P.: 1960, 'Le principe de causalité et le second principe de la thermodynamique', *J. Physique* **21**, 681–684.

Terrell, J.: 1959, 'Invisibility of the Lorentz Contraction', *Phys. Rev.* **116**, 1041–1245.

Tiomno, J.: 1955, 'Mass Reversal and the Universal Interaction', *Il Nuovo Cimento* **1**, 226–232.

Tolman, R. C.: 1974, *Relativity, Thermodynamics and Cosmology*, Clarendon Press, Oxford.

Tomonaga, S. I.: 1946, 'On a Relativistically Invariant Formulation of the Quantum Theory of Wave Fields', *Prog. Theor. Physics* **1**, 27—42.

Treder, H. J.: 1971, 'The Einstein-Bohr Box Experiment' in *Perspectives in Quantum Theory* (W. Yourgrau and A. van der Merve, eds), MIT Press, Cambridge, Mass., pp. 17—24.

Tribus, M.: 1963, 'Information Theory and Thermodynamics' in *Boelter Anniversary Volume*, McGraw-Hill Book Company, pp. 348—368.

Umezawa, H. and Visconti, A.: 1956, 'Commutation Relations and Relativistic Wave Equations', *Nuclear Physics* **1**, 348—354.

Vargas, J. G.: 1984, 'Revised Robertson's Test of Special Relativity', *Found. Physics* **14**, 625—652.

Vessiot, E.: 1909, 'Essai sur la propagation des ondes', *Ann. l'Ecole Normale Supérieure* **45**, 405—448.

Voigt, W.: 1887, 'Theorie des Lichtes für bewegten Medien', *Göttingen Nachrichten*, No. 8, 177—237.

Waals, J. D. van der: 1911, 'Uber die Erklärung der Naturgesetze auf statistisch-mechanischer Grundlage', *Physik. Z.* **12**, 547—549.

Wald, A.: 1950, *Statistical Decision Functions*, Wiley, New York.

Watanabe, S.: 1951, 'Reversibility of Quantum Electrodynamics', *Phys. Rev.* **5**, 1008—1025.

Watanabe, S.: 1955, 'Symmetry of Physical Laws', *Rev. Mod. Physics* **27**, 26—39, 40—70 and 179—186.

Weinberg, S.: 1977, *The First Three Minutes*, Basic Books, New York.

Weinstock, R.: 1982, 'Dismantling a Centuries-Old Myth: Newton's Principia and Inverse Square Orbits', *Amer. J. Physics* **50**, 610—617.

Weiszäcker, C. F. von: 1931, 'Ortsbestimmung eines Elektrons durch ein Mikroskop', *Z. Physik* **70**, 114—130.

Weiszäcker, C. F. von: 1939, 'Der zweite Hauptsatz und der Underschied von Vergangenheit und Zukunft', *Ann. Physik* **36**, 281—291. Reprinted in: 1971, *Die Einheit des Natur*, Carl Hausen Verlag, München, pp. 172—182.

Werner, S. A., Colella, R. Overhauser, A. W. and Eagen, C. F.: 1975, 'Observation of the Phase Shift of a Neutron due to Precession in a Magnetic Field', *Phys. Rev. Letters* **35**, 1053—1055.

Weyl, H.: 1949, *Philosophy of Mathematics and Natural Science*, Princeton Univ. Press, Princeton, N.J.

Weyssenhof, J.: 1947, 'Further Contributions to the Dynamics of Spin Particles', *Acta Physica Polonica* **9**, 26—45.

Weyssenhof, J. and Raabe, A.: 1947, 'Relativistic Dynamics of Spin Fluids and Spin Particles', *Acta Physica Polonica* **9**, 7—18.

Wheaton, B. R.: 1983, *The Tiger and the Shark*, Cambridge Univ. Press, Cambridge, U.K.

Wheeler, J. A. and Feynman, R. P.: 1945, 'Interaction with the Absorber as the Mechanism of Radiation', *Rev. Mod. Physics* **17**, 157—181.

Wheeler, J. A. and Feynman, R. P.: 1949, 'Classical Electrodynamics in Terms of Direct Interparticle Action', *Rev. Mod. Physics* **21**, 425—433.

Wheeler, J. A.: 1978, 'The Past and the Delayed Choice Double Slit Experiment' in R. Marlow (ed.), *Mathematical Foundations of Quantum Mechanics*, Academic Press, New York.

Whitehead, A. N.: 1929, *Process and Reality*, Cambridge Univ. Press, Cambridge, U.K.

Wiener, N.: 1958, *Cybernetics*, Hermann, Paris.

Wigner, E. P.: 1932, 'Uber die Operation der Zeitumkehr in der Quantenmechanik', *Göttinger Nachrichten* **31**, 546—559.

Wigner, E. P.: 1967, *Symmetries and Reflections*, MIT Press, Cambridge, Mass.

Wilson, A. R., Lowe, J. and Butt, D. K.: 1976, 'Measurement of the Relative Planes of Polarization of Annihilation Quanta as a Function of Separation Distance', *J. Physics* **G2**, 613—623.

Wu, C. S. and Shaknov, I. S.: 1950, 'The Angular Correlation of Scattered Annililation Radiation', *Phys. Rev.* **77**, 136.

Wu, C. S. *et al.*: 1957, 'Experimental Test of Parity Conservation in Beta Decay', *Phys. Rev.* **105**, 1413—1415.

Wu, T. Y. and Rivier, D.: 1961, 'On the Time Arrow and the Theory of Irreversible Processes', *Helv. Phys. Acta* **34**, 661—674.

Yanase, M. M.: 1956, 'Reversibilität und Irreversibilität in der Physik', *Ann. Japan Assoc. Philosophy of Science* **1**, 131—149.

Yilmaz, H.: 1972, *Two New Derivations of the Lorentz Transformation* (mimeographed).

Zermelo, E.: 1896, 'Über einem Satz der Dynamik und die mechanische Wärmetheorie', *Ann. Physik und Chemie* **57**, 485—494.

Zurek, W. H.: 1981, 'Pointer Basis of Quantum Mechanics: Into What Mixture does the Wave Packet Collapse?', *Phys. Rev.* **D24**, 1516—1525.

ADDED IN PROOF

ON PAULING'S AND PERUTZ'S 1987 CRITICISMS OF SCHRÖDINGER'S BOOK 'WHAT IS LIFE'

At the 1987 Schrödinger Centenary Conference both Pauling (1987, pp. 225—233) and Perutz (1987, pp. 234—251) severely critized Schrödinger's 1944 book *What Is Life.* According to them living systems do not feed upon negentropy (as Schrödinger says) but on Helhmoltz's free energy. What they are both overlooking (and what Schrödinger did not make clear) is the point discussed by Brillouin (1967, Chapter 18) under the title 'Writing, printing and reading'.

Let me quote Perutz (1987, p. 243): "A well ordered configuration of atoms in a single molecule of an enzyme catalyst can direct the formation of an ordered sterospecific compound at the rate of 10^3—10^5 molecules a second, thus creating order from disorder at the ultimate expense of solar energy". All right. The point is, however, that this multiplication of information by fast printing does tap the negentropy cascade "from the hot sun to the cold earth" — as Boltzmann (quoted by Perutz, p. 241) stated in 1886.

By including the Sun's negentropy source in the picture one shows, with Brillouin, that the overall negentropy goes down.

Additional remarks are as follows:

1. The Helmholtz free energy F of a thermodynamic system embedded in an environment at constant temperature Θ is a state function defined as $F = U + \Theta N$, with U and N denoting the internal energy and negentropy. Thus, via $\mathrm{d}F = \mathrm{d}U + \Theta\,\mathrm{d}N$, $\mathrm{d}U$ and $\mathrm{d}N$ are coupled. To the classicists $\mathrm{d}N$ was ancillary to $\mathrm{d}U$ but (as seen from Pauling's and Perutz's remarks) it is $\mathrm{d}F$ that is ancillary to $\mathrm{d}N$ in biochemistry.

2. An enzyme is a polypeptide specialized in duplication. The 'structural negentropy' of these biomolecules amounts to "superastronomic numbers of bits" (Hoyle and Wickramasinghe, 1983, Chapter 2). In comparison the structural negentropy of the macro-skeleton of an eohippus, alluded to p. 114, is ridiculously small. To paraphrase

Eddington: "Do not ask me if I believe that these wonderful micro-skeletons are born from chance and necessity".

REMARK ON NONSEPARABILITY AT THE MACROLEVEL

One often reads in quantum textbooks that "the pointer of a macroscopic measuring device is never seen in a superposition of states". That this statement needs some qualification is shown by the following example.

A birefringent crystal separates an incident light beam in two orthogonally polarized beams, which one can afterwards separate widely. These beams can be interpreted as the two macroscopically distant divisions of a dial. That the belonging of a photon to one beam or the other cannot be ascertained is tantamount to saying that the macroscopic pointer is in a superposition of its two possible states. As each of these states is uniquely correlated with one of the orthogonal polarizations, it is a faithful macroscopic representation of the quantal state — faithful to the point of reproducing the 'non-separation'.

In order (so to speak) to bring the polarization reading into focus, one must 'move the eyepiece' so as to lose knowledge of the phase difference; for example, one can intersect both beams by a photographic plate. Thus, when 'appropriately focused', the pointer is no more 'seen in a superposition of states'.

Thus a measurement is a severance procedure in which the phase relations are lost to the observer. In this sense it is an approximation, leading from the quantal to the macro-world. However, far from being a step away from 'super-fluity' and towards 'realism', it implies a loss of underlying information, following a free decision of the observer.

The underlying information is not *essentially* lost, however, as it can be retrieved in principle by including the measuring device into the quantal description — that is, by respecting the 'nonseparability'.

This illustrates how 'illusory' is the so called 'reality' of the world as we perceive and conceive it macroscopically. And, as it happens, this supports Bergson's (1907) contention that *homo sapiens* truly is a *homo faber* (Chapter 2) who (equipped so to speak with a 'machete') acts very much as the one who cut the Gordian knot — that is, 'arbitrarily severs for mastery' (Chapter 3).

NOTE TO PAGE 283

Consider in this respect the three equivalent expressions

$$\langle\Psi|v\Phi\rangle=\langle\Psi v|\Phi\rangle=\langle\Psi|v|\Phi\rangle$$

of the transition amplitude between a preparation $|\Phi\rangle$ and a measurement $|\Psi\rangle$, with U denoting the unitary evolution operator. The first one, projecting the retarded preparation upon the measurement, expresses the familiar 'collapse' concept. The second one, projecting the advanced measurement upon the preparation, expresses what has been called in Section 4.3.18 "retrocollapse". The third one is a symmetric "collapse and retrocollapse" concept.

This is a concise formalization of Fock's (1948) and Watanabe's (1955) connection between retarded or advanced waves, and statistical prediction or retrodiction, respectively.

It is worth noting that CPT-*invariance of the transition amplitude is a corollary to these twin symmetries.*

So, the evolution operator U connects the 'preparation representation' and the 'measurement representation' of a system, thus providing what has been called, in Section 4.7.6, a 'dramatization of the quantal chance game'.

ON AN INTRIGUING PAPER BY ALBERT, AHARONOV·AND D'AMATO

In a recent (1983) paper entitled 'Curious New Statistical Prediction of Quantum Mechanics' Albert, Aharonov and d'Amato (AAA) express, and uphold, views which are (as they admit) at variance with both the generally accepted quantum mechanical paradigm and the algebraic reversibility of transition amplitudes. As the questions they raise (and, in my opinion, answer inadequately) are central, and are germane to those just examined, a brief discussion of them is apposite.

Quoting AAA: "Consider a quantum mechanical system [prepared] at time t_i . . . in the state $|A = a\rangle$ . . . and . . . measured at time $t_f > t_i$ . . . in the state $|B = b\rangle$. What do these results imply about . . . other experiments that *might* [my italics] be carried out within the interval $t_i < t < t_f$?"

Assuming for simplicity that the system evolves freely, and denoting as C the observable alluded to, AAA write down

$$P(c_j) = \frac{|\langle C = c_j | A = a \rangle|^2 \cdot |\langle C = c_j | B = b \rangle|^2}{\Sigma |\langle C = c_j | A = a \rangle|^2 \cdot |\langle C = C_j | B = b \rangle|^2}$$

as the probability that, *if* measured, (their italics) C comes out as c_j.

Then, remarking that "$P(a) = P(b) = 1$ whatever a and b" they draw the staggering conclusion that, between times t_i and t_f, the system "must have definite values of *both A* and *B* whether *or not A* and *B* ... commute" (their italics).

Well, the very fact that $P(a) = P(b) = 1$ even if $a \neq b$ most certainly means that *two different probabilities are subsumed under the same symbol* — which is thus insufficiently explicit. These are *two conditional probabilities, the one predictive,* $P(C = c_j, B = b | A = a)$, *the other retrodictive,* $P(A = a, C = c_j | B = b)$.

An example will help clarifying the matter. Suppose that a low intensity photon beam issuing from a laser A is prepared in a linear polarization state $|A\rangle$ and, before its reception in a photodetector, is measured in a linear polarization state $|B\rangle$, the orientations $A = a$ and $B = b$ of the two polarizers being neither parallel nor orthogonal to each other. Suppose also that a birefringent crystal C is inserted between A and B, the length of which is such that a zero phase shift (molulo $2n\pi$) takes place. The transition amplitude $\langle A | B \rangle$ remains unchanged, and the crystal C can be arbitrarily rotated. This takes care of AAA's "complete orthogonal set" C.

Can we then say, following AAA, that, while travelling inside the crystal C, an individual photon does "have" both of the linear polarizations a and b? Of course not. This photon is in a superposition of virtual states $|C\rangle$; in Miller and Wheeler's (1983) wording, it is a "smoky dragon".

Confirmation is obtained by modifying slightly the thought experiment, and using as C a simple linear polarizer. Then, the transition amplitude $\langle A | B \rangle$ is modified, except if $c_j = a$ or $c_j = b$.

What appears, in AAA's paper, as a formal self-contradiction, is also latent as such in the writings of many other authors. As AAA put it "So far as the past is concerned, the quantal formalism *requires* [their italics] that [the uncertainty relations] be violated". Two illegitimate procedures led to this statement. 1. *Counterfactual thinking*; 2. *Reification of wave retardation.* In contrast, all phantasms dissolve if two demands are obeyed: 1. Consider as actual *only* those states that are *actually prepared or measured*; 2. Dismiss the concept of an 'evolving state vector', and *stick to the* (intrinsically reversible) *transition amplitude concept.*

If, instead of the quantal, the classical, prequantal probability calculus were at stake, the formula 'corresponding' to AAA's would be

$$P(c_j) = \frac{(C = c_j | A = a)(C = c_j | B = b)}{\Sigma(C = c_j | A = a)(C = c_j | b = b)}$$

with the parentheses denoting conditional probabilities. There c_j refers to any 'real hidden state' compatible with the preparation $|A = a\rangle$ and the measurement $|B = b\rangle$.

It is usual to conceptualize this in terms of a fully retarded causality, but it could be done just as well in terms of a fully advanced causality. What the discussion of AAA's paper clearly shows is that causality *must* be conceived as arrowless at the quantal level — with then, *inevitably*, the far reaching implication pointed to by Wigner (1967, pp. 171—184) and others, and discussed by Schmidt (1982).

INDEX OF NAMES

INDEX OF SUBJECTS

BOSTON STUDIES IN THE PHILOSOPHY OF SCIENCE

Editors:

ROBERT S. COHEN and MARX W. WARTOFSKY

(Boston University)

1. Marx W. Wartofsky (ed.), *Proceedings of the Boston Colloquium for the Philosophy of Science 1961–1962.* 1963.
2. Robert S. Cohen and Marx W. Wartofsky (eds.), *In Honor of Philipp Frank.* 1965.
3. Robert S. Cohen and Marx W. Wartofsky (eds.), *Proceedings of the Boston Colloquium for the Philosophy of Science 1964–1966. In Memory of Norwood Russell Hanson.* 1967.
4. Robert S. Cohen and Marx W. Wartofsky (eds.), *Proceedings of the Boston Colloquium for the Philosophy of Science 1966–1968.* 1969.
5. Robert S. Cohen and Marx W. Wartofsky (eds.), *Proceedings of the Boston Colloquium for the Philosophy of Science 1966–1968.* 1969.
6. Robert S. Cohen and Raymond J. Seeger (eds.), *Ernst Mach: Physicist and Philosopher.* 1970.
7. Milic Capek, *Bergson and Modern Physics.* 1971.
8. Roger C. Buck and Robert S. Cohen (eds.), *PSA 1970. In Memory of Rudolf Carnap.* 1971.
9. A. A. Zinov'ev, *Foundations of the Logical Theory of Scientific Knowledge (Complex Logic).* (Revised and enlarged English edition with an appendix by G. A. Smirnov, E. A. Sidorenka, A. M. Fedina, and L. A. Bobrova.) 1973.
10. Ladislav Tondl, *Scientific Procedures.* 1973.
11. R. J. Seeger and Robert S. Cohen (eds.), *Philosophical Foundations of Science.* 1974.
12. Adolf Grünbaum, *Philosophical Problems of Space and Time.* (Second, enlarged edition.) 1973.
13. Robert S. Cohen and Marx W. Wartofsky (eds.), *Logical and Epistemological Studies in Contemporary Physics.* 1973.
14. Robert S. Cohen and Marx W. Wartofsky (eds.), *Methodological and Historical Essays in the Natural and Social Sciences. Proceedings of the Boston Colloquium for the Philosophy of Science 1969–1972.* 1974.
15. Robert S. Cohen, J. J. Stachel, and Marx W. Wartofsky (eds.), *For Dirk Struik. Scientific, Historical and Political Essays in Honor of Dirk Struik.* 1974.
16. Norman Geschwind, *Selected Papers on Language and the Brain.* 1974.
17. B. G. Kuznetsov, *Reason and Being: Studies in Classical Rationalism and Non-Classical Science.* (forthcoming).
18. Peter Mittelstaedt, *Philosophical Problems of Modern Physics.* 1976.
19. Henry Mehlberg, *Time, Causality, and the Quantum Theory* (2 vols.). 1980.
20. Kenneth F. Schaffner and Robert S. Cohen (eds.), *Proceedings of the 1972 Biennial Meeting, Philosophy of Science Association.* 1974.
21. R. S. Cohen and J. J. Stachel (eds.), *Selected Papers of Léon Rosenfeld.* 1978.
22. Milic Čapek (ed.), *The Concepts of Space and Time. Their Structure and Their Development.* 1976.

23. Marjorie Grene, *The Understanding of Nature. Essays in the Philosophy of Biology.* 1974.
24. Don Ihde, *Technics and Praxis. A Philosophy of Technology.* 1978.
25. Jaakko Hintikka and Unto Remes. *The Method of Analysis. Its Geometrical Origin and Its General Significance.* 1974.
26. John Emery Murdoch and Edith Dudley Sylla, *The Cultural Context of Medieval Learning.* 1975.
27. Marjorie Grene and Everett Mendelsohn (eds.), *Topics in the Philosophy of Biology.* 1976.
28. Joseph Agassi, *Science in Flux.* 1975.
29. Jerzy J. Wiatr (ed.), *Polish Essays in the Methodology of the Social Sciences.* 1979.
30. Peter Janich, *Protophysics of Time.* 1985.
31. Robert S. Cohen and Marx W. Wartofsky (eds.), *Language, Logic, and Method.* 1983.
32. R. S. Cohen, C. A. Hooker, A. C. Michalos, and J. W. van Evra (eds.), *PSA 1974: Proceedings of the 1974 Biennial Meeting of the Philosophy of Science Association.* 1976.
33. Gerald Holton and William Blanpied (eds.), *Science and Its Public: The Changing Relationship.* 1976.
34. Mirko D. Grmek (ed.), *On Scientific Discovery.* 1980.
35. Stefan Amsterdamski, *Between Experience and Metaphysics. Philosophical Problems of the Evolution of Science.* 1975.
36. Mihailo Marković and Gajo Petrović (eds.), *Praxis. Yugoslav Essays in the Philosophy and Methodology of the Social Sciences.* 1979.
37. Hermann von Helmholtz, *Epistemological Writings. The Paul Hertz/Moritz Schlick Centenary Edition of 1921 with Notes and Commentary by the Editors.* (Newly translated by Malcolm F. Lowe. Edited, with an Introduction and Bibliography, by Robert S. Cohen and Yehuda Elkana.) 1977.
38. R. M. Martin, *Pragmatics, Truth, and Language.* 1979.
39. R. S. Cohen, P. K. Feyerabend, and M. W. Wartofsky (eds.), *Essays in Memory of Imre Lakatos.* 1976.
40. B. M. Kedrov and V. Sadovsky. *Current Soviet Studies in the Philosophy of Science*
41. M. Raphael, *Theorie des Geistigen Schaffens auf Marxistischer Grundlage*
42. Humberto R. Maturana and Francisco J. Varela, *Autopoiesis and Cognition. The Realization of the Living.* 1980.
43. A. Kasher (ed.), *Language in Focus: Foundations, Methods and Systems. Essays Dedicated to Yehoshua Bar-Hillel.* 1976.
44. Trân Duc Thao, *Investigations into the Origin of Language and Consciousness.* (Translated by Daniel J. Herman and Robert L. Armstrong; edited by Carolyn R. Fawcett and Robert S. Cohen.) 1984.
45. A. Ishmimoto (ed.), *Japanese Studies in the History and Philosophy of Science*
46. *Peter L. Kapitza, Experiment, Theory, Practice.* 1980.
47. Maria L. Dalla Chiara (ed.), *Italian Studies in the Philosophy of Science.* 1980.
48. Marx W. Wartofsky, *Models: Representation and the Scientific Understanding.* 1979.
49. Trân Duc Thao, *Phenomenology and Dialectical Materialism.* 1985.
50. Yehuda Fried and Joseph Agassi, *Paranoia: A Study in Diagnosis.* 1976.
51. Kurt H. Wolff, *Surrender and Catch: Experience and Inquiry Today.* 1976.

52. Karel Kosík, *Dialectics of the Concrete*. 1976.
53. Nelson Goodman, *The Structure of Appearance*. (Third edition.) 1977.
54. Herbert A. Simon, *Models of Discovery and Other Topics in the Methods of Science*. 1977.
55. Morris Lazerowitz, *The Language of Philosophy. Freud and Wittgenstein*. 1977.
56. Thomas Nickles (ed.), *Scientific Discovery, Logic, and Rationality*. 1980.
57. Joseph Margolis, *Persons and Minds. The Prospects of Nonreductive Materialism*. 1977.
58. G. Radnitzky and G. Andersson (eds.), *Progress and Rationality in Science*, 1978, x + 416 pp.
59. Gerard Radnitzky and Gunnar Andersson (eds.), *The Structure and Development of Science*. 1979.
60. Thomas Nickles (ed.), *Scientific Discovery: Case Studies*. 1980.
61. Maurice A. Finocchiaro, *Galileo and the Art of Reasoning*. 1980.
62. William A. Wallace, *Prelude to Galileo*. 1981.
63. Friedrich Rapp, *Analytical Philosophy of Technology*. 1981.
64. Robert S. Cohen and Marx W. Wartofsky (eds.), *Hegel and the Sciences*. 1984.
65. Joseph Agassi, *Science and Society*. 1981.
66. Ladislav Tondl, *Problems of Semantics*. 1981.
67. Joseph Agassi and Robert S. Cohen (eds.), *Scientific Philosophy Today*. 1982.
68. Wuadysuaw Krajewski (ed.), *Polish Essays in the Philosophy of the Natural Sciences*. 1982.
69. James H. Fetzer, *Scientific Knowledge*. 1981.
70. Stephen Grossberg, *Studies of Mind and Brain*. 1982.
71. Robert S. Cohen and Marx W. Wartofsky (eds.), *Epistemology, Methodology, and the Social Sciences*. 1983.
72. Karel Berka, *Measurement*. 1983.
73. G. L. Pandit, *The Structure and Growth of Scientific Knowledge*. 1983.
74. A. A. Zinov'ev, *Logical Physics*. 1983.
75. Gilles-Gaston Granger, *Formal Thought and the Sciences of Man*. 1983.
76. R. S. Cohen and L. Laudan (eds.), *Physics, Philosophy and Psychoanalysis*. 1983.
77. G. Böhme et al., *Finalization in Science*, ed. by W. Schäfer. 1983.
78. D. Shapere, *Reason and the Search for Knowledge*. 1983.
79. G. Andersson, *Rationality in Science and Politics*. 1984.
80. P. T. Durbin and F. Rapp, *Philosophy and Technology*. 1984.
81. M. Marković, *Dialectical Theory of Meaning*. 1984.
82. R. S. Cohen and M. W. Wartofsky, *Physical Sciences and History of Physics*. 1984.
83. E. Meyerson, *The Relativistic Deduction*. 1985.
84. R. S. Cohen and M. W. Wartofsky, *Methodology, Metaphysics and the History of Sciences*. 1984.
85. György Tamás, *The Logic of Categories*. 1985.
86. Sergio L. de C. Fernandes, *Foundations of Objective Knowledge*. 1985.
87. Robert S. Cohen and Thomas Schnelle (eds.), *Cognition and Fact*. 1985.
88. Gideon Freudenthal, *Atom and Individual in the Age of Newton*. 1985.
89. A. Donagan, A. N. Perovich, Jr., and M. V. Wedin (eds.), *Human Nature and Natural Knowledge*. 1985.
90. C. Mitcham and A. Huning (eds.), *Philosophy and Technology II*. 1986.
91. M. Grene and D. Nails (eds.), *Spinoza and the Sciences*. 1986.

92. S. P. Turner, *The Search for a Methodology of Social Science*. 1986.
93. I. C. Jarvie, *Thinking About Society: Theory and Practice*. 1986.
94. Edna Ullmann-Margalit (ed.), *The Kaleidoscope of Science*. 1986.
95. Edna Ullmann-Margalit (ed.), *The Prism of Science*. 1986.
96. G. Markus, *Language and Production*. 1986.
97. F. Amrine, F. J. Zucker, and H. Wheeler (eds.), *Goethe and the Sciences: A Reappraisal*. 1987.
98. Joseph C. Pitt and Marcella Pera (eds.), *Rational Changes in Science*. 1987.
99. O. Costa de Beauregard, *Time, the Physical Magnitude*. 1987.
100. Abner Shimony and Debra Nails (eds.), *Naturalistic Epistemology: A Symposium of Two Decades*. 1987.